Edward Tanjore Corwin

Corwin Genealogy in the United States

Edward Tanjore Corwin

Corwin Genealogy in the United States

ISBN/EAN: 9783337330194

Printed in Europe, USA, Canada, Australia, Japan

Cover: Foto ©berggeist007 / pixelio.de

More available books at **www.hansebooks.com**

The Corwin Genealogy

(CURWIN, CURWEN, CORWINE)

IN THE

UNITED STATES.

BY

EDWARD TANJORE CORWIN,

MILLSTONE, N. J.

THE GLORY OF CHILDREN ARE THEIR FATHERS.—PROVERBS, 17 : 6.

NEW-YORK:
S. W. GREEN, PRINTER, 16 AND 18 JACOB STREET.
—
1872.

DESCENDANTS OF MATTHIAS CORWIN. (Six Generations.)

Matthias Corwin, settled at Ipswich, Mass., 1634; removed to Long Island, 1640; died, 1658.

Theophilus,
1630–92.

- Mehetable.
- Mary.
- Phebe.
- Bethia.
- Martha.

John, 1630–1702.

- **Mehetable.**
  - **John. D. 1640.**
    - Jonathan.
    - Mehetable.
    - Amaziah, (Md.)
  - **Thomas.**
- **Mary.**
  - **Timothy, 1720.**
    - Timothy.
    - Sarah.
    - Amaziah.
- **Theophilus, 1675–1769.**
  - **Theophilus, 1700.**
    - Jonathan.
    - David.
    - Mary.
    - George.
  - **Huldah.**
  - **Samuel, 1710–62.**
    - Asa.
    - David.
    - Samuel. — Henry.
    - Henry. — Henry.
    - Benjamin.
- **Bethia.**
- **Jedediah.**
  - **Daniel.**
  - **John.**
- **David, 1673–1759.**
  - **Daniel, 1690–1747.**
    - Silas.
    - Lucas.
    - Michal.
    - Mary.
    - Jedediah.
    - Peintiah.
    - Nathan.
    - Edward.
  - **Simeon, 1710.**
    - Theophilus.
    - Rachel.
    - Jemima.
  - **Henry, 1700–85.**
    - Gilbert.
    - Gershom.
    - Naomi.
  - **Rev. Jacob, 1710–82.**
    - Sarah.
    - Mary.
    - Matthias.
- **Matthias, 1676–1769.**
  - **Jeremiah.**
  - **Jesse, 1700.**
    - Amy.
    - Jesse.
    - Israel.
- **Mary.**
  - **Samuel, 1706–84.**
    - Nathaniel.
    - Stephen.
    - Ezra.
    - James.
    - Mary.
    - Phebe.
    - Samuel.
  - **Jedediah.**
    - John.
- **Abigail.**
  - **David, 1710–80.**
    - Eli, 1757–1833.
    - Phineas.
    - Annie.
    - Joseph.
    - Joshua.
    - David. — 5 —
- **John, 1663–1729.**
- **Hannah.**
- **John, 1705–55.**
  - Benjamin, 1690–1721.
  - Sarah. Died 1738.
  - Elizabeth.
  - Hester.
    - William, 1744–1818.
    - Sarah.
    - Elizabeth.
    - James, 1741–91.
    - John, 1735–1817.
- **Rebecca.**
- **Sarah.**

See annexed paper.

DESCENDANTS OF GEORGE CURWIN. (SIX GENERATIONS.)

George Curwin, settled at Salem, Mass., 1638; died 1683.

Abigail.

John.

Elizabeth.

Bartholomew.
(New-Jersey, 1717.)

Lucy.

Hannah.

Samuel.

Richard,
Died young.

George.
(Kentucky.)
(Ohio.)

Margaret.

Jonathan.

Sarah.

Jonathan.
Died young.

Elizabeth.

William,
Died young.

John.

Joseph.

Samuel.

Descendants in
Baltimore, and
perhaps South-
Carolina.
(See Supplement.)

Keziah.
Naomi.
(Canada.)

George.
Abigail.
Rebecca.
Amy.
Samuel.
Lydia.
Rachel.
(New-Jersey.)

Hannah.

Elizabeth.

Rev. George,

Jonathan,
Died young.

Samuel,
(Author of the Journal.)
No ch.

Jonathan.
Died young.

George.

Penelope.

Susannah.

George.

John,

Margaret,

Anna.

Jonathan.
Died young.

Margaret.

George. Sarah. Mehetable.
Mar. Richard Ward.
(Salem, Mass.)

Daniel.
Martha.
Elizabeth.
Mehetable.
Sarah.
Samuel C.
George C.

Samuel Curwen, (6.)
(See p. 195.)

Samuel Corwine,
settled in Essex Co.,
Mass., 1650.

Mar. Mary Thomas Smith.

Thomas Curwin,
a Quaker,
in Boston, 1676.

John Curwen,
of Keswick, Eng.,
settled at Philadelphia, 1784.

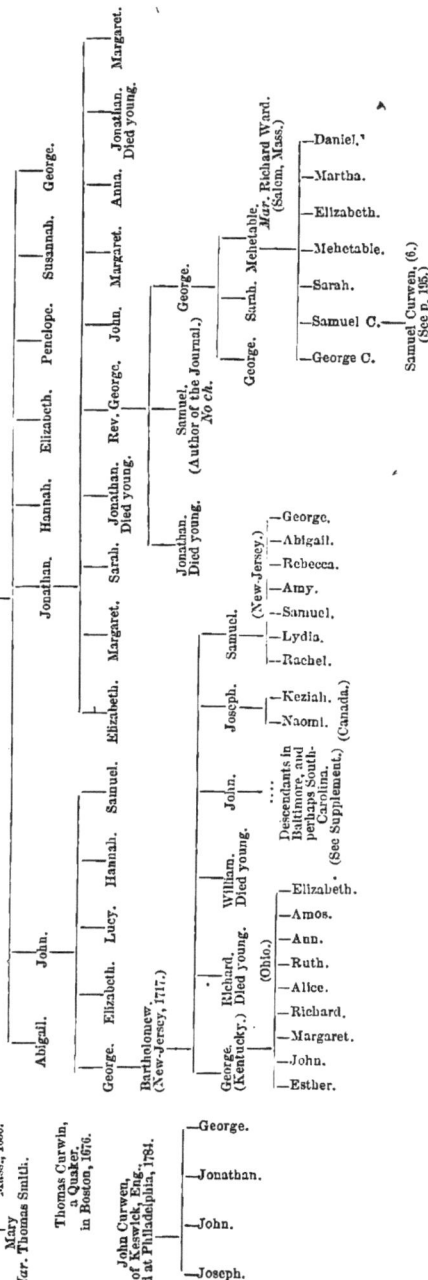

George.
Jonathan.
John.
Joseph.

Elizabeth.
Amos.
Ann.
Ruth.
Alice.
Richard.
Margaret.
John.
Esther.

Malcolm Corwin, of London, Eng.,
settled in New-York City, 1844.

Charles H. Curwen, of Workington, Eng.
New-Orleans, La., 1868.

Janos Hunyadi Corvinus

born 1376 died 1456, 1443-1456 -
Regent & Governor of Hungary -

made Vaywood or Standertwained or Wladislaus,
defeated the Turks under Amurath on the
plains of Wallachia, conquired Nissa, and
gained a signal victory before Sophia, 1443.
Was defeated at the little Isle of Varna from lack
of food, the battle raged for three days, and
three nights without result, On the fourth day
Hunyadi charging with his horse put to flight the
Janisaries, The Sultan was flying. But two
bishops, jealous of Hunyadi, left their positions
contrary to his orders, and pursued the flying
enemy. This turned the day against him.
Hunyadi and a few followers escaped. Hunyadi
was made Governor of Hungary. In 1452, the
Captive Prince Lladislaus Posthumus, was
released and Hunyadi resigned, Was made
Chief. General. The Sultan invaded Hungary
with 150,000 men, and in 1456, appeared
before Belgrade. The Hungarian King by
advice of Hunyadi fled to Vienna, Hunyadi
quickly raised a force of 10.000 men, and
hastened to the relief of the beleagured city,
After a siege of two weeks, Hunyadi compelled
the haughty Ottoman to retreat to Adrianople,
leaving 24000 dead on the field of battle.
This battle decided the fate of Europe.
The great fatigue brought on Hunyadi a
sickness, of which he died a month later,
September 10. 1456. He left two sons,
Lladislaus, who was cruelly murdered
by King Ladislaus, and Matthias.

Matthias Corvinus,
born 1441. died 1491.
King of Hungary - 1458-1491-
Married first, the daughter of
Podiebrad, married secondly,
Beatrice of Naples,

The Crown of Hungary had been elective
since 1301. The royal elections were often
scenes of wild confusion. In 1458 when
Matthias Corvinus was elected to the throne,
he was proclaimed by 40,000 soldiers
standing on the frozen Danube, beneath
the terraced heights of Buda.

By his first wife King Matthias had
two sons.

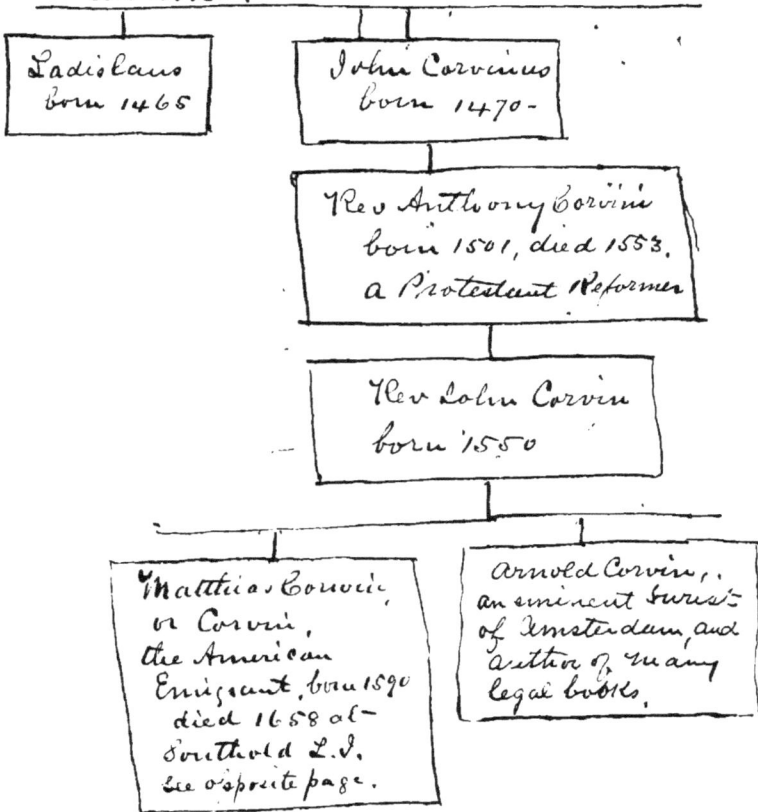

```
┌──────────────┐        ┌──────────────────┐
│ Ladislaus    │        │ John Corvinus    │
│ born 1465    │        │ born 1470-       │
└──────────────┘        └──────────────────┘
```

```
┌──────────────────────────┐
│ Rev. Anthony Corvini     │
│ born 1501, died 1553.    │
│ a Protestant Reformer    │
└──────────────────────────┘
```

```
┌──────────────────────────┐
│ Rev. John Corvin         │
│ born 1550                │
└──────────────────────────┘
```

```
┌────────────────────────┐   ┌──────────────────────┐
│ Matthias Corvin,       │   │ Arnold Corvin,       │
│ or Corvin,             │   │ an eminent jurist    │
│ the American           │   │ of Amsterdam, and    │
│ Emigrant, born 1590    │   │ author of many       │
│ died 1658 at           │   │ legal books.         │
│ Southold L.I.          │   │                      │
│ See opposite page.     │   │                      │
└────────────────────────┘   └──────────────────────┘
```

Matthias Corwin, born 1590.

Settled at Ipswich, Mass., 1633; removed to Southold,
Long Island, 1640; died 1658; married in England,
Margaret Martin.

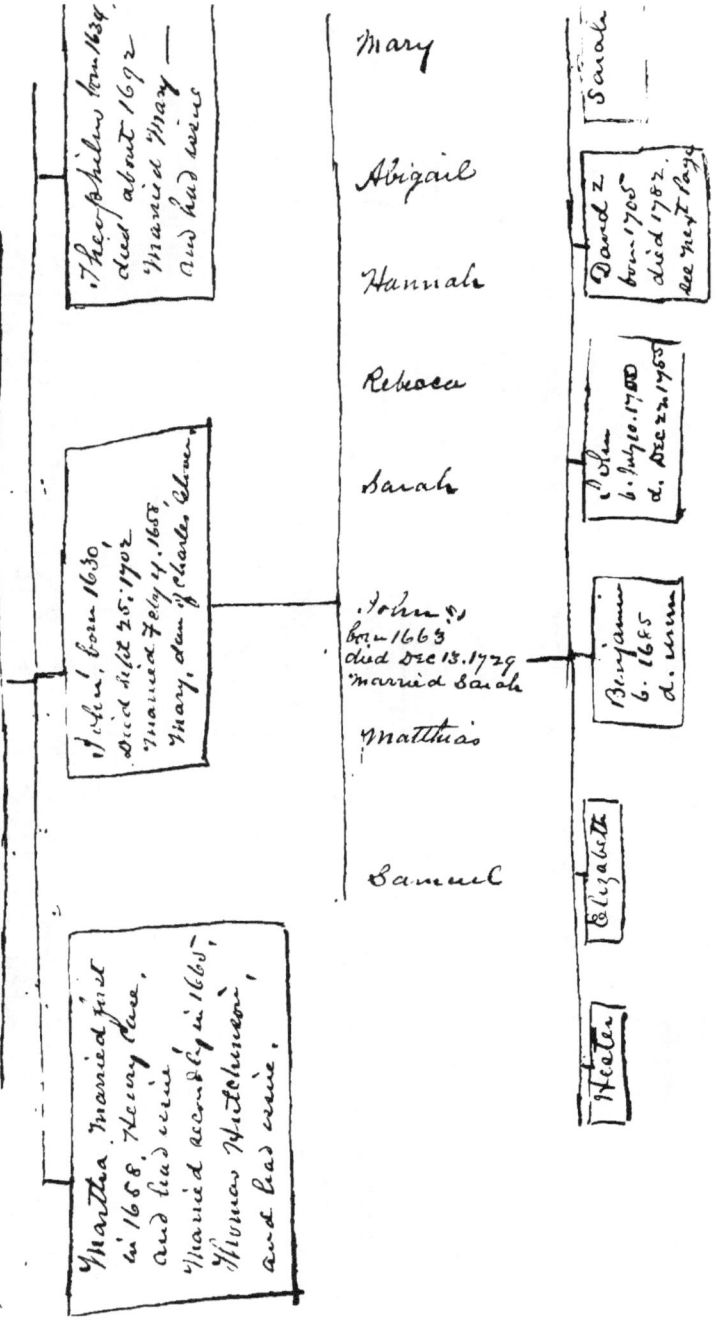

Theophilus born 1639,
died about 1692.
Married Mary —
and had issue

John, born 1630,
died Sept 25, 1702.
Married Feby 4, 1658,
Mary, dau. of Charles Glover

Martha, married first
in 1658, Henry Case,
and had issue;
married secondly in 1665,
Thomas Hutchinson,
and had issue.

Mary

Abigail

Hannah

Rebeca

Sarah

John 2d
born 1663
died Dec 13. 1729
married Sarah

Matthias

Samuel

David 2,
born 1708,
died 1782.
See next page

John
b July 20. 1700
d. Dec 22, 1765

Benjamin
b. 1665
d. unm

Elizabeth

Heala

Sarah

David Corwin 2
(2ms of Lothan 2)

born 1705, died 1762, married 1730 Deborah Wells
who was born 1717, and died Nov. 24. 1798. His David
Corwin moved to Orange Cou'ty, New York, probably at
the opening of the Revolutionary War.

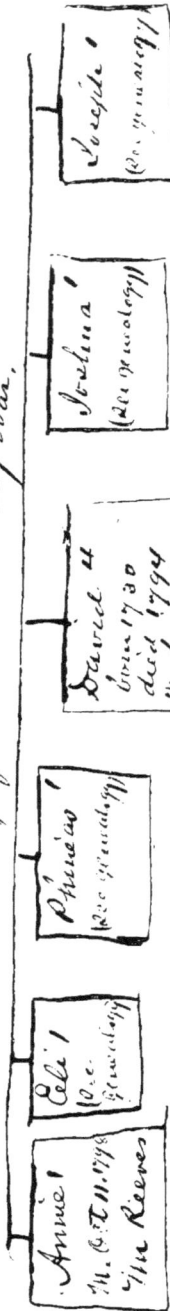

Anne 1	Eli 1	Phineas 1	David 4	Joshua 1	Joseph 1
m. Oct 11. 1798	(see genealogy)	(see genealogy)	born 1730	(see genealogy)	(see genealogy)
Jno Reeves			died 1794		
			m. Feb. 20. 1757		
			Mary Wells		
			m. secondly		
			Abigail Brown		

— Deborah 1
b. 1756
m. Ephraim
Everett

— Mehitable
died young

— Abigail 2
b. Dec 9. 1775
m. March 9 1796
Walter Everett

— Mary 12
m. James H.
Reeves

— Daniel 4
b. 1759. d. 1829
m. Anna Hulse

— David 6
b. 1760 d. 1802
m 1. Corey, m 2. Mc ?
m 3. Burchon
— Jesse 3
b. Sept 12. 1763 d. ?
m. Mrs Walker

— Eli 3
b. Jan 27. 1765
d. June 1. 1826

— Joseph 9
b. 1781, d. May 4
m. Hannah
Finch

— Elisha 1. b. Sept
d. May 20. 1866
m. May 24. 1808
Christiana Sm
— Nebat 1 b. 1783
m. Mary Howard

— Phineas 2
b. 1784 d young

— Naboth D. 1
b. Feb. 13. 1787

b. 1782

832.

g. 1831

Julia Ann / b. 1814 / M. Abram / Hennitta
— Helen, Henrietta.

Gabriel / b. 1808 / M. Jane Murray
— Louisa

1st. Eliza 2 / married / Gabriel / Horton
— Julia, Henry, Abigail, Sarah, Caroline, Mary, Adelia, Charles, Harriet

Mary 26 / b. March 29.1810 / M. Nov 13. 1633 / William A. Thomson / she died. Feby 23.1663
— Samuel, William, Joseph, Anna Mary, Emma, Elizabeth, Edward, Clara.

Sarah Jane 32 / born 1802 / M. George A. Miller
— Sylvanus K., Mary, Charles A.

Clara Thomson
b. Jany 1. 1848
d. March 20. 1850

Elizabeth Thomson
b. July 15. 1843
d. Feby 19. 1844

Emma Thomson
b. Aug 27. 1841 - d. June
m. Feby 15. 1883.   1891
Robert C. Shaw.

Anna Mary Thomson
b. Jany 27. 1840
m. Sept 24. 1873
Samuel T. Clark

Samuel Thomson
b. Feby 22. 1835
d. May 6. 1891

William Thomson
b. Sept 11. 1836
m. May 13. 1863
Julia L. Clearwater

Joseph Thomson
b. July 21. 1838
m. Nov 4. 1867
Jane L. Amidon

Edward Thomson
b. Jany 11. 1845
m. 1. Oct 5. 1870
Sarah T. Shanks
m. 2 Sept 29. 1874
Emma Talbot

Mary Corwin 25 (Joseph 9)
born March 28. 1810
died Feby 23. 1893.
Married Nov 13. 1833.
William Archibald Thomson.
born May 29. 1808.

Lillian Genevieve
b. Aug 30.1876.
Clara Estella
b. Oct 9.1880.
(Brooklyn, N.Y.)

Josephine
b. Sept 1.1870
Amidon
b. Oct 30.1878

Nettie
b. April 30.1864
(Rhinebeck N.Y.)
(N.Y. City)
m. May 13 1891-

Mary Conner
b. June 29.1874
Wm A. Thomas
b. Feby 26.1876
(Sing Sing N.Y.)

William Thompson
Croston L. Sisson

TABLE OF CONTENTS.

———•●•———

PREFACE.

THE CORWIN GENEALOGY herewith given to the public is the result of the slow accumulation of material during many years. It only recently, however, became possible to arrange the earlier generations of the family of Matthias Corwin, of Long Island, in a sure and satisfactory manner. This was accomplished by the publication of Moore's *Indexes of Southold*, a few years since. A circular letter was then issued, and sent to every person connected with the family, whose address was ascertained, asking for such additional information as was necessary to complete the work. Probably few American families of the Corwin (Curwen or Corwine) name will experience much difficulty, with the material here presented, in making their own family record complete, if it be not already so in the present volume.

The work indeed contains the records of several Corwin families, originating from different American ancestors. The writer took it for granted at first, as several New-England authors have done, that the Salem and Long Island Corwins were probably connected. Under this supposition, he investigated, as opportunity permitted, the history of the English Curwens, of Cumberland County, whose *Arms* George Curwen, of Salem, Mass., had brought with him to America, in 1638. Yet it seemed a little remarkable that, if the Salem and Long Island families were related, all trace of such relationship should have been lost on both sides, though living not very distant from each other. But in 1859, when Appleton's *New American Cyclopedia* was publishing, it was noticed in a sketch of the Hon. Thomas Corwin, found therein, that the family was referred to an Hungarian origin. The article was written by the Hon. Anthony H. Dunlevy, of Ohio. (See *Lucinda Corwin*, 4, p. 142.) About the same time, a volume of speeches of Governor Corwin was published, to which was prefixed a memoir of him, by Isaac Strohm, of Ohio. Herein again, still more definitely, the same fact respecting the family's origin was stated. (*Corwin's Speeches*, p. 8.) The writer addressed Governor Corwin on the subject, and received the following reply:

LEBANON, OHIO, September 5, 1859.

DEAR SIR: When the publishers of the *Encyclopædia* advised me of their wish to insert in that work a brief sketch of my life, I was in haste to leave home, and handed their note to a friend, requesting him to correspond with the publishers, and give them, in his own way, what they wanted.

This gentleman had known me and my family intimately for fifty years. The result was the article to which you refer. He had given some attention to our family history, and I had not. There is in my hands, amongst much of such lumber, several letters showing our connection with the family of the Hungarian Corvinus. Somebody's history of Connecticut is referred to, I remember; and at the time I read these communications, their account of the matter struck me as quite plausible. I could never bring myself to feel interest enough in the subject to withdraw me from necessary labor long enough to make such researches as to enable me to form even a plausible guess as to the persons who might have been at work, for ten centuries back, in the laudable effort to bring me, *nolens volens*, into this breathing world on the 29th day of July, (a most uncomfortable time of the year,) in the year of grace 1794. I have seen the book you refer to.* I read a page or two of it, and it was captured by an aged aunt, who carried it away, and I have not seen it now for about five years. It was sent to me by a Mr. Ward, then of Staten Island, and I suppose it is the same you propose to send me. There was a likeness of Judge Curwen in the volume I had. Your researches† sent me, for which I thank you, are curious enough, and must have cost great labor and much time. . . . (But) I have resolved not to engage in that Battle of the Books. Nevertheless, had I the leisure, I dare say I should take as much delight in such inquiries as yourself. But my tastes have never had fair play. The actual affairs around me and upon me have driven me like a slave, through a very busy and very unprofitable life thus far. My paternal grandfather's name was Jesse; his father's name was William.‡ So says the family legend with us, and I believe the records so say.

I should like to know you personally. Can you tell me how I can find you, and when? I am again in that turbid water, politics. If you come to Washington next winter, you will find me amongst the monsters, big and little, that swim in that sea of troubles. Truly yours,

E. T. CORWIN.   THOMAS CORWIN.

Such statements from so distinguished an authority seemed worthy of further investigation. It was soon ascertained that there was a wide-spread tradition in the family concerning an Hungarian extraction. This tradition, indeed, did not extend to all branches; but it existed in various quarters. The writer has subsequently frequently heard of it in conversation, and there have been a number of references to it in letters. He has, therefore, felt himself obliged, though reluctantly, to relinquish his former cherished opinions of the relation of the Long Island family to the English Curwens. The results of his investigations in this matter,§ as well as the facts con-

* Curwen's *Journal and Letters.*
† The *English Curwen Genealogy.* (See Appendix II.)
‡ But see Jesse, 1, p. 113, and William, 1, p. 224.
§ Curiously enough, long after this volume was ready for the press, and two thirds of it actually printed, (April, 1872,) the writer received a note from George R. Curwen, Esq., of Salem, Mass., stating that an interesting record had just turned up in Salem, written upon

cerning several later immigrants to America, of the Corwin or Curwen name, are presented in the INTRODUCTION.

The peculiar difficulties in the preparation of an American genealogy have been fully experienced in this work. The families are scattered from ocean to ocean, and from lake to gulf. They are often without any records of their remote origin, and "The Old Family Bible" * has often been carried

the fly-leaf of an old book, formerly the property of Rev. George Curwin, (4,) p. 79, (being inscribed Georgii Curwin Liber, Anno 1715,) in his own handwriting, as follows:

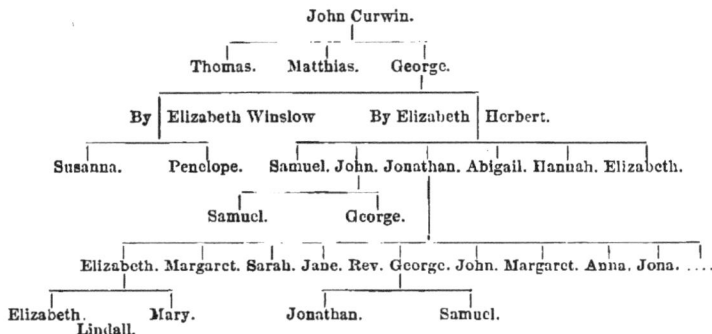

John Curwin.

Thomas.   Matthias.   George.

By | Elizabeth Winslow   By Elizabeth | Herbert.

Susanna.   Penelope.   Samuel. John. Jonathan. Abigail. Hannah. Elizabeth.

Samuel.   George.

Elizabeth. Margaret. Sarah. Jane. Rev. George. John. Margaret. Anna. Jona. ....

Elizabeth.   Mary.   Jonathan.   Samuel.
Lindall.

Now, this " tree," without a word of explanation more than the names arranged as above, makes Thomas Curwin, the Quaker, Matthias Corwin, of Long Island, and George Curwin, of Salem, three original immigrants, though at different times, to be brothers, and the sons of a John Curwin, of England. Thomas and George may have been brothers, but the remarks already made, and the general tenor of the Introduction, will apparently quite for-bid the placing of Matthias here. Probably this arrangement by Rev. George Curwin was a *supposition*, upon which he was beginning to work out a Curwin Genealogy; but his death soon after, (1717,) prevented him from either verifying or disproving the same. It seems impossible to allow such a record to overturn all the testimony which the reader will find respecting another origin of Matthias Corwin. We hope, however, that the truth may be elucidated. The writer would simply observe that the family of Matthias Corwin, on Long Island, have never possessed (so far as his knowledge extends) the Curwen Arms, which the family of George, both in the East and in the West, have cherished; that the early gene-rations of these families are not known to have ever visited or even corresponded with each other. The name of Matthias might have become known to Rev. George, from the neigh-boring Ipswich records where it yet remains. Rev. George was but two years old when his grandfather died. Matthias moved from Ipswich to Long Island about the time of the birth of the first George's children, 1640; while Thomas does not, as far as known, appear in this country till forty or fifty years after the arrivals of George and Matthias.

* Says a writer in the *New-Brunswick Review*, 1855, Among time-honored customs, there is none more touching or instructive than that which employs the blank leaves of the "Old Family Bible" for genealogical records. In that consecrated place, between the Old and New Testaments, thereby acknowledging that true religion is the basis of the pure domestic relation, we place the records of our births, marriages, and deaths. And this is highly appropriate, as the Bible is the family record in general of all the children of Adam. It tells of the divine origin of our race, of the true aim of life, of our physical and moral destiny, and of our high dignity in the scale of animated and rational being. It tells how the lost image of God may be restored, and with it, every other capacity invigorated anew. For this very position of our "family records" seems to indicate that, while we trace our pedi-gree up to Adam, though not without many long breaks, we yet love to place it as near as possible to the genealogy of our Lord Jesus Christ. For while we acknowledge our relation

off by daughters, and thus lost to the family name. Others, as is some·times the case, care very little for such pursuits. Concerning the Salem family, extensive researches had already been made, and published in Curwen's *Journal and Letters*, and in the volumes of the Massachusetts Historical Society. The new material of the western branches of this family, under the name of Corwine, is here for the first time published and connected with the Salem ancestral stock. The Rev. Jacob Corwin, it is said, prepared a genealogy of the Long Island Corwins about the opening of this century, but his papers were burned.

The increase of the several families, classified according to generations, sexes, and surnames, is represented by the following tables:

FAMILY OF MATTHIAS CORWIN, (APPROXIMATELY COMPLETE.)

	CORWINS.		OTHER SURNAMES.		
Generations.	Male.	Female.	Male.	Female.	Total.
1	1	..	..	..	1
2	2	1	..	..	3
3	6	23	3+	1	33
4	16	2+	4+	12	34
5	62	21+	6+	10+	99
6	136	101	2+	4+	243
7	326	307	96+	136+	865
8	415	349	362	280+	1406
9	438	143	153	176	610
10	10	15	28	26	79
Total.....	1112	962	654	645	3373

FAMILY OF GEORGE CURWIN, (INCOMPLETE.)

1	1	..	..	..	1
2	3	5	..	..	8
3	6	9	1	1	17
4	5	..	..	4	9
5	7	2	..	..	9
6	9	16	1	..	26
7	22	14	4	3	43
8	37	26	3	3	69
9	38	40	15	16	109
Total.....	128	111	23	27	292

A dozen, perhaps, of the sixth generation still survive, while a few of the tenth generation have come on the stage. The eighth is now in its prime. It will be noticed that few names of the female descendants of the fourth and fifth generations have been ascertained.

The Corwin families have been mostly engaged in agriculture, that great and unfailing source of national wealth. Not a few in each generation have also been pioneers on the ever-receding border, forcing the forests and help-

to our first father, according to the flesh, yet as Christians, we value still more highly the new relation to a higher and better nature which now may be established, through faith, between us and the Second Adam, who is the Lord from heaven. For the first left us the legacy of sin and death; the Second waits to bestow on us the blessings of righteousness and eternal life.

ing to open up new States for the growing nation. Many have occupied
positions in the various mechanical arts, in commerce, and in navigation.
Yet the learned professions and the legislative halls of several States and of
Congress have not been without representatives from this family. Many of
the youth, also, are now engaged in seeking a liberal education, and pre-
paring for increased usefulness. There have also been in every generation
many devoted Christians, acting either as members or officers in almost
all the various denominations.

The subject of genealogy is becoming increasingly interesting in America as
time removes us further from the original settlers. It seems a misfortune that
any of the old ante-revolutionary families of New-England should care so little
for their fathers as to allow themselves to forget the stern virtues and noble
courage which pervaded their souls. They were men of profound religious
convictions, who, when they settled these shores, consecrated themselves
and their children, to the latest generation, to the cause of truth and human
liberty. They left homes which they loved because they were God's chosen
ones, animated by his spirit with grand and holy principles, which are now
beginning to be universally recognized as just, and which are diffusing their
sacred leavening influences throughout the world. Should not each family,
then, especially if descended from the Puritans, warmly cherish the memory
of their fathers, and be stimulated by their history to a similar zeal for piety
and liberty? The object of genealogy to an American should not be, indeed, chiefly to show what men call a noble rank. Mere hereditary rank,
in itself considered, is certainly an illusion. But hereditary virtue, though,
alas! it may sometimes leap over a generation, gives a divine patent of the
truest nobility, better than all the blazonry of heralds' colleges. For while
the Puritans were not without their faults, at which unbelief or bigotry
loves to sneer, yet, all things considered, they were as noble a race of men
as ever trod the earth. "The principles of New-England," says De Tocque-
ville, an unbiased foreign writer, "spread at first to the neighboring States;
then they passed successively to the more distant ones, and at length they
imbued the whole confederation." We see this now fulfilled on a still
grander scale in the principles of universal liberty and political equality,
which have at length become the common doctrine of our land. In all the
battles, then, which are yet to be fought with error and corruption, let the
descendants of New-England, wherever they have made their homes, be
found on the side of Truth and Virtue, the unflinching and ever loyal cham-
pions of Right.

The writer would take this opportunity of thanking all who have in any
way assisted him in the preparation of this work. He is under special obligations to Abel Corwin, of New-York City, who many years ago prepared
the record of the descendants of David Corwin, No. 2; to Charles B.
Moore, Esq., of New-York City, except for whose *Index of Southold* this
work would hardly have been completed, and who has also most kindly
aided him by correspondence; to Robert Corwin, Esq., of Dayton, Ohio,

for much material concerning the Ohio Corwins; to Hon. Amos B. Corwine, of New-Rochelle, N. Y., for very full details concerning the Ohio
Corwines; and to George R. Curwen, of Salem, Massachusetts, for frequent
assistance in reference to the Salem Curwens, and for the plates of Judge
Curwen, and the Curwen Arms; also partly for the lithograph of Workington Hall. He is also indebted to Mr. Comly, of Cincinnati, Ohio, who
allowed him to procure impressions of the steel plate, in his possession, of
the Hon. Thomas Corwin.

INTRODUCTION.

———•♦•———

THERE are several families of the name of Corwin, Corwine, or Curwen in the United States, the majority of whom are, no doubt, of English descent; while one of them, according to tradition and many corroborative circumstances, is said to be of remote Hungarian origin, though also, perhaps, for a time naturalized in England. This latter family is descended from MATTHIAS CORWIN, who was the first of this name in America. He is found settled at Ipswich, Massachusetts, in 1634, whence he removed to Southold, Long Island, in 1640. He came, it is said, from Warwick, England.

In 1638, GEORGE CURWIN (or CORWIN) came from Northampton, England, and settled at Salem, Massachusetts. In 1652, we find a SAMUEL CORWINE (or CURWIN) settled in Essex and Old Norfolk, in the same State. In 1676, THOMAS CURWIN and, perhaps, WILLIAM CURWIN, Quakers, are found at Boston. There then appear no new settlers of the name until 1784, when JOHN CURWEN, of Keswick, Cumberland, England, settled in Philadelphia or vicinity. Again, in 1844, MALCOLM CORWIN, of London, settled in New-York City. About 1868, CHARLES H. CURWEN, of Workington, Cumberland, England, went to Denver, Colorado, and subsequently removed to New-Orleans. (See these names in their appropriate places.)

THE NAME CORWIN.

This name, as used by the family of traditional Hungarian descent, is of Latin origin, being simply the Anglicized form of the Latin adjective *Corvinus*, derived from *Corvus, a raven;* but as used by the families of pure English descent, Corwin, or Corwine, is a corruption of *Curwen*, which latter form is itself a corruption of *Cul-wen*, a compound Celtic word, *Cul-wen*, meaning a *white cowl* or *hood*, such as monks used to wear. Neither of these names has any connection with the Welsh *Corwen*, (*Caer-wen, Corwenna,* or *Cor-owain*,) with which Corwin has sometimes been confounded The name has assumed various orthographies at different times and in different branches of these families, as follows :

The Continental Name.	The English Name.
Corvinus, (Latin form.)	Culwen, (Celtic or Saxon form.)
Corvin, (German.)	Curwen, (English.)
Corwin, "	Curwyn, "
Corwin, (English and American.)	Curwenne, ⎫
	Corin, ⎪
	Curran, ⎬ (Irish orthographies.)
	Corwine. ⎭

Currin, ⎧ (Erroneous orthographies, ori-	Curwin, ⎫	
Curran, ⎨ ginating from the traditional	Corwin, ⎪	
Curwin, ⎪ Hungarian pronunciation.)	Curwen, ⎬ (American orthographies.)	
⎩ See note, p. xv.	Corwine, ⎭	

Thus, these families have sometimes fallen upon the same orthography, although etymologically the names are very different. The Long Island family have never used *Curwen* or *Corwine*, but in a few instances *Curwin*, which was a tendency toward the true orthography from the Hungarian pronunciation of Corwin, which was Currin. On the other hand, the English *Curwen* name is found spelled Curwyn and Corwine, even in Great Britain; and in Ireland has been corrupted also into Corin and Curran.* In America, the two forms, Curwen and Corwine, are now used.

THE ORIGIN OF THE FAMILY OF MATTHIAS CORWIN.

Matthias Corwin is found settled at Ipswich, Massachusetts, in 1634,† whence, in 1640, he removed to Southold, Long Island. No original documents now exist, so far as the writer is aware, respecting the particular European origin of this family, though he has heard of such documents as having only recently existed. Although the name of Corvinus is very ancient, and of flattering celebrity in Roman history, all claims will be cheerfully waived as to any connection with that old patrician family. It may, however, be observed, that Marcus Valerius Messala Corvinus,‡ in connection with Augustus, was the last of the Roman consuls under the Republic, and a most excellent man, in an age of abounding corruption. Now, the Italian historians declare that the village of Corvinus, in Hungary,§ was thus named by one of the Greek emperors, out of respect to this worthy Roman. The modern surname Corvinus (Corvin, Corwin) is probably

* It is also known that the names of some recent Irish families in the United States, such as Kirwin, Curran, (or other names of similar sound,) are sometimes recorded in our city directories as Corwin, Corvin, Curvin, and Curwin. The writer has met with several such cases. They are generally Romanists in religion. Some of them, however, may be offshoots of the English Curwens, in very early times, or the names may have originated from other sources. Bernard Corvin and his cousin, Frank Corvin, came about 1850 to Philadelphia, Pa., from Ireland, but ultimately from Denmark. They may be of the continental Corvini. A few additional orthographies to the above lists, never used by the families, but simple clerical blunders in the records, are omitted, such as Curwinne, Curwine, Carwine.

† He probably came over a few years earlier, or about 1630.

‡ For the origin of the name Corvinus, and a few extract about the Romans of that name, see Appendix A.

§ Then in Hungary, now in Wallachia, Turkey, on the river Danube.

derived from this village. Corvinus is now a common name in Austria and Germany.*

Now, from this village, say some writers, sprang, in the fourteenth century, John Hunyadi, who was surnamed Corvinus. He was a famous warrior, and firmly held the Turks in check when they were designing to overrun all Europe. He was the commander at the celebrated battle of Belgrade, (1456,) when the Moslems left 24,000 of their men dead on the battle-field. He was honored with the title of the hero of Christendom.†

He was also the father of the still more celebrated Matthias Corvinus, under whose wise and liberal statesmanship Hungary reached the acme of her greatness. The proverb yet remains, "Matthias is dead and justice is gone!" He was elected to the high position, which he held for thirty-three years, (1458–1491,) by the votes of the nobles. They hoped that the ability of the father would not be wanting in the son, and they were not disappointed. Like his father, he was the firm bulwark of Europe against the Mohammedans. He was also a renowned champion of literature.‡

Now, tradition § reports that Matthias Corwin, the American immigrant, was not only of ultimate Hungarian extraction, but was also a descendant of this excellent and worthy man. Some branches of the family did not possess this tradition, while others have warmly cherished it. But the writer has not personally met with any original documentary or historical evidence by which he could indubitably verify the tradition. The mere identity of names, the American *Matthias Corwin*, and the Hungarian *Matthias Corvinus*, though striking, is not sufficient in itself. Two hundred and forty years have passed away since the family immigrated to America. Manuscripts of a merely private character would hardly survive so long, although such manuscripts are said ‖ to have existed recently on Long Island. These were, of course, unintelligible to Americans, and therefore, perhaps, not properly valued, or they may have been burned with Rev. Jacob Corwin's papers. It is also, probably, impossible to find in America the necessary works of reference to trace particularly and satisfactorily the successive links for the century which intervened between the death of the Hungarian Matthias, 1491, and the birth of the American Matthias, about 1590. Notwithstanding, we do know that persons of the name of Corvinus, or Corvin, are found scattered through Germany, reaching even to Hanover and Holland, during the sixteenth century, and that one named Matthias Corwin, descended, according to tradition, from the Hungarian

* The writer was informed of this fact by fellow-students, who fled their country in the Kossuth rebellion of 1848. The fact is also known from other sources.
† See Appendix B. ‡ See Appendix C.
§ See Corwin's Speeches, Dayton, Ohio, 1859, p. 8. In the memoir of Governor Corwin, prefixed to the above volume, the editor says, "The Corwin family originally immigrated to the American Colonies from Hungary, and, it is said, are traceable to the celebrated Matthias Corvinus of that country."
‖ David Corwin, of Brooklyn. See also Rev. Jacob Corwin's name, who prepared a genealogy about 1800, and the letter of Hon. Thomas Corwin, in the Preface. Also Appendix G.

Corvini, is found consorting with the English Puritans about 1630. It is also a well-known fact that the early American pronunciation of the Long Island *Corwin* name was Currin, and this is identical with the Hungarian pronunciation to this day of the Corvin name.*

Now, an American, while earnestly believing in the American political doctrines, may justly desire, as a matter of laudable curiosity or as a stimulus to duty, to know his origin and what circumstances brought him to his present condition and home. He may be justly inquisitive concerning the antecedents of his ancestors, especially if they have been on the side of freedom, education, and humanity. Otherwise, even if of royal extraction, he could derive no honor therefrom as an American, but might rather blush if obliged to show his pedigree. But Matthias Corvinus, of Hungary, though not without his faults, stands ever on the page of history as a bright and shining light, mitigating the persecutions which the Romish hierarchy enforced in his dominions, while he was a most zealous advocate of learning, and a pattern of wisdom and of kindness. It would not, therefore, be unpleasant or disgraceful to point back, if possible, to such an ancestor. But the writer has been unable to prove it, although tradition affirms it.

Taking it for granted, however, that at least the Hungarian extraction is true, the reasons of the exodus of the family from Hungary were, no doubt, partly religious and partly political. The name of the excellent Corvinus would naturally excite jealousy, while the later Corvini, with whose lives the writer has become somewhat acquainted, were, with few exceptions perhaps, Protestants. It is also well known that the Reformers in Hungary and Bohemia accomplished a great work before the reformation of Luther. Multitudes forsook the Roman Church, and, in the periodical intervals of restored papal power, those who escaped death fled their country. Very large numbers of Hungarians are known to have scattered, at these times of persecution, through Germany, Holland, and England.† Hungary and England also entered into peculiarly intimate relations in the early part of the sixteenth century, through the intermarriage of the royal families,‡ which, by the increased intercourse which thence sprung up, would greatly facilitate migrations of Hungarians to the latter country.

* In the Hungarian and Latin languages, the letter *w* is altogether wanting, (as in most of the languages of Europe,) the letter *v* or *u* (originally the same) taking its place. But this *v* or *u*, in Hungarian, is pronounced like the English *w*; while the *o* of that language is pronounced like the French *eu*, or the German *u*. (See Art. *Hungarian Language*, in Appletons' *Cyclopedia*, vol. ix., p. 365.) Therefore, the name *Côrvin*, in the language of Hungary, was pronounced *Curwin*, or *Curwn*, or less accurately, *Currin*, or *Currun*. The form Curvinus is also found. Now, Currin was the popular pronunciation of Corwin, in America, down to a very recent time, and in some retired districts is not yet extinct. Plenty of documents, such as wills and deeds, attest the same. See Currin, Curran, Curwin, in Index. Savage, in his *Genealogical Dictionary*, says, "Often this name (*Curwin*) is written Corwin; sometimes Currin, to conform to sound." He supposed that this family was identical with the Curwen family of Salem. So also several other New-England writers on these subjects.

† See Ullman's *Reformers before the Reformation*, Gillett's *Life of John Huss*, and D'Aubigné's *Hist. Reformation*, v. p. 415. Also, Appendix D.

‡ The daughter of James I. married a Bohemian prince.

Now, it is well known that Matthias Corvinus had two sons, Ladislaus and John. The latter was born about 1470, and was yet living in 1540. He was placed under the tutorship of Anthony Bonfinius,* of Lucca, (Loccum,) Italy. In the next generation appears a Rev. Anthony Corvin, who studies at Loccum, and the circumstances would seem to suggest the probability of his being a son of John, and named after Bonfinius. He was born in 1501, and died in Hanover in 1553. He was driven from the cloister of Loccum in 1523, on account of his favor to the reformed doctrines. He went to Wittenberg, and became a celebrated preacher, reformer, and author.† During this same period, in 1540, there was a papal legate at the Council of Trent, by the name of Corvinùs.‡ Skipping over the period of one generation, a John Corvinus again appears, born about 1560, who had a son, Arnold Corvinus, born about 1590. This John Corvinus may be identical with Rev. John Corvinus,§ who was a preacher at Dantzic (northeastern Prussia) in 1621. The writer has been able to obtain but little information about this man, and would only add that the chronology is entirely suited to meet the facts, supposing he may have been the father of the American Matthias, as he was born about thirty years before the latter. If this be so, then Arnold Corvinus, the eminent jurist, may have been a brother of the American Matthias. But nothing positive is yet known on this subject.

In 1649, Arnold Corvinus published, at Amsterdam, Holland, *Digests of the Law, in Aphorisms.* Such a work required a man of remarkable legal learning. He had the degree of *Juris utriusque Doctor*, or doctor of both laws, the canon and the civil.‖ In his days, also, there appeared a Corvinus in the Synod of the Reformed Churches, held at Dordrecht, (Dort,) Holland, in 1618, 1619.¶ Again, omitting a generation, we find a John Augustus Corvinus,** an engraver, who was born in 1686, and died in 1738.

---

* Bonfinius published a life of Matthias and a genealogy of the Corvini, dedicating the latter to John. (See Article Bonfinius, in Chambers's *Cyclopædia*.) Bonfinius is, perhaps, the same as Anthony Bonvise, a merchant of Luke, (Lucca,) Italy. He was living in London in 1526 and 1535, being referred to in letters of Sir Thomas More, who calls him "the most beautiful and choice of all his friends." (C. B. Moore.)

† His life was written in 1749, by Baring, in the German language. See Appendix E.

‡ See Smith's *Tables of Ecclesiastical History.*

§ See Mosheim's *Ecclesiastical History.* (vol. iii. p. 388,) in which it is said, "In 1621, John Rathman, a pious and not unlearned minister of the Gospel at Dantzic, (north-eastern Prussia,) was thought by John Corvinus, his colleague, and by many others, to derogate from the majesty and efficacy of the Holy Scriptures." For in 1621, Rathman, in a German work on "Christ's Gracious Kingdom," said that, in addition to the written word of God, God must give an internal word to the soul before a man could experience regeneration. Corvinus and his associates contended that this was "mystical," etc.

‖ The title of this work, a copy of which the writer has, is, *Arnoldi Ioh. Fil. Corvini, J.U.D. Digesta per Aphorismos strictim explicata. Editio secunda auctior. Amsterodami, apud Ludov. Elzevirium. MDCXLIX. Tribonius. Theophilus.*

¶ See *History Synod of Dort.* This may be the John Corvinus above alluded to.

** See Appendix F. It might also be remarked in passing that Pope Innocent IV., in 1245, and Saint Louis, in 1248, sent mendicant friars as missionaries among the Mongols, who had, in 1226, invaded Europe, under Ghengis Khan. And that, in 1289, John de Monte Corvino

He, of course, has nothing to do with the Matthias Corwin who came to America a half-century before his birth.

Thus, we find Corvini scattered from Hungary to the coasts of Holland during the century and a half succeeding the death of the renowned Matthias, (1491–1649.) We also know, from the general histories of the reformation period, that numbers of Hungarians fled to England. We know, in addition, that the descendants of Matthias Corwin, who came over with the Puritans, about 1630, claim to be of Hungarian origin, pointing even to the good and learned Matthias of Hungary as their ancestor. The writer here leaves this branch of the subject to those who may care to prosecute it further.*

There are a few traditions, also, which refer the Long Island Corwins to a German and even to a Welsh origin. (See Jesse, 1.) The tradition of the German origin can easily be explained, as many of the Corvini lived in Germany.† For a similar reason, they might be called English. The idea of a Welsh origin, perhaps, arose from noticing the town called Corwen, in Merionethshire, Wales. Upon inquiry, by letter, of the parish clergyman of that place, as to the origin of the name of that town, and whether any Corwin families now lived there, or ever had, the following courteous reply was received:

The Rectory, Corwen,‡ Wales, }
December 29, 1870. }

Rev. and dear Sir: I have much pleasure in answering your letter about the supposed meaning of the word Corwen. The received opinion

translated the New Testament and Psalms into the Tartar language. (Appleton's Cyclopædia, Art. Foreign Missions, vol. xi. p. 569.)

* The writer was also assured, several years since, by a member of the family, that Kossuth, when in this country, in 1848, alluded in some of his speeches to the emigration of descendants of Matthias Corvinus to America. In the same year there was also one of those not uncommon excitements in some portions of the family about a European estate awaiting the heirs of Corvinus. It subserved, at least, the good purpose of reviving and strengthening fast-dying traditions. See some letters on the subject in Appendix G.

† Daniel Corwin, of Howells, Orange Co., N. Y., informed the writer that a woman of the name of Corwin, ninety years of age, informed him, (while on a visit to Southold, about 1810–20,) that the Corwins were of German descent. She herself was a native of Germany.

‡ The Celtic Caer, or Car, is also said to mean camp, or fort. (Webster's Dictionary, p. 1626.)

Corwen, Merionethshire, Wales, is a town, and the seat of a poor-law union, in the Parish of Corwen, and in the Hundred of Edernion. It is situated on the right bank of the Dee. Population in 1851, 2069. The living is a sinecure rectory in the archdeaconry of Montgomery and diocese of St. Asaph. Corwen poor-law union contains 15 parishes and townships, and a population of 15,409. It is a small, neat town. Its situation above the river imparts to it a pleasant appearance, and also contributes to its celebrity. The town is supplied with water. Corwen is regarded with interest by the Welsh, as the scene of victories over the English—one in 1165, by Owen Gwynedd, over Henry II., and the other by Owen Glyndwr, over Henry IV. In the vicinity of Corwen, a British or Welsh post existed. It consisted of a circular wall, one and a half miles in circumference, on the summit of a steep hill, and a circular habitation, now in ruins, within the inclosure. The parish church, a neat cruciform building, stands in a picturesque situation, immediately at the foot of a rocky precipice, forming part of the Berwyn mountains. The Calvinistic and Wesleyan Methodists, Baptists, and Independents, also here exist. There is a parochial school of about fifty

here is, that Corwen is so called from the Welsh words *Caer-wen*—that is, white stone, descriptive of the rocks at the back of the town; or from *Cor-wenna*, a daughter of Brannas, a Welsh prince who lived in the neighborhood; or from *Cor-owain*—that is, Owen Glyndwr's choir. No family of the name of Corwin live anywhere in these parts.

I am, Rev. and dear sir, your faithful servant,

The Rev. E. T. Corwin. William Richardson.

.

MATTHIAS CORWIN, THE AMERICAN IMMIGRANT.

Among the earliest settlers of the Massachusetts Bay Colony, we find a Matthias Corwin, (pronounced Currin.) The *Commoner's Record*, at Ipswich, yet preserved, says, "Given and granted to Matthias Currin, two acres of land, lying unto his house, on the east end thereof, to him, his heirs, and assigns," etc. This was in 1634, and seems to be a second grant of land. The name in the same records is also spelled Curwin, and they note, concerning him, that he finally removed to Southold, Long Island.* The first grant was probably the fifty or two hundred acres given to all the first settlers.† It is said that he came from Warwick, England.‡

These settlers of Ipswich came under the lead and governorship of John Winthrop. That the circumstances of that settlement may be understood, permit a condensed statement to be given from Bancroft's *History of the United States:*

Rev. Mr. White, a minister of Dorchester, in the south of England, a Puritan, but not a separatist, breathed into the attempted settlement at Cape Ann (1624) a higher principle than the mere desire of gain. Roger Conant obtained the agency of the adventure, (1625.) The attempt at Cape Ann failed. But Conant, confiding in the active friendship of White, made choice of Salem as a convenient place of refuge for the exiles for religion, and he and his companions remained as the sentinels of Puritanism on Massachusetts Bay.

The desire of a plantation was now ripening in the mind of White and his associates in the south-west of England. About the same time, some friends in Lincolnshire fell into discourse about New-England, (1627;) and

.

children, and also a reading-room. Corwen possesses an endowed hospital for widows of clergymen. (See Parry's *Cambrian Mirror;* Cliffe's *Book of North Wales;* *Land we Live in,* vol. iii.; Zell's *Encyclopedia.*)

* These facts are also all presented in Felt's *History of Ipswich.* The writer visited Ipswich in July, 1870.

† Each adventurer who took £50 stock in the Massachusetts Company was to receive 200 acres, or 50 acres if he took no stock, but simply transported himself and family. Fifty additional acres, however, were given for every person, additional to the family, whom any of the immigrants brought over.

‡ This tradition was brought to light in the excitement, in 1848, about the Austrian estate. The writer is not aware of the ultimate authority of the tradition, and can not pronounce either for or against it. (See letter of Irena Husted, in Appendix G.)

from London, Lincolnshire, and the *west-country*, men of fortune and religious zeal, merchants and country gentlemen, offered the help of their purses to advance the glory of God, (1628,) by planting a colony of the best of their countrymen on the shores of New-England. To facilitate the grant of a charter from the crown, they sought the concurrence of a Council of Plymouth, (England,) for New-England; they were befriended in their application by the *Earl of Warwick*, and obtained the approbation of Sir Ferdinando Gorges; and on March 19th, 1628, the Council conveyed to Sir Henry Roswell, Sir John Young, Thomas Southcoat, John Humphrey, John Endicott, and Simon Whetcomb, a belt of land extending three miles south of the River Charles, and three miles north of every part of the River Merrimack, from the Atlantic to the Pacific Ocean. The grantees associated to themselves Sir Richard Saltonstall, Isaac Johnson, Matthew Cradock, Increase Nowell, Richard Bellingham, *Theophilus Eaton*, William Pynchon, and others; of whom nearly all united religious zeal with a capacity for vigorous action. Endicott was made governor. He arrived in September, 1628, and united his own party and those who had formerly been planted there into one body, amounting to fifty or sixty persons altogether. With these he founded Salém. Thomas Walford, a blacksmith, already dwelt at Charlestown, and William Blackstone, an Episcopal clergyman, a courteous recluse, lived on the opposite peninsula, (Boston,) while Samuel Maverick, himself a prelatist, son of a pious Nonconformist minister, of the west of England, was at East-Boston; all within the bounds of the new Massachusetts Bay Colony.

After the departure of the emigrant ship from England, (1628,) the company, counseled by White, an eminent lawyer, and supported by Lord Dorchester, obtained from the king a confirmation of their grant. On the second of March, 1629, a new offer of "Boston men," that promised good to the plantation, was accepted; and on the fourth of the same month, the broad seal of England was put to the letters-patent of Massachusetts Bay. The freedom of Puritan worship was the purpose and the result of the colony.

The company was authorized to transport to its American territory any persons, whether English or *foreigners*, who would go willingly, would become lieges of the English king, and who were not forbidden.

In April, 1629, the new embarkation was far advanced. The company was directed to make plentiful provision of godly ministers, as the propagation of the Gospel was their professed aim in settling. One of these was Samuel Skelton, of Clare Hall, Cambridge, under whose preaching Endicott had sat. Another was Francis Higginson,* of Jesus College, Cam-

---

* Revs. Higginson, Skelton, and Bright came over to America in 1629, (being silenced ministers in England,) and with them came over sundry honest and well-affected people, in several ships, which were engaged to transport planters to New-England, all of whom arrived alive and safe at Naumkeag, (Salem,) Mass., intending to settle a plantation there. (Hubbard's Hist. New-Eng., 112.)

Higginson came over in ship Talbot, (Captain Thomas Beecher,) and he is called the father and pattern of the New-England clergy. His relation of the voyage is printed in

XX   INTRODUCTION.

bridge. Deprived of his parish in Leicester for nonconformity, he received the invitation to conduct the emigrants as a call from heaven. Two other ministers were added as missionaries to the Indians.

The party included also six shipwrights, and an experienced surveyor, who, with Samuel Sharpe, master gunner of ordnance, was to drill the company at appointed times. A great store of horses, cattle, and goats was put on ship-board. Higginson called his children, as the ship was receding from Land's End, and bade them look, for the last time, on their native country, not as the scene of sufferings from intolerance, but as the home of their fathers and the dwelling-place of their friends. They did not say, "Farewell, Babylon ! Farewell, Rome!" but " *Farewell, dear England !*" " On the Sabbath, they added preaching twice, and catechising; and twice they faithfully kept solemn fasts. The passage was quiet and Christian-like ; for even the ship-master and his religious company set their eight and twelve o'clock watches with singing a psalm, and with prayer that was not read out of a book."

In the last days of June, 1629, the little band of two hundred arrived in Salem. The old and new planters numbered about three hundred, besides women and children. They believed themselves to be chosen emissaries of God, the favorites of heaven, selected to light in the wilderness the beacon of pure religion. Mr. Skelton was chosen pastor, and Mr. Higginson, teacher, and were consecrated by the imposition of hands of some of the flock. Mr. Higginson wrote a glowing description of New-England, which went through three editions in a few months, and many letters were sent back to friends by the colonists. While former attempts to colonize, from motives of gain, had failed, this effort could not, since their object was purity of religion. They purposed to form a peculiar government, and to colonize only *the best.*

In July, 1629, Matthew Cradock, the governor of the company, proposed to transfer the government of the company to the emigrants. This led several persons of worth and quality, wealthy commoners, zealous Puritans, to think about casting their destinies with the American colonists. Twelve men, of large fortunes and liberal culture, on the 26th of August, 1629, among whom were John Winthrop, Isaac Johnson, Thomas Dudley, and Richard Saltonstall, offered to bind themselves to each other, that if the government of the colony were thus transferred to America, they would emigrate. On the 29th, the transfer took place. Thus the commercial corporation became the germ of a commonwealth. John Winthrop was chosen governor. Of the stock of the company, nine tenths was sunk in the first year. Cradock furnished two ships. The Eagle was purchased, and its name changed to Arbella.* From the resources of the emigrants,

Hutchinson's Collection of Papers. 1630. London. Neal's Puritans, i. 300, 301. Harper's edition, N. Y. 1843.

* Mr. Drake, in his early *History of New-England*, gives the passenger-lists of about fifty or sixty ships which brought some of the early settlers to New-England; but the name of Matthias Corwin is not found among those yet published. We understand that additional passenger-lists of the Puritan emigrants are about to be published in England.

seventeen vessels, during 1630, brought over about a thousand souls. About seven hundred others, not conformists, yet not separatists, many of them men of high endowments and large fortunes; scholars well versed in the learning of the times; clergymen, who ranked among the best educated of the realm, embarked with Winthrop in eleven ships, bearing with them the charter of their liberties. They arrived in June, 1630, at Salem, and afterward settled at Boston. The *west-country men, who, before leaving England, had organized their church, with Maverick and Warham for minis- ters, and who, in a few years, were to take part in calling into being the commonwealth of Connecticut,* were found at Nastasket, where they had landed just before the end of May. Some of the emigrants remained in Salem; some founded Lynn. William Coddington, of Boston, England, settled on the peninsula, since called Boston. Malden, Watertown, Rox- bury were also founded. Ludlow and Rossiter, with the men from the *west of England,* settled South-Boston.

On the 5th of February, 1631, the ship Lyon, from Bristol, arrived, laden with provisions, and twenty passengers. In 1631, only ninety came over. In 1632, only two hundred and fifty. In April, 1631, Roger Williams began to preach at Salem. In the same month, the Indians of Connecticut invited a colony to their river. In July, 1633, the ship Griffin brought over two hundred passengers. Among them was Haynes, Cotton, and Hooker. In 1634, the bay of Massachusetts was thronged with squadrons.[*]

Now, some of these settlers came over from London, Lincolnshire, and the *west-country,* about 1630. These persons are expressly said to be those who afterward took part in the settlement of Connecticut, whither Matthias Corwin went in 1640, though these *west*-countrymen at first remained for a while in South-Boston. The *west-country* might 'refer to Warwick, especially as the Earl of Warwick was a special friend of the enterprise. The company was also to convey either Englishmen or for- eigners. There is, therefore, some probability that Matthias Corwin came over as early as 1630, while we find his name among the inhabitants of Ipswich in 1634. His name is not in the imperfect list of the first ten [†] who went there in March, 1633; but in 1634, he received a second grant of land. He was a freeman of the place.

John Winthrop, Jr., purchased Agawam, (the Indian name of Ipswich,) of the Indian chief Masconnomet, for £20, in 1638. The numbers of this settlement were soon increased by additions from abroad. Rev. Nathaniel Ward,[‡] persecuted by Bishop Laud, arrived in June, 1634, in which year "fourteen great ships arrived at Boston and Salem." The passengers of at least two of these ships were expressly designed for Agawam. They

---

[*] See Bancroft's interesting account in his Hist. U. S., vol. i., 339–382. Palfrey, in his *History of New-England,* gives a still more minute account. Vol. i., chaps. viii.-xv.

[†] The original list gives the names of Mr. John Winthrop, Jr., Mr. William Clark, Robert Coles, Thomas Howlet, John Biggs, John Gage, Thomas Hardy, William Perkins, Mr. John Thorndike, William Serjeant. Felt's Ipswich, p. 11.

[‡] A very excellent memoir of Ward was published by John Ward Dean in 1868. Munsell, Albany.

arrived in August, and at once changed the name of the place to Ipswich,* because of the great kindness which they had received at Ipswich, England, when embarking for their distant home. Rev. Mr. Ward was chosen preacher, and Rev. Mr. Parker,† who had already been preaching there, was chosen teacher. The next year, the latter removed to Newbury, and Revs. Thomas Bracey and John Norton took his place. In 1636, Mr. Ward resigned on account of ill health, but continued to reside in Ipswich; and Rev. Nathaniel Rogers succeeded him in February, 1637.

Ipswich was, at this time, one of the most intellectual and refined towns in the colony. An unusual proportion of the people were persons of wealth and education. Four members of the colonial government resided here, namely, Bellingham, Saltonstall, Bradstreet, and Symonds. Ipswich was also one of the most beautiful of the sea-board towns, resembling the charming rural scenery of Dorsetshire. Mr. Ward, on returning to England in 1646, says that in the twelve years of his residence there, he had only heard one oath, had only seen one man drunk, and had only heard of three bad women.

Cotton Mather, in 1638, at the time of Mr. Rogers's ordination at Ipswich, says, "Here was a renowned church, consisting mostly of such illuminated Christians, that their pastors in the exercise of their ministry might, in the language of Jerome, 'Perceive that they had not disciples, so much as judges.'" Johnson remarks, 1646, "The peopling of this town is by men of good rank and quality, many of them having the yearly revenue of much land in England before they came to this wilderness. But their estates being employed for Christ, and left in bank, they are well content till Christ shall be pleased to restore *it* again to them or theirs, which, in all reason, should be out of the prelates' hands in England."

They had been so greatly persecuted by the prelates that they were forced to emigrate. Some of these pilgrims were merchants. Many of them were among the most estimable of their parishioners, whom the clergymen brought over with them from England.‡

Ipswich increased rapidly in wealth. In 1633, the town was assessed only £8, but the next year £50. After this, however, they were yet obliged to use bullets for farthings. Persons were not allowed to use tobacco under a penalty of 2s. 6d. §

But no sooner was a settlement formed, than it began to send out emigrants to form new ones. This process began in Ipswich in 1635. In 1641, several families in Lime and Ipswich, "having proposed to inhabit

---

* The town of Ipswich, England, derives its name from the river Gipping, and was, in the time of the Saxons, called *Gippeswic*, now corrupted into Ipswich. This town suffered from persecution during the reformation. Its history was written by John Glyde, Jr., and published at Ipswich, England, 1850.

† In May, 1634, Rev. Thomas Parker, of Wiltshire, England, became a resident of Ipswich. He preached much that the "emigrants had come over for good reasons, and that God would multiply them, as he did the children of Israel." He remained only one year. (Felt's Ipswich, 216.)

‡ Felt's Ipswich, p. 38.   § Felt's Ipswich, pp. 40, 69, 72.

Long Island, their leaders are called before the General Court, and persuaded from proceeding any further, because it would strengthen the Dutch, whom Winthrop called 'doubtful neighbors.'" Matthias Corwin had left in the preceding year.[*]

In 1637, the Pequot war occurred in Connecticut, causing much suffering to the first settlers there, who had gone from Massachusetts. As early as 1631, Seguin, the Indian Sagamore of the Connecticut Valley, requested Governor Winthrop to send a colony thither. Individuals from Devonshire, Dorsetshire, and Somersetshire, England, had located at Dorchester, Mass., in 1630, under Rev. John Warham as pastor, and Rev. John Maverick as teacher. Roger Ludlow and Henry Wolcott were in this company. Sir Richard Saltonstall's people, the same year, settled at Watertown, Mass., under the care of Rev. Mr. Philips. In 1632, a congregation, under the lead of Rev. Thomas Hooker, of Chelmsford, Essex, England, settled at Cambridge, Mass., and were joined by certain others from Weymouth, Mass. These parties applied, as early as 1634, for permission to remove to Connecticut; but their request was denied. The next year, their renewed request was reluctantly granted.

Thus Hartford, Windsor, and Wethersfield were settled in the spring of 1636. But the most distinguished company of emigrants that ever came to New-England, arrived in Boston, from London, July 26, 1637. Their pastors and leaders were Rev. John Davenport, a preacher of London; Governor Theophilus Eaton, a wealthy merchant, and others. They proceeded to Connecticut, and founded New-Haven, April 18th, 1638. In January, 1639, they framed a written constitution, the first example of such a thing in history. "There, by the influence of Davenport, it was resolved that the Scriptures were the only perfect rule of a commonwealth. A committee of twelve was selected to choose seven men qualified for the foundation-work of organizing the government. Eaton, Davenport, and five others were 'the seven pillars' for the new house of wisdom in the wilderness. As neighboring towns were planted, each was likewise a house of wisdom, resting on its seven pillars, and aspiring to be illuminated by the Eternal Light. The pleasant villages spread along the Sound, and on the opposite shore of Long Island." [†] About this time, or shortly after, Matthias Corwin left Ipswich, and came to New-Haven.

At New-Haven arrived soon after Rev. John Youngs, a minister from Hingham, Norfolkshire, England, with a part of his congregation. On October 21st, 1640, he reorganized his church, and removed to Southold, [‡]

[*] One very direct tradition says he went first to Saybrook, in Connecticut, about the time that the Rev. Mr. Hooker settled Hartford. Felt says (p. 72) that it is unknown what became of many of these emigrants from Ipswich, though he tells what became of Corwin. So also Farmer's Register.

[†] Bancroft's U. S. i. 404. In Felt's Ipswich, p. 55, we read that Rev. Mr. Davenport also directed the brethren at Ipswich to nominate eleven of their most godly men for church pillars, out of which the seven pillars of wisdom were to be chosen.

[‡] Although Rev. Mr. Youngs and his colony came from Hingham, Norfolkshire, yet their

Long Island, where Captain Howe and other Englishmen had purchased of the Indians a large tract of land. * This little colony remained in union with New-Haven, though they reluctantly yielded to the rule that none but church members should hold office. Says Trumbull in his *History of Connecticut*, vol. i., p. 119, "Some of the principal· men (of Southold) were Rev. Mr. Youngs, William Wells, Barnabas Horton, Thomas Mapes, and Matthias Corwin." † The other parties who made up this Southold colony were Peter Hallock, Richard Terry, Robert Akerly, Jacob Corey, John Conkline, Isaac Arnold, and John Budd. ‡

Says Hinman in his *Settlers of Connecticut*, p. 726, "Matthias Corwin was one of the leading men of Southold in its first settlement. . . . It was first called Yennecock. . . . Many of the first planters came with Rev. John Young, from Hingham, Norfolkshire, England. Mr. Young stood at the head of the civil and religious affairs,§ aided by Corwin, Wells, Tuthill, Horton, and others of his church. The name of *Corwin* was not strictly a Connecticut name, but only at the time Southold was under the jurisdiction of the Connecticut colony." We see from the above facts that Matthias Corwin took part in the settlement of at least two towns in New-England, namely, Ipswich and Southold, and perhaps three. For it is not certain whether the reference in Hollister's work,‖ which places him among the founders of New-Haven, refers to New-Haven proper or to Southold, which was then in the colony of New-Haven. He had lived about six years at Ipswich, and he spent the eight remaining years of his life at Southold. His will may be found in the town records of that place. He was also, at times, a director of town affairs. In a description of his pro-

new town was called Southold, and their county Suffolk, no doubt after Southwold, Suffolkshire, England. The band is said to have sailed from Yarmouth, Norfolkshire.

The following works are on Southwold, Suffolk, England :

Suffolk, Southwold, and its Vicinity, Ancient and Modern. By Robert Wake, M.R.C.S.L. With maps and plates. Thick octavo. Yarmouth, 1839. 7s. 6d.

Another : *Suffolk, Southwold, and its Vicinity ; an Historical, Antiquarian, and Picturesque Guide for the especial use of Visitors.* With plate. Small octavo. 1844. 1s. 6d.

These works are noticed in Mr. James Coleman's catalogue of books, 22 High street, Bloomsbury, London, W. C.

* They made this purchase in behalf of Connecticut. The tract extended from the eastern part of Oyster Bay to the western part of Holmes Bay, and to the middle of the great plain. It lay on the north side of the island, and extended about half-way across. (Trumbull's Conn., vol. i. p. 119.)

† This passage from Trumbull is quoted by Barber in his *History of New-Haven*, and by Lambert and Hinman in their histories of Connecticut.

‡ John Ketcham became a citizen of Southold in 1648. He also was from Ipswich, like Matthias Corwin. (Moore's Southold, p. 253.)

§ The following are the pastors at Southold and Mattituck until the Revolution : Rev. John Youngs was pastor at Southold, 1640–72 ; Rev. Joshua Hobart, from Boston, 1674–1717 ; Rev. Benjamin Woolsey, 1720–36 ; Rev. James Davenport, 1738–46 ; Rev. William Troop, 1748–56 ; Rev. John Storrs, 1763–76 ; again, 1782–87. At Mattituck, Rev. Joseph Lamb, 1717–49 ; Rev. Joseph Parks was pastor, 1752–56 ; Rev. Nehemiah Barker, 1756–72 ; Rev. Jesse Ives, 1772–3 ; Rev. John Davenport, 1775–7 ; Rev. Benjamin Goldsmith, 1777–1810.

‖ In *The First Planters of New-Haven*, by Hollister, vol. i., p. 506, occurs the name of "Curwin."

perty three years before his death, (1655,) no less than nineteen plots of land are described as belonging to him, situated in Southold, on the northern shore, on Tom's Creek, toward the north-west and the north-east, at Oyster Pond, toward the south-west, at Peehaconnicke River, and at Corchack, (Cutchogue.) His will mentions John,* Martha, and Theophilus as his children, all of whom seem to have been of age at the writing of his will in 1658. Hence they were probably born at Ipswich, before 1637, or possibly some or all of them in England still earlier. The families of the two sons were large, embracing seven or eight children each, all of whom continued to reside at Southold or immediate vicinity. The names of John's children are positively known by his will, which has also been found.† The other names therefore of the third generation, which are recorded in the census list of 1698, must belong to the family of Theophilus. Most of these removed eight or ten miles west of Southold, to Mattituck.

In the fourth generation removals from the Island began to be made, though to a very limited extent, until the breaking out of the Revolution. Before that event, however, Amaziah, 1, had removed to Maryland, about 1750 ; Jesse, 2, to Connecticut, about 1760 ; Theophilus, 4, to Orange Co., N. Y., about 1760 ; Gilbert, 1, to Rockland Co., N. Y., about 1768 ; while David, 2, in his old age, accompanied his children to Orange Co., N. Y., about the opening of the war. But with the Revolution removals became frequent. That really broke up the family on Long Island.

In April, 1775, a meeting was held at Southold, to secure the signatures of those who would support Congress. The list is preserved and printed in the Calendar of Revolutionary Papers. In May, the paper was carried around to get the signatures of those not present at the meeting. About 223, in the little town of Southold, L. I., agreed to support Congress, while only 40 declined. Among those who signed were most of the Corwins. (See *Index, Revolution.*)

After the battle of Long Island, 1776, great consternation seized the people of Suffolk County. The American army being obliged to abandon the island, the more prominent Whigs of Suffolk County fled across the sound to Connecticut, carrying with them what they could, leaving their houses and farms to the enemy. The convention aided the removals. Many of these joined the American army. Some crossed over to the Hudson River and settled in Orange County, N. Y., while others afterward returned to the Island. (Prime's Long Island, p. 65. Onderdonk's *Revolutionary Incidents in Kings and Suffolk Counties, L. I.*)

Among those who never returned were James, 1, who settled near Middletown, Orange Co., N. Y. ; William, 1, his brother, and a Benjamin, who settled near Chester, Morris Co., N. J. ; Joshua and Eli, with their father

---

* Several of the inhabitants of Southold consented to be made free of this colony, unless any thing appear to interrupt the same. Among these was John Corwin, 1662. (*Col. Rec. of Ct.*)

† This was found by Charles B. Moore, Esq., of New-York, and kindly furnished to the writer.

David, who settled near Scotchtown, Orange Co., N. Y., while their brother Phineas removed to Central New-York; Stephen went to Essex Co., N. J., near Springfield, and subsequently to Ohio; while Jesse, already in Connecticut, had, about 1767, removed to the vicinity of Flanders, Morris Co., N. J., and about the opening of the war proceeded to Fayette Co., Pa., and in 1789 to Bourbon Co., Ky.

Since this first general scattering, the migrations have continued in every direction for a century, but the reader is referred to the particular names for further information. New-York State has always been the chief home of the family, especially the counties of Suffolk, Orange, Sullivan, Ulster, Cayuga, and other counties in Central New-York. New-Jersey has had the next share, perhaps, especially Morris County. Ohio stands next, until now fully three fourths of the States have members of this family for citizens. They are found in each of the New-England States, excepting Maine, and in all the others, except, perhaps, Utah, Nevada, Delaware, and West-Virginia.* (See *Index*.)

Not a few college graduates, clergymen, lawyers, and doctors are also found in this family record. Judges are here, and legislative members in various States. One has been a governor, a member of the President's cabinet, and an ambassador to another nation, while others have been members of Congress. (See *Index*.)

THE ORIGIN OF THE FAMILY OF GEORGE CURWEN.

Captain George Curwen came from Northampton, England, and settled at Salem, Massachusetts, in 1638. He was descended from the ancient Curwen family † of Workington, ‡ Cumberland, England, bringing over with him memorials of such descent, such as a seal with the Curwen arms, etc. The English family have written their name Curwen since about 1433. Before that time, it was Culwen, which name was assumed about 1140, from a Scotch source. Hutchinson, in his *History of Cumberland County*, gives the records of this family back to the reign of Elthelred, about A.D. 870. John de Talbois is the earliest ancestor named, and his descendant Gospatric, of the fifth generation, was the first Lord of Workington. He was thus named from his maternal grandfather. Gospatric's son, Thomas, received by gift, from one Rowland, the lordship of Culwen.

---

* The name has been incorporated in the geography of the country, as follows:

> Corwin, a township of Ida Co., Iowa. Population, 200.
> Corwin, a village of Warren Co., O.
> Corwinville, a village of Lorain Co., O.
> Corwin, a railroad station in Morris Co., N. J.
> Curwinville, a village in Clearfield Co., Pa. Population, 455.
> Corwin, a village in ―― Co., O.
> New-Corwin, Highland Co., O.

One of the steamers of the United States marine is called "Corwin," after Governor Corwin.

† For a sketch of this English family, see Appendix II.   ‡ See Appendix I.

CURWEN HOUSE, ENGLAND

Thomas died in 1152, and *his* eldest son, Thomas, before his father. In the mean time Patric, the second son, had received from his father the lordship of Culwen, and married the heiress of Culwen. In a cartulary (or church record) of St. Begh's, in the Earl of Oxford's library, "Thomas, son of Gospatric, gives, among other benefactions," to Patric, his son, "*Salinum in Culwen*," or a salt-pit in Culwen. Here Patric expected to make his home. But his elder brother dying without issue, Patric removed to Workington Hall, retaining, however, his new name, Patric de Culwen, which his posterity (lords of Workington) changed into Curwen.[*]

The origin and meaning of Culwen is somewhat under dispute. Some think that it comes from Culdees. Mr. Toland says, "That the old Culdees were a sort of *lay religious*, who had the power of electing their own bishops or superintendents, and that they were so named from their original Irish or ancient Scottish word *ceilede*, signifying *separated or espoused to God*." [†] But George Buchanan, well versed in Irish lore, explains the word by *Dei Cultores*, or worshipers of God.

The short story of these monks is, that they were of the Irish rule, carried into Scotland by St. Columba, and thence dispersed into the northern parts of England. They were so named from the black habit [‡] which they wore; for *Culdee*[§] signifies as plainly *a black monk*, from the color of his hood or cowl, as *Culwen* signifies *a white one*, from the white cowl worn. John Leland mentions two rivers, illustrative of the meaning of these words, *Clar-duy*, or black clar, and *Clar-wen*, or white clar.[‖]

This Culwen was on the sea-coast in Galloway, (county Kirkcudbright, Scotland,) and had its name from a neighboring rock, which was thought to resemble a white monk.

This Curwen family, in the direct line of eldest sons, has several times failed, when some younger branch succeeded. A little less than a century ago, with the death of Sir Henry Curwen, a single daughter, Isabella, was

---

[*] It is sometimes found in the form of Curwenne. Hugh Curwen, Archbishop of Dublin, wrote it also Corin. His nephew, Philip, the converted friar, wrote it Corwine. The celebrated Irish barrister wrote it Curran. (See Appendix II.)

[†] Hutchinson's Cumberland, ii. p. 144.

[‡] The dark-attired Culdees
    Were Albyn's earliest priests of God,
    Ere yet an island of her seas
    By foot of Saxon monk was trod.
                    Campbell.

[§] The word *Culdee* is explained in many different ways. The least satisfactory is its derivation from the Latin *Cultores Dei*, or worshipers of God. Culdee was, no doubt, Latinized into Colidei. *Cul*, however, may be etymologically connected with the Latin *cucullus*, a hood or cap, and with the provincial German *Kogel, gugle*, and Anglo-Saxon *cuhle, cugle, cugele*, all of the same signification, namely, a cowl. Ebrard gives Kile De, a man of God. Braun gives *Gille* De, a servant of God. A later authority gives Cuildich, (as used by the Celts,) a secluded corner, referring to the life of a recluse. Culdee became at length a general term for the whole Celtic church. These people had a simple Bible Christianity, and seem to have been models of piety in every respect. They anticipated Augustine in introducing Christianity into Britain. Their history has been frequently written. (See Herzog's, McClintock's, and other cyclopedias on *Culdee* and *Iona*.)

[‖] Hutchinson's Cumberland, ii. p. 144.

left heiress of all her father's vast estate. She married John Christian, Esq., of the Isle of Man, who, upon receiving a conveyance of the estate in 1790, assumed the arms and name of the Curwens, and has left a numerous family. The present head of the family, in England, is Edward Stanley Curwen.* There were also Curwens in Yorkshire, England, in 1612 and 1665. Eight names are mentioned in a list of pedigrees contained in William Paver's *Consolidated Visitations of Yorkshire*, being taken in the above years. (*N. E. Ant. and Gen. Reg.* xi. p. 265.)

GEORGE CURWIN, THE AMERICAN IMMIGRANT.

Captain George Curwin or Corwin,† who was born in England, December 10, 1610, settled at Salem, Massachusetts, in 1638. His father's name has recently been discovered to be John, (see Preface, p. 3,) but the links uniting them with the Workington family have not yet been ascertained. But Sir Christopher Curwen, who died in 1491, besides his eldest son, Sir Thomas, had by a second wife five sons, namely, John and William, who were priests, Robert, Edward, and George. Their mother was Anne, daughter of Sir Roger Pennington, of Pennington and Mancaster. The dates of their births would, therefore, be about 1450–70. Now, Robert was a merchant of Workington, and traded largely with Virginia, (that is, America,) and possibly some of his descendants may have settled there.‡ The name of Rev. Thomas Curwen is found in Bridges's county history of Northampton, in the century preceding the emigration of George to Ame-

---

* Burke, in his Cyclopædia of Heraldry, gives no less than five families of the name of Curwen, and one of the name of Curwin, with their arms, as follows :

(1) Curwen, of Workington, County Cumberland.
    Ar.* fretty gu. a chief az. Crest—a unicorn's head erased arg. armed or. Motto, Si je n'estoy.
(2) Curwen, of Cumerton, County York.
    Ar. fretty gu. a chief escallop ar.
(3) Curwen, Ar. fretty sa. a chief or.
(4) Curwen, Ar. fretty gu. or a chief az. three escallops or.
(5) Curwen, Ar. fretty of six and a chief az.
(6) Curwin, Ar. fretty sa. a chief az.

† Captain George, the first settler of the name at Salem, generally wrote his surname Corwin, but always used the arms of the Curwens, of Workington Hall. His son, John, wrote Corwin and Curwin, while his brother, Jonathan, wrote Corwin. George, son of this John, wrote Corwin and Curwen. Rev. George, son of Jonathan, wrote Corwin and Curwin; while Rev. George's sons, Samuel and George, wrote Corwin during their minority, but afterward Curwen, to which latter orthography their branch of the family have adhered ; but the descendants of John have adhered to Corwin or Corwine. (*George R. Curwen*, Salem, Mass.)

In N. E. Hist. Reg. ix. 368, the name of George, the first settler, is once Corwine.

‡ This is a suggestion of Rev. Alfred F. Curwen, now of Harrington, England, (son of Edward Stanley Curwen,) who communicated the above names, which are not found in the county histories.

* A brief explanation of these heraldic terms may be seen in Webster's Dict., p. 1712.

rica, and who may be George's grandfather.* George died in 1685, at Salem, leaving a large estate. [GEORGE, (1).]

He came to America, like Matthias Corwin, under the auspices of the Massachusetts Bay Colony. He and his descendants were united in marriage with some of the principal families of the colony. The branch of his family which descended from his son Jonathan became extinct, as to male heirs, about the beginning of this century, when Samuel Curwen Ward, (grandson of Richard Ward and Mehetable Curwen,) at the request of Samuel Curwen, (a great-uncle, who died in 1802,) assumed the name of Curwen, by an act of the Legislature of Massachusetts, and his descendants are now the only individuals bearing this name (it is believed) in New-England.

The descendants of his son John removed to Hunterdon County, New-Jersey, in 1717, and some of them remain there, or in the neighboring counties of Mercer and Somerset, to this day. At the close of the Revolution, a portion of this New-Jersey family removed to Mason County, Kentucky, where they met descendants of the family of Matthias Corwin, who had shortly preceded them. Thence they scattered to Ohio, and many other States. This family has furnished several judges, lawyers, and clergymen. (*See Index.*) The house which they occupied for many years in Salem, (yet standing,) was built by Roger Williams. In it he was tried, and in it, subsequently, some of the witch-trials were held. Samuel Curwen, a descendant of this family, was the well-known loyalist of the American Revolution, whose Journal and Letters have now been for thirty years before the public, and which give us the best account we have of men and habits, in England, during the American Revolution. It has gone through four editions, and is now out of print. [SAMUEL, (2).]

In the Broad Street Cemetery, at Salem, Mass., are found the following inscriptions on three sides of a small monument:

Here lie the remains of George Curwin, Esq., born in England, 1610, died at Salem, 1685; his widow, Elizabeth, only daughter of Governor Edward Winslow, and many of their descendants.	The tomb of Samuel Curwen, born in 1773, when the remains of his ancestors were removed to it, from a vault in the family ground.	Here lie the remains of Richard Ward, Esq., born 1741, d. 1824; his wife, Mehetable, daughter of George Curwin, Esq., and many of their descendants.

Josiah Winslow, son of Governor Winslow, mentions his sister, Elizabeth Corwin, in his will, 1675, (yet preserved,) giving her a pocket-watch, and asks Captain George Corwin to assist his executrix.

* "In the church of Luffwick was a chauntry for two priests, founded in 1498, by the last will of Edward, Earl of Wiltshire, and endowed with the manor of Culworth, and other lands in Bedfordshire. These lands were settled on THOMAS CURWEN and John Ross, the first chaplains, and their successors, by Robert Whittlebury and Thomas Montagn, the earl's feoffes. In 1535, 26 of Henry VIII., the profits arising to each chaplain were rated at 9*l*. 13*s*. 8*d*." This was the Hundred of Luffwick, (or Lowick?) in Northamptonshire. Each county was subdivided into sections called hundreds, comprising a hundred families or freemen. A chauntry (or chantry) was an endowed chapel, where one or more priests daily

In the Winthrop papers, John Curwen is mentioned by Lucie Downinge, in a letter, January 17th, 1661-2, p. 38.

In 1648, she mentions a Mr. Curwins, p. 55.

In 1670, Elizabeth Stone mentions a Mr. and Mrs. Curen, p. 103. (Mass. Hist. Coll., vol. i. Fifth series.)

Mrs. Anne Bradstreet, in 1712, gave, by will, to Mrs. Margaret Corwin, bedding, etc. (*N. E. Ant. and Gen. Reg.*, vol. xiii. 230.)

M. E. Curwen wrote *Sketches of the Campaign in Northern Mexico in 1846 and 1847.* New-York, 1853. 8vo. Maps.

Maskell E. Curwen wrote a *Sketch of the History of Dayton*, Ohio. (See his name printed, by mistake, Marshall, p. 148.)

A few years ago, (1867 or 1868,) appeared the following book notice. This Harry Curwen is apparently an Englishman :

" *French Love Songs, and other Poems*, selected and translated by HARRY CURWEN, (Carleton,) is a reprint of an excellent series of English translations from several of the most celebrated French poets, including Charles Baudelaire, Alfred de Musset, André Chenier, Victor Hugo, Gautier, Lamartine, Béranger, and others of less prominent note. The volume will have the charm of novelty to most readers in this country, whom it may initiate into a knowledge of a branch of French literature the best portions of which, as here exhibited, are equal in merit to the infamy of the worst specimens."

A certain George Corwin transcribed, in the year 1700, a copy of *Compendium Physicæ*, written by Rev. Charles Morton. This manuscript is now in possession of the Massachusetts Historical Society. (See Proceedings of 1868, p. 59.)

George R. Curwen,* Esq., of Salem, has several relics of the family, and portraits of early members, as well as the original voluminous manuscript (only partially published) of Curwen's Journal and Letters.

The following are the portraits :

1. Captain George Curwen, (1 ;) nearly full-length.
2. His daughter Hannah, (1,) wife of Hon. William Brown, Jr. ; half-length.
3. Rev. George † Curwen, (4 ;) half-length.
4. Hon. Samuel Curwen, (2,) son of Rev. George ; half-length.
5. Mrs. Abigail Curwen, wife of Hon. Samuel, (2 ;) half-length.

---

sung or said mass for the souls of the donors, or such as they appointed. (Bridges's *History of Northamptonshire, England*, vol. ii. p. 248.)

The fact that Thomas Curwen was a Roman priest does not necessarily include the fact of celibacy, as " The larger and better portion of the clergy had wives in the reign of Henry the First, and this with the monarch's approval." (Lyttleton's Life of Henry II., vol. iii. p. 42, 328.) Wilkins's Concilia, vol. iii. p. 277, says. "Priests continued to marry. to some extent. as late as the fifteenth century." (Extract from Palfrey's New-England, i. p. 102.)

* He corrected several errors in the Curwen genealogy, as formerly published in the N. E. Ant. and Gen. Reg., vol. ii. p. 50.

† A sermon on the death of Rev. George Curwin may be found in the Boston Public Library. (See in catalogue, Barnard, J. Shelf 17. No. 165.)

6. Mrs. Sarah Curwen, widow of George, (5,) the son of Rev. George, (4;) cabinet size.
7. Richard Ward, Esq., who married Mehetable Curwen, (1;) cabinet size.
8. Mehetable Curwen, (1,) daughter of George, (5,) and wife of Richard Ward ; cabinet size.
9. Sarah Curwen, sister of the last ; half-length.
10. Samuel Curwen Ward, (5,) son of Richard Ward and Mehetable Curwen, (1;) half-length.
11. The same ; cabinet size.
12. Daniel Ward, (1,) son of Richard Ward and Mehetable Curwen, (1;) cabinet size.
13. Richard Ward, (,) son of Richard Ward and Mehetable Curwen, (1;) cabinet size.
14. Mrs. Priscilla Barr Curwen, widow of Samuel Curwen, (6;) half-size.

The silver-headed cane and lace bands of Captain George Curwen, (1,) are also yet in existence, and in possession of George R. Curwen, of Salem, Mass. The ring in which Captain George Curwen, (1,) was painted can not be traced later than July 28th, 1802, when it was in possession of Daniel Ward, son of Mehetable Curwen, (1.) Mr. George R. Curwen, however, has an impression which was made from it. Other smaller relics also exist. Several volumes of Curwen papers, bound, are also in the Worcester Historical Library, some of them dating back to the first George.

SAMUEL CORWINE.

A person of this name is found settled in Essex and Old Norfolk, Massachusetts, in 1652. He may have been of the same English Curwen family as George Curwin, of Salem. He is known to have had a daughter Mary, but probably no sons. See his name and that of his daughter in their respective places. His name is also spelled Curwin.

THOMAS CURWIN.

He and his wife, Alice Curwin, were Quakers, and were in Boston, Massachusetts, in 1676. They were taken out of a church there, and whipped for their faith ! In the same year, a William* and Alice Curwin, Quakers, visited Shelter Island, Long Island, and New-York, having come from New-England. The laws were not then so severe against the Quakers as they had been twenty years before, but Thomas Curwin and his wife, Alice, seem to have suffered,† being imprisoned three days, and then whipped and

---

* From Onderdonk's Scrap-Book, found in New-York Historical Society's Library. This William may be a mistake for Thomas.

† In the Essex Institute Library, at Salem, Mass., there is a volume entitled " Truth and Innocency defended against falsehood and envy, and the martyrs for Jesus, and sufferers for his sake, vindicated ; in answer to Cotton Mather, (a Priest of Boston,) his calumnies, lyes,

set at liberty.   Alice Curwin appears to have written an account of her sufferings.   It is not known what became of them.   Probably they returned to England.

JOHN CURWEN.

He was of Keswick,* Cumberland, England, and, no doubt, of the ancient Curwen family of that county.   But, though an Englishman, he sympathized with the principles of the American revolution; and accordingly, after the war, in 1784, he emigrated to Philadelphia, Pennsylvania.   His son, John Curwen, bought land, and laid out the town of Curwinsville, Clearfield County, Pa., though the family have no interest there at present.

MALCOLM CORWIN.

He came from London to New-York in 1844, and died in a few years. He left no sons, but two daughters.

CHARLES H. CURWEN.

He is one of the younger sons of Edward Stanley Curwen, the present head of the Curwen family, at Workington, England.   He went to Denver, Colorado, about 1868, on business pursuits, and has subsequently removed to New-Orleans, Louisiana.

and abuses of the people called Quakers, in his late Church History of New England, with remarks and observations on several passages in the same, and his confessions to the just judgment of God on them. By John Whiting, London. Printed and sold by T. Sowle, in White Hart Court, in Gracious street, 1702."

The above is bound up in a volume entitled "New-England Judged, not by Man's, but the Spirit of the Lord, and the Sum sealed up of New-England's Persecutions. Being a brief relation of the sufferings of the People called Quakers, in those parts of America, from the beginning of the 5th month, 1656, (the time of their first arrival at Boston, from England,) to the latter end of the 10th month, 1660; Wherein the Cruel Whippings and Scourgings, Bonds and Imprisonments, Beatings and Chainings, starvings and huntings, fines and confiscations of estates, Burning in the hand, cutting off ears, Order of sale for Bondmen and Bondwomen, Banishment upon pain of death, and putting to death of those people, are shortly touched, with a relation of the manner, and some of the other most material proceedings, and a judgment thereon. In answer, To a certain Printed Paper, Intituled a Declaration of the General Court of the Massachusetts, holden at Boston the 18th Octob', 1658, apologizing for the same. By George Bishop, London. Printed in the year 1661, and now reprinted, 1702 | 3."

On page 101 is the following:

"Cotton Mather might well say, book v. page 95, c. 1, They have seen many instances upon which God might say to them, when I would have healed New-England, then its iniquity was but the more discovered, and did they not accordingly, not long after, (or in the year 1676,) come and forcibly drive Tho. Curwin and Alice, his wife, etc., out of their meeting, at Boston, all along the street, until they came to the prison or house of correction, whereinto they thrust them, which was of service to the truth, for many people, rich and poor, came to look upon them, and some were convinced; it being a time of great tribulation, their hearts failed for fear, and the third day of their imprisonment, they brought them down to the whipping-post; but the presence of the Lord was manifest, and bare them over their cruelty, and they could not but magnify the name of the Lord, and declare of his wondrous works, at which the Heathen were astonished, and shook their heads; and the next day they were set at liberty, and went to their meeting again, which was peaceable; for you saw its like would not do."

A marginal note says, "See Alice Curwin's relation of her labors, travels, and sufferings, p. 5." (Communicated by George R. Curwen, of Salem, Mass.)

* Keswick is described in Hutchinson's Cumberland, vol. ii. p. 153.

THE LANDING OF THE PILGRIMS.

THE breaking waves dashed high
  On a stern and rock-bound coast,
And the woods against a stormy sky
  Their giant branches tossed ;
And the heavy night hung dark
  The hills and waters o'er,
When a band of exiles moored their bark
  On the wild New-England shore.

Not as the conquerors come,
  They, the true-hearted, came ;
Not with the roll of the stirring drum,
  Or the trumpet that sings of fame ;
Not as the flying come,
  In silence and in fear—
They shook the depths of the desert's gloom,
  With their hymns of lofty cheer.

Amidst the storm they sang,
  And the stars heard, and the sea ;
And the sounding aisles of the dim woods rang
  To the ANTHEM OF THE FREE.
The ocean eagle soared
  From his nest by the white wave's foam,
And the rocking pines of the forest roared—
  This was their welcome home.

What sought they thus afar ?
  Bright jewels of the mine ?
The wealth of seas, the spoils of war ?
  They sought a faith's pure shrine.
Ay, call it holy ground,
  The spot where first they trod ;
They have left unstained what there they found—
  FREEDOM TO WORSHIP GOD.

MRS. HEMANS.

ABBREVIATIONS AND EXPLANATIONS.

M. Married.
b. Born.
d. Died.
Ch. Children.
Unm. Unmarried.

The arrangement is in the alphabetical order of the Christian names of all the descendants of the original Corwin immigrants, whatever the surname may be. When the Christian name is printed in SMALL CAPITALS, the surname is Corwin. Other surnames are attached to the Christian name.

The small *superior* number, (as John⁴,) indicates the number of generations from the original immigrant, including also the latter.

The parent's name immediately follows, *italicized* and in a (parenthesis.)

The simple figures, 1, 2, 3, etc., connect together the descendants of Matthias Corwin, of Long Island.

The figures in parentheses, (1,) (2,) (3,) etc., connect together the descendants of George Curwin, of Salem.

The small letters, (a,) (b,) (c.) connect together the descendants of John Curwen, the immigrant of 1784.

The Roman letters, (I,) etc., refer respectively to Samuel Corwine, and the capitals, (A,) (B,) to Malcolm Corwin, and their descendants, who were very few.

Thomas and William Curwen left no descendants, so far as is known.

The fractional numbers refer to names which were received after the book was ready for the press. See also corrections and additions at the end of the volume.

The definitions of the Christian names are given as far as ascertained, unless they were surnames of other families.

The residences or places of birth are frequently omitted, when a reference to the father's name would be sufficiently definite.

The index gives intermarriages, descendants of other surnames, localities of residence, etc., etc.

By the arrangement of this genealogy, the ancestry or posterity of any person of the family can be at once, and without difficulty, traced to the earliest or latest generation.

Copies of this work will be sent to any address, post-paid, on receipt of price, $3.

Address

Rev. E. T. CORWIN,
Millstone, Somerset Co., N. J.

THE CORWIN GENEALOGY,

ARRANGED IN ALPHABETICAL ORDER

OF

CHRISTIAN NAMES.

———•••———

Aaron. (Hebrew,) lofty, inspired.

1 AARON,[6] (*Gershom*, 2 ?) b. ——, 1800–20.
M. Elmira ——.
In 1851, sold land in Hannibal, Cayuga Co., (now Oswego Co.,) N. Y.

(1) *AARON HOUGHTON,[7] [*Amos*, (1,)] b. ——, 1780–1800, d. in Philadelphia,
Pa. June, 1830. Unmarried.
  He was an artist of considerable repute, migrating from Maysville, Mason
Co., Ky., to Cincinnati, O., at an early age. He at once took rank among
the very first of his profession. His portraits were noted for truthfulness
to the original, and to this day are highly prized by the families in Cincinnati, who are their fortunate possessors. He started for Italy in the early
part of 1829, to perfect his art; but upon his arrival in London, his delicate
health, which had become much impaired by the long voyage, induced him
to return to America, which he barely reached before his death.

(2) *AARON HOUGHTON,[8] [*Joab II.*, (1,)] b. ——, 1825–35, d. young.

Abbiette, (Latin,) little Abby.

1 ABBIETTE,[7] (*Benjamin*, 8,) b. 1790–1800.
M. —— Hansen.
Ch. Ann E., Frederick B.   (*Long Island*, N. Y.)

Abby, see Abigail.

1 ABBY,[7] (*Jeremiah*, 2,) b. June 23d, 1792, d. April 13th, 1838. Unmarried.

2 ABBY CHARLOTTE,[8] (*Horton*, 1,) b. July 8th, 1846.
M. Charles H. Burton, August 25th, 1868.
Ch. Edith M.

3 Abby Taylor,[8] (*David J. Taylor*, 30,) b. ——, 1857 ?

* Corwine.

4 Abbie F. Dunham,[10] (*Alida M. Durkee*, 1,) b. February 14th, 1865.

(*Vt.*)

(1) * Abby,[7] [*Richard*, (2,)] b. ——, 1780–90, d. young.

(2) * Abby,[8] [*Samuel*, (6½,)] b. July 27th, 1804.
    M. George Saxon.
    No Ch.   (*Omega, Pike Co., Ohio.*)

Abel, (Hebrew,) transitoriness.

1 Abel,[6] (*John*, 6,) b. March 21st, 1762, d. October 8th, 1808.
    M. Ruth Hedges, b. about 1761, d. September 20th, 1821.
    Ch. Abel, Phila, Seth, Hudson, Nathaniel, Frederick, Mary, George,
    Grover, Buel. Will at Riverhead, Suffolk Co., N. Y. Was consta-
    ble in Riverhead, 1792.

2 Abel,[7] (*Eli II.*, 2,) b. December 7th, 1805.
    M. Mary, dau. of John Poillon, August 21st, 1827. She d. February 9th,
    1870, and was bur. at Cypress Hill Cem., L. I.
    No Ch.   (*New-York City.*)

3 Abel,[7] (*Abel*, 1,) b. ——, 1785 ? d. ——.
    M.
    Ch. Abel, Jane, Sarah A., Elizabeth, Phila, Irad.

4 Abel Dixon,[8] (*Nathaniel*, 6,) b. June 31st, 1834, d. August 22d, 1839.

5 Abel,[*] (*Abel*, 3,) b. ——, 1805–15.
    M.
    Ch. Bartholomew ?

6 Abel W. Watkins,[8] (*Marietta Woodruff*, 2,) b. 1820–30, d. 1855.
    M. Mary Louisa ——.
    Ch.

Abiah, (Hebrew,) whose father is Jehovah.

1 Abiah,[9] (*Jedediah*, 2,) b. ——, 1760–80.

Abigail or Abby, (Hebrew,) whose father is joy.

1 Abigail,[3] (*John*, 1,) b. ——, 1660–70. Not married in 1690, when her
    father's will was written, nor in 1698, when her name yet appears as
    Abigail Corwin. (*Doc. Hist. N. Y.* i. 450.)

2 Abigail,[6] (*David*, 4,) b. December 9th, 1775, d. December 9th, 1825.
    M. Walter Everett, March 9th, 1795. (He b. July 9th, 1772, d. July
    27th, 1848. He *M.* (2) Phebe Case, July 23d, 1827. His father, Ephraim
    Everett, was b. 1741, d. 1834.)
    Ch. Lewis Hudson, Harriet, Sarah, Walter Collins.

3 Abigail,[6] (*Eli*, 1,) b. November 10th, 1780, d. February 4th, 1852.
    M. Isaac Little, April 24th, 1800. (He b. February 16th, 1766, d.
    March 6th, 1827.)
    Ch. Lewis, Fanny, Horace, Cynthia, Jane, John, Harriet, James, Edwin.

* Corwine.

4 ABIGAIL,[7] (*Daniel*, 4,) b. ——, 1790–1800, d. ——, 1836. Unmarried.
Will. (*Walkill, N. Y.*)

5 ABIGAIL,[7] (*John*, 9,) b. January 30th, 1788, living.
M. (1) John Hubbard, b. ——, 1764, d. ——, 1825. (2) William Wickham. (*Aquebogue, L. I.*)

6 ABIGAIL ANNA,[7] (*William B.*, 23), b. March 27th, 1847.

6½ ABIGAIL,[7] (*Thomas*, 2,) b. ——, 1815–25.

7 ABIGAIL,[7] (*Stephen*, 2,) b. ——, 1790–1800.
M. Cornelius Wilcox.

8 ABIGAIL,[8] (*David*, 16,) b. February 9th, 1837, d. ——, 1852.

9 ABIGAIL,[8] (*Noah*, 2,) b. April 20th, 1842.
M. Levanda Wakely, March 28th, 1864.
Ch. Corwin, b. October 21st, 1869.

10 ABIGAIL,[8] (*Jabin*, 1,) b. August 12th, 1822, d. 1840. (*Vermont.*)

11 Abigail Case,[6] (*Benjamin Case*, 27,) b. ——, 1745, d. June 25th, 1752.

12 Abigail Everett,[7] (*Deborah Corwin*, 1,) b. ——, 1810–20.

13 Abigail Horton,[8] (*Eliza Corwin*, 2,) b. ——, 1830–40.

14 Abigail Ann Woodruff,[8] (*Silas Woodruff*, 15,) b. ——, 1825–30.
M. William Trumbull. (*Newburgh, N. Y.*)

15 Abigail Jane Little,[8] (*Harry Little*, 28,) b. December 9th, 1824, d.
October 6th, 1840.

(1) *ABIGAIL,[2] [*George*, (1),] b. August 1st, 1637, in England.
M. E. Hawthorne, August 28th, 1663. *M.* (2) Hon. James Russell.

(2) †ABIGAIL,[7] [*Samuel*, (4),] b. December 23d, 1788, d. October 2d, 1821?
M. Harris Van Crosswick, (or Harrison Crossthwaite,) April 1818.

(3) †ABIGAIL,[8] [*George*, (9),] b.
M. Silas Bray. (*North-Branch, N. J.*)

(4) †ABIGAIL,[6] [*Samuel*, (3),] b. ——, 1765, d. ——, 1857.
M. John Vanderveer, June 14th, 1806. No *Ch.*

(5) †ABIGAIL,[7] [*John*, (3),] b. ——, d. young.

Abner, (Hebrew,) father of light.

1 ABNER,[6] (*Joshua*, 1,) b. March 3d, 1760, d. ——, 1838.
M. Sarah Overton, February 28th, 1782. (*Aquebogue Recs.*)
Ch. Seth, John, Joshua, David, Jemima.

1½ ABNER,[7] (*Barnabas*, 1.)
M. Ellen — – ·.

From 1819–1829, he was of Lansing, Tompkins Co., N. Y. In 1830, sold

---

* Curwen. † Corwine.

land to the Ref. Dutch Church of Ithaca, N. Y. The writer's memoranda have also an Abraham, son of Barnabas, probably an error.

2 ABNER,[6] (*Benjamin*, 2.)   (*Morris Co., N. J.*)

3 ABNER,[7] (*William*, 4,) b.   d.
   M. Mary Corwin, (37,) January 18th, 1818.

4 ABNER,[7] (*William*, 4,) b. January 13th, 1800.
   M. Mrs. Polly Youngs, dau. of Jeremiah Corwin. *No Ch.*
                                      (*Upper Aquebogue, L. I.*)

Abraham, (Hebrew,) father of a multitude.

1 ABRAHAM,[7] (*John*, 7,) b. ——, 1790–1810.
   M. Clarissa Smith.
   Ch. Ann E., Abraham, Mary.

An Abraham Corwin bought land of Daniel Corwin in Genoa, Cayuga Co., N. Y., in 1827.

2 ABRAHAM,[8] (*John*, 16,) b. February 4th, 1830, d. December 8th, 1848.

3 ABRAHAM,[8] (*Abraham*, 1,) b. ——, 1820–30, d. young.

4 Abraham Jaques,[8] (*Julia Corwin*, 3,) b. ——, 1820–30.

Ada, or Edith, (Old German,) happiness. Ada, (Hebrew,) beauty, or ornament.

1 Ada Irene Crans,[8] (*Cynthia Little*, 7,) b. ——, ·d. August 5th, 1850.

2 Ada Easton,[9] (*Hannah E. Corwin*, 16,) b. February 23d, 1855.

3 Ada Horton,[9] (*Henry Horton*, 29,) b. ——, 1855–65.
4 *Ada B. Call[9] (W ≈ E. Call 90½) b. Dec 6. 1859, m May 2, 1889. George Strong*

Adaline, (Old German,) of noble birth, a princess.

1 ADALINE,[7] (——.)
Related to an Oliver Corwin.

2 ADALINE,[8] (*Benjamin W.*, 11,)·b. December 13th, 1834.
   M. Lewis Seybolt, September 20th, 1853.
   Ch. Edward L.

3 ADALINE,[8] (*Isaac*, 4,) b. December 26th, 1822, d. November 15th, 1857.
   M. Thomas Chattle, February 28th, 1842. (*See Julia* |9|)

4 ADALINE,[8] (*William O.*, 29.)
   M. (1) —— Benjamin, (2).

5 Adaline Terry,[8] (*Charlotte Corwin*, 5,) b. ——, 1815–30.
   M. John McNally.
   Ch.   (*Newburgh, Orange Co., N. Y.*)

6 Addie M. Reynolds,[9] (*Mary A. Smith*, 146,) b. August 3d, 1860.
                                                 (*Vermont.*)

7 ADDIE,[9] (*David*, 24,) b.

Adam, (Hebrew,) red-earth, man.

1 ADAM,[7] (*Jason*, 1,) b. July 23d, 1809.
   M. Melinda G. Owen, November 25th, 1833. She d. February 22d, 1863.

Ch. Samuel, Mary.

2 Adam Selah Howell,[8] (*Susan Corwin,* 1.)

3. Adam Blinn,[8] (*Lucinda Thatcher,* 5,) b. February 26th, 1825.
M. Cath. Derrick, January, 1850.
Ch. Martha, Amos, Susan, Lucy, Jacob, Ellen.

4 Adam Crans,[8] (*Cynthia Little,* 9.)

1 ADDISON MOORE,[8] (*Jabez,* 3,) b. ——, 1826, d. ——, 1871.

2 ADDISON LINWOOD,[9] (*Egbert C.,* 1,) b. June 14th, 1858.·

3 Addison Harlow,[8] (*Susan C. Corwin,* 2,) b. ——, 1840–50.

Adelaide, same as Adaline.

1 ADELAIDE,[8] (*Nathan II.,* 3,) b. November 1st, 1846. Teacher in Wells Seminary, Aurora, N. Y., 1870.

2 ADELAIDE,[8] (*Matthias,* 8,) b.

3 ADELAIDE,[8] (*Joshua C.,* $8\frac{1}{2}$,) b.
M. —— Decker.
Ch.

4 Adelaide Mary Colegrove,[9] (*Rev. Clinton Colegrove,* 1,) b. October 23d, 1853.

1 Adelbert Smith,[9] (*Loren Smith,* 1,) b. ——, 1850–55.
(*St. Lawrence Co., N. Y.*)

Adelia, same as Adaline.

1 Adelia Horton,[8] (*Eliza Corwin,* 2,) b. ——, 1830–40.
M. William Porter. *Lemuel C.*
Ch. Horton.

2 ADELIA[8] (*Cortlandt,* 1,) b.
M. —— Bancroft.
Three Ch. (*White Water, Wis.*)

Adell, same as Adaline.

1 Adell Little,[8] (*John A. Little,* 69,) b. ——, 1850.

1 Adelmer Knox,[8] (*Julia Corwin,* 7,) b. July 6th, 1846.

Albert, (Old German,) Illustrious.

1 ALBERT,[7] (*John II.,* 8,) b. March 4th, 1835, d. March 4th, 1837.

2 ALBERT P.,[7] (*Thomas,* 2,) b. July 15th, 1818.
M. Mary E. King, May 13th, 1841. She b. September 14th, 1818.
Ch. James N., Mary I., Edith G., Herbert R.

3 ALBERT LEWIS,[8] (*Alsop,* 1,) b. ——, 1810–20.

4 ALBERT II.[8] (*Isaac,* 4,) b. September 17th, 1837.
M. Addie Ludlum, April 30th, 1862.

5 ALBERT,[8] (*Henry,* 11,) b. April 18th, 1837.

M. Elizabeth Miller, January, 1870.
Ch. Albert G.   (*Peconic, L. I.*)

6 ALBERT,[8] (*Gabriel,* 2.)

6½ ALBERT GENEVIEVE,[9] (*Albert,* 5,) b. ——, 1870.

7 Albert B. Porter,[7] (*Elizabeth Corwin,* 8,) b. ——, 1820–30.

8 Albert Terry,[7] (*Keturah Corwin,* 1,) b. ——, 1780–90.

9 Albert Chase,[8] (*Bethia Corwin,* 5,)

10 Albert Swina,[8] (*Bethia Hobart,* 11,) b. August —, 1830.
M. Henrietta Sickles.
Ch. Marietta, Frank, Byron.   (*Wisconsin.*)

11 Albert B. Campbell,[8] (*Daniel Campbell,* 34,) b. April 9th, 1846.

12 Albert E. Millspaugh,[8] (*John M. Millspaugh,* 72,) b. April 7th, 1844.

13 Albert Baldwin,[8] (*Jane Corwin,* 1,) b. ——, in Goshen, N. Y.
                                        (*Starved to death at Andersonville.*)

14 Albert Mullock,[8] (*Mary Corwin,* 22.)

—— 1 Alburtis Burdette Sperry,[8] (*Mary Corwin,* 40,) b. August 21st. 1831.
M. Mary Hopkins.

1 ALETTA,[7] (*William,* 4,) b. ——, 1802–23 ?
M. Cyril Howell.

1½ ALETTA,[8] (*Uriah,* 1,) b.   (*Greenport, L. I.*)

2 Aletta Terry,[7] (*Keturah Corwin,* 1,) b —— , 1780–90.
M. —— Alsop.   (*New-York City.*)

3 Aletta Mapes,[9] (*Mary Corwin,* 49.)

Alexander, (Greek,) a defender of men.

1 ALEXANDER,[6] (——,) b. ——, 1790–1800, d. ——, 1850.
M. Sarah Bigelow.
Ch. Hector, De Witt C., William, James T.   (*Hamptonburgh, N. Y.*)

2 Alexander Thatcher,[7] (*Jemima Corwin,* 1,) b. about 1799.
                                                       (*Valparaiso, Ind.*)

Alfred, (Old German,) good counselor.

1 ALFRED,[7] (*Phineas,* 3,) b. September 14th, 1821, d. August 1st, 1853.
                                                       (*Hancock, Ill.*)

2 ALFRED EGBERT,[7] (*Silas,* 2,) b. in Orange Co., N. Y., June 30th, 1812.
M. (1) Sarah M. Roe, November 28th, 1840.   She d. 1846.   (2) Elizabeth F. Foster, March 25th, 1847.
Ch. Ann A.
He moved to Susquehanna Co., Pa., 1838.   Carpenter and farmer.
                                                       (*Montrose, Pa.*)

— 3 ALFRED H.,[8] (*John,* 16,) b. April 5th, 1818.

4 ALFRED,[9] (*William,* 34.)

5 Alfred Higby,[8] (*Harriet Corwin*, 2,) b. December ——, 1858.

6 Alfred Hulse,[8] (*Lucetta Corwin*, 1.)

7 Alfred Smith,[8] (*Sarah Corwin*, 24.)

8 Alfred Brown,[8] (*Elizabeth Woodhull*, 45,) b. September 6th, 1823.
M. Louisa Bartell. (*Milltown, N. J.*)

Alice, same as Adaline.

1 ALICE E.,[6] (*James*, 20,) b. July 12th, 1855.

2 ALICE M.,[8] (*Hervey*, 1,) b. September 26th, 1854.

2½ ALICE MAY,[9] (*James*, 16,) b. May 5th, 1856, d. January 8th, 1861.

3 ALICE M.,[9] (*William II.*, 38,) b. April 29th, 1854, d. November 26th, 1859.

4 ALICE ELIZABETH,[9] (*Silas G.*, 12,) b. April 4th, 1852, d. October 8th, 1854.

5 ALICE G.,[9] (*William*, 39,) b. ——, 1855–65.

6 ALICE V.,[9] (*Hubbard*, 2,) b. March 22d, 1843.
M. Benjamin F. Howell, January 8th, 1863.

7 Alice Julia Millspaugh,[8] (*John M. Millspaugh*, 72,) b. September 15th, 1848.

8 Alice Vanelia Armstrong,[8] (*Charlotte Millspaugh*, 10,) b. October 20th, 1848. (*Booneville, Ind.*)

9 Alice K. Chamberlin,[9] (*Jennette Corwin*, 1,) b. September 6th, 1866.

10 Alice D. Smith,[9] (*Sarah Corwin*, 40.)

11 Alice Robinson,[10] (*Hannah Corwin*, 22.)

12 Alice E. Colegrove,[9] (*Rev. Clinton Colegrove*, 1,) b. October 14th, 1861.

13 Alice Jewett,[9] (*Mary F. Colegrove*, 141½,) b. about 1859.

(1) *ALICE,[6] [*George*, (6),] b. January 18th, 1752.
M. Peter, Bellos.
Ch. Elizabeth, Rebecca, John, Amos, Peter, Isaac.

(2) *ALICE,[7] [*Samuel*, (4),] b. August 4th, 1798.
M. William Howell, April, 1818.

(3) †ALICE,[7] [*John*, (3),] b. December 29th, 1776, d. July 13th, 1867, at Titusville, N. J.
Bought land in Hopewell, Hunterdon Co., N. J., 1828.

(4) †ALICE,[9] [*George*, (11½),] b. ——, 1850. (*Carthage, Mo.*)
(Alice Curwin, a Quaker; see Thomas Curwin, at the end of list of Thomases.)

1 Alida Maria Durkee,[9] (*Philena Corwin*, 2,) b. April 27th, 1843.
M. Elbridge F. Dunham, October 16th, 1860.
Ch. Charles F., Grace E., Abbie F., Flora W.
(*Randolph, Orange Co., Vt.*)

* Curwen. † Corwine.

Allen, or Alan, (Slavonic,) a hound ; (Celtic,) harmony.

1 ALLEN WICKHAM,[9] (*John E.*, 45,) b. June 18th, 1870.

2 Allen E. Baker,[10] (*Herbert Baker*, 2,) b. Nov. 1st, 1869.

1 ALMINA,[9] (*Ezra*, 3,) b. ———, 1831.
   M. Hiram Coff.   (*Owasco, Mich.*)

1 Almodan Osman,[4] (*Sarah Corwin*, 1,) b. ———, 1700–1710.

Alonzo, or Alphonso, (Old German,) all-ready, willing.

1 ALONZO,[7] (*Stephen*, 7½,) b. Feb. 11th, 1812.   (*Dunkirk, N. Y.*)
   M. Sarah Pease, January 31st, 1833.
   Ch. William A., Charles.

2 ALONZO B.,[9] (*De Forest*, 1,) b. Aug. 27th, 1855.

3 ALONZO J.,[6] (*James*, 20,) b. April 8th, 1858.

Alma, (Latin,) nourishing.

1 ALMA IRENE,[6] (*Daniel*, 11,) b. ———, 1811.
   M. Horace Tryon.

1 Almeda Chase,[9] (*Phebe Corwin*, 8½.)

1 ALMERIN,[8] (*Pollydore B.*, 1,) b. Oct. 6th, 1827.
   M. Huyler Cleveland.   (*Owego, N. Y.*)

1 ALSOP,[7] (*Joshua*, 2,) b. ———, 1788, d. ———, 1858.
   M.
   Ch. Richard, Joshua, Albert Lewis, David, Benjamin, William.
                                                  (*Montrose, Pa.*

2 ALSOP HOLLETT,[7] (*Richard*, 1,) b. ——, 1790 ? d. 1851 ?
   M. Julia Ann ———.
   Ch. Hannah A., Alsop H., William, Joseph S., Mary A., Miriam.
                         (*Southampton and Jamesport, L. I.*)

3 ALSOP LEWIS,[8] (*Joshua*, 6,) b. Nov. 11th, 1819.
   M. Elizabeth ———.
   Ch. Selah ? Harvey ?   (*Elmira, N. Y.*)

4. ALSOP HOLLETT,[8] (*Alsop H.*, 2.)
   M. Aletta Corwin.   (*New-York City.*)

Alva, Alvan, or Alvah, (Hebrew,) iniquity.

1 ALVA,[6] (*Henry*, 10,) b. Jan. 3d, 1798, d. April 12th, 1865.
   M. Mehetable Jennings.   She b. Nov. 10th, 1802, d. Mar. 31st, 1864.
   Ch. Eliza, Mary A., Emma, John H., Oliver, Ann M., Theodore C.
                                        (*Upper Aquebogue, L. I.*)

1 Alvan Squires,[8] (*Patty Corwin*, 1.)

1 Alvirda Novilla Millspaugh,[9] (*Benj. F. Millspaugh*, 31,) b. Jan. 8th, 1849.

Amanda, (Latin,) worthy to be loved.

1 AMANDA,[7] (*David*, 8.)

2 AMANDA,[7] (*Silas*, 5,) b. Jan. 24th, 1817.
M. Alexander Galloway, March 29th, 1837.   He d. Feb. 1867.
Ch. Jerush, Mary, Silas, John, Frank, Silas.   (*Morristown, N. J.*)

3 AMANDA,[7] (*Nathan*, 5,) b. ——, 1825-30.

4 AMANDA,[8] (*Ebenezer*, 3,) b. April 25, 1842.
M. Benjamin Welsh.

5 Amanda J. Brown,[8] (*Silas C. Brown*, 16.)

6 Amanda Halstead,[8] (*Cynthia Hobart*, 10,) b. Nov. 24th, 1828, d. Nov. 28th, 1850.
M. Ab. H. Jones, Oct. 28th, 1847.
Ch. Cynthia.

7. Amanda Chappel,[8] (*Charlotte Hobart*, 9,) b. Feb. 1847.
M. Daniel Fuller, Feb. 1864.

8. Amanda F. Smith,[9] (*George Smith*, 74,) b. ——, 1853-55.
(*St. Lawrence Co., N. Y.*)

Amarantha, (Greek,) unfading.

1 Amarantha Brown,[8] (*Silas C. Brown*, 16.)

Amaziah, (Hebrew,) whom Jehovah strengthens.

1 AMAZIAH,[5] (*Timothy*, 1,) bapt. at Mattituck, Jan. 31st, 1762, d. Oct. 2d, 1841.
M. Joanna Brown, April 3d, 1785, (Aquebogue Rees.) She was b. ——, 1759; d. Aug. 6th, 1842.
Ch. Timothy, Ebenezer, Samuel, Amaziah W., Joanna.   In 1779, after the battle of Long Island, he went to Lyme, Ct.   In April, 1780, he was permitted to visit Long Island, with others, for supplies.
(*Mattituck, L. I.*)

2 AMAZIAH,[5] (*John*, 4,) b. 1730 ?   (*Snow Hill, Worcester Co., Md.*, 1752.)

3 AMAZIAH WEBB,[6] (*Amaziah*, 1,) b. ——, 1799, d. Aug. 19th, 1838.
M. Abby Aldrich.
Ch. Harriet J., Joanna E., Mary.

Ambrose, (Greek,) immortal, divine.

1 AMBROSE, F.[6] (——) b. 1750-60 ?
M.
Ch. Ambrose.

2 AMBROSE,[7] (*Ambrose F.*, 1,) b. ——, 1780-90 ?   (*Suffolk Co., L. I.*, 1820.)
He sold land to his father in 1820 ; he bought land of Dan. Holt, 1819.

1 Amidon Thomson ?[Joseph Thomson 27] b Oct 30 1878

Amelia, (Old German,) busy, energetic.

1 AMELIA,[7] (*Jemuel*, 1,) b. May 19th, 1834.

2 AMELIA,[7] (*Asa*, 2,) b. ——, 1780-1800.

3 AMELIA,[7] (*Matthias*, 5,) b. March 9th, 1804.
 M. Rev. Lewis Osborn, May 30th, 1827.
 Ch. Ann, William, Sophia, Francis, Matthias, Patience, John, Mary.
                (*Plymouth, Ill.*)

5 AMELIA HALLECK,[8] (*Nathaniel*, 7,) b. October 30th, 1837.
 M. J. Albert Wells, September 10th, 1860.
 Two Ch.   (*Southold, L. I.*)

6 AMELIA A.,[9] (*David*, 24,) b.

7 Amelia Osborn,[9] (*Matthias Osborn*, 12,) b. September 7th, 1857.

(1) * AMMON B.,[8] [Amos, (1½).]
                  (*Broadwell, Ill.*)
     Amos, (Hebrew,) strong, courageous.

1 AMOS,[8] (*John*, 30,) b. ——, 1822.
 M. Eliza Jane Chase.
 Ch. Perry W., Charles, Henry, and Mary.
        (*Jerusalem, N. Y., and Tioga, Pa.*)

2 Amos Thatcher,[9] (*Jesse Thatcher*, 10,) b. June 30th, 1829.

3 Amos Thatcher Blinn,[9] (*Lucinda Thatcher*, 5,) b. August 9th, 1829.
 M. Lucy Paring, November 6th, 1854.
 Ch. Willard.

4 Amos Blinn,[9] (*Adam Blinn*, 3,) b. ——, 1853 ?

(1) * AMOS,[6] [*George*, (6,)] b. in Amwell, Hunterdon Co., N. J., about 1756.
 M. Sarah Houghton.
 Ch. Joab H., Richard, William, John, Clarissa, Aaron H.

He removed to settle in Mason Co., Ky., 1788, though we find him selling land at Hopewell, N. J., in 1792. He bought land of his father in same place in 1780. (*Hunterdon Co. Records.*) He was related to the celebrated Colonel Joab Houghton, of revolutionary fame. The news of the battle of Lexington was received in Hopewell on the Sabbath, during service. Colonel Houghton, immediately after the benediction, called the male members of the congregation together, and made a stirring appeal to them in behalf of independence, and then and there commenced recruiting a company. A letter of Amos and Sarah Corwine, from Mason Co., Ky., to his brother, John, dated April 20th, 1800, is still preserved.

(1½) * AMOS,[7] [*Richard*, (2,)] b. November 22d, 1791, d. about 1857.
 M. (1) Mary Merrill, 1812. She died February 2d, 1836. (2) Elizabeth Milday, July 1st, 1837. She now lives at Lincoln, Illinois.
 Ch. Elizabeth, Susan, George, Richard, Benjamin F., Ammon B., William A., Lewis C., John H., Sarah.

He removed from Kentucky to Pike Co., Ohio, in 1828, and to Illinois in 1855.

He was a generous, noble-hearted man. He won distinction by his liberality and talents. His capacity to accumulate wealth was materially

         * Corwine.

affected by his unbounded liberality and his public spirit. He was a
public speaker of no inconsiderable merit. He took an active part in all
public affairs. He was commissioner of the county, and a member of the
State Legislature. He was not an aspirant for public honors. Had he
been, he would have secured almost any position he might have coveted.

(2) * Amos Breckenridge,³ [*Joab II.*, (1),] b. November, 1816.

M. Caroline Augusta, daughter of Captain Simeon H. Ackerman,† in
New-York City, on March 27th, 1854.

Ch. William R., Harriet L., Ernestine, Kate R. R., Elizabeth, Amos A.,
Frederick M., Carrie A.

(*New-Rochelle, N. Y.*)

He was engaged with his brother, Samuel L. Corwine, for several years
in the publication of the *Yazoo Banner*, at Benton, Mississippi, before and
after the presidential campaigns of 1840 and 1844. When the Mexican
war broke out, he volunteered, in capacity of Lieutenant of Company A,
in Colonel Jefferson Davis's First Mississippi Rifles, in which he served the
full term of his enlistment, and returned home with his company. They
were in action at the battle of Buena Vista, and he was wounded by an
escapet ball in the left side. Colonel Davis complimented him, in his report to General Taylor, for having remained on the field in discharge of
duty after being wounded.

After his return to the United States, he and his brother Samuel engaged in the publication of *The Cincinnati Daily Chronicle.* In December, 1849, he was commissioned to Panama, New-Granada, as United
States Consul, by President Taylor, and was continued in the same office
during the whole of Fillmore's administration. In the spring of 1856,
President Pierce commissioned him, as special agent of the United States
to Panama, to investigate the riot or massacre which had occurred there
in April, resulting in the death of some Americans, and the destruction
of their property. His report was the basis of the convention, soon after-
terward concluded, between the United States and New-Granada, for the
adjustment of the claims of American citizens against New-Granada, for
damages suffered in the riot. Mr. C. had already been the means of
suppressing several riots there, during his consulship. In July, 1856, he
was reappointed Consul at Panama, and continued there till the fall of
1861, when he returned home, and was at once offered the command of a
regiment at Cincinnati, which was ready to take the field against the re-
bels, and which desired his services. But owing to important private en-

---

* Corwine.

† Captain Ackerman was born in Portsmouth, New-Hampshire, his ancestors having come
from England. (Probably, remotely, from Germany or Holland, as the name is German.—
E. T. C.) He followed the sea during the most of his life, commanding vessels plying between
New-York and Liverpool. At the commencement of the rebellion he built a steamer, which
he chartered to the government, which afterward also purchased her. He amassed a
handsome fortune. He died in 1868. His wife's maiden name was Greer. She was a native
of Bristol, R. I., and died at New-Rochelle, N. Y., in December, 1865. Her ancestors were
from England. Her grandfather, on her maternal side, George Thorpe, was an early Governor at Savannah, Ga.

gagements, mostly of a fiduciary nature, he was obliged to decline the offer.

Governor Marcy, Secretary of State under President Pierce, submitted to Mr. C. the first draught of the treaty between the United States and New-Granada, for suggestions or modifications, and which were incorporated in the treaty. The President offered him, in 1856, a mission to Bogota, to negotiate a new postal convention with that country, but the offer was declined.

Mr. C. is now engaged, in company with his brother, Richard M. Corwine, in the prosecution of claims before the joint commission of the United States and Mexico.

(3) * AMOS BRECKENRIDGE,[9] [*Samuel*, (12),] b. December 3d, 1857, d. April 14th, 1860.

(4) * AMOS ACKERMAN,[9] [*Amos B.*, (2),] b. at New-Rochelle, August 13th, 1866.

(5) Amos Bellos,[7] [*Alice Corwine*, (1),] b. ——, in New Jersey.

Amy, (Latin,) beloved.

1 AMY,[5] (*Jesse*, 1,) b. ——, 1730–40, on Long Island.

(1) * AMY,[6] [*Samuel*, (3),] b. ——, 1755 ? d. ——.   Unmarried.

Andrew, (German,) strong, manly.

1 ANDREW KING CHANDLER,[8] (*Howard*, 1,) b. September 26th, 1855.

2 ANDREW J.,[8] (*Noah*, 2,) b. August 3d, 1830.
M. Louisa Wakely, November 5th, 1856.
Ch. Ophelia, Lewis A., Anna M.

3 Andrew Burton Wetherby,[8] (*Eliza Corwin*, 3,) b. January 24th, 1838, d. January 17th, 1841.

4 Andrew T. Fairchild,[9] (*Eliza E. Corwin*, 10.)

Angeline, (Greek,) angelic, lovely.

1 ANGELINE,[8] (*Hervey*, 1,) b. July 1st, 1845, d. 1859.

2 Angeline Hulse,[8] (*Lucetta Corwin*, 1,) b. ——.
M. —— Terwilliger.

3 Angeline Mullock,[8] (*Mary Corwin*, 22.)

Ann, Annie, Anna, or Hannah, (Hebrew,) grace.

1 ANNIE,[5] (*David*, 2,) b. ——, 1740–50.
M. —— Reeve.
An Anna Corwin married Benjamin Reeve October 11th, 1798.
                    (*Aquebogue Records.*)

2 ANN,[6] (*Thomas*, 1,) b. ——, 1790–1800.
M. Timothy Gregg.

3 ANNIE,[6] (*Joshua*, 1,) b. March 12th, 1767, d. April 22d, 1858.
                    Corwine.

M. David Penney, October 14th, 1783.   He b, February 11th, 1762, d.
April 29th, 1834.

Ch. Esther, David, Deborah, Huldah, Anna, Jesse.

4 ANNA⁶, (*David*, 5,) b. October 20th, 1804.
M. Brown, son of William and Laura Upright, February 28th, 1829.
Ch. Maria E., William.

4½ ANNA,⁶ (*Jedediah*, 2,) baptized at Mattituck, February 3d, 1754.
An Anna Corwin married Youngs Wells, August 30th, 1778; perhaps
this one.   (*Aquebogue Records.*)

5 ANNIE,⁷ (*Stephen*, 2,) b. ——, 1790–1800.
M. Samuel Price.   (*Tioga Valley, Pa.*)

6 ANNIE,⁷ (*James*, 3,) b. ——, 1800–10.   (*Camillus, N. Y.*)

7 ANNA,⁷ (*Daniel*, 4,) b. ——, 1780–1800.
M. Nathaniel Mapes.
Ch. Harrison, Isaiah.   (*Shawangunk, N. Y.*)

8 ANNA,⁷ (*Joseph*, 4,) b. July 28th, 1792, d. October 22d, 1832.
M. —— Wheat.

9 ANN ELIZA,⁷ (*James*, 6,) b. February 15th, 1810, d. ——. Unmarried.
                                         (*New-York City.*)

10 ANN ELIZA,⁷ (*John*, 14,) b. January 1st, 1817. Unmarried.
                                         (*New-York City.*)

10⅓ ANN, (*Rev. Joseph*, 5,) b. about 1804, d. about 1864.
M. Daniel Wells.

11 ANN M.,⁸ (*Isaac*, 4,) b. February 11th, 1824.
M. William E. Millspaugh, November 21st, 1849.

                                         (*Middletown, N. Y.*)

12 ANN ELIZA,⁸ (*Abraham*, 1,) b. ——, 1820–30.
M. —— Brennan.   (*San Francisco, Cal.*)

13 ANN,⁸ (*Benjamin H.*, 10,) b. ——, 1840–50.

14 ANN,⁸ (*Benjamin*, 15,) b. ——.   (*Ohio.*)

15 ANN ELIZA,⁹ (*Salem*, 1.)

16 ANNIE,⁸ (*George S.*, 12,) b. March 12th, 1837, d. June, 1862.

17 ANNA MARIA,⁸ (*Gabriel*, 4,) b. at Middletown, N. Y., June 28th, 1819,
d. January 22d, 1842.
M. Rev. Horatio Tompkins, August 12th, 1838, in Greenfield, Luzerne
Co., Pa.   He was born April 14th, 1809.
Ch. Maria L., Julia A.   (*Shunk, Pa.*)

18 ANN AUGUSTA,⁸ (*Benjamin*, 7,) b. ——, 1825–30.
M. —— Garrabrant.

19 ANN ELIZA,⁸ (*Peter*, 3,) b. ——, 1810–20.
M. David Ammerman.

                                 (*Chester, N. J. and Long Island.*)

20 ANNIE,[6] (*Harrison*, 1.)

21 ANN KATE,[8] (*Daniel*, 16,) b. ——, 1826.
     M. Valentine Losee.
     Ch. Susan, Josephine.   (*Brooklyn, N. Y.*)

22 ANN,[8] (*Henry*, 10,) b. July 6th, 1809, d. ——.
     M. David Tuthill, December 22d, 1826.
     Ch. Mary A., Huldey, Halsey, Elizabeth.

23 ANN MARIA,[8] (*Daniel*, 11,) b. ——, 1809.
     M. Joseph Knapp.

24 ANN AUGUSTA,[8] (*Richard*, 7,) b. ——, 1815-25.

25 ANN,[8] (*Lewis W.*, 1,) b. ——, 1825-35.
     M. Vernon Churchill.   (*Geneva, N. Y.*)

26 ANN G.,[8] (*John*, 22,) b. February 10th, 1815.
     M. James ——.   (*New-York City.*)

26½ ANN AMELIA,[8] (*Alfred E.*, 2,) b. ——, 1846, d. 1864.

27 ANNA MARIA,[8] (*William*, 7,) b. July 22d, 1859.

27½ ANN ELIZA,[8] (*Thomas*, 5,) b. ——, 1851.

28 ANNIE,[9] (*Charles N.*, 15½,) b. July 10th, 1861.

28½ ANNA M.,[9] (*Andrew*, 2,) b. September 27th, 1866.

28½ ANN MEHETABLE,[9] (*Alva*, 1,) b. Nov. 3d, 1827.
     M. James B. Storer, Feb., 1867.   (*Brooklyn, N. Y.*)

29 Anna Campbell,[7] (*Sarah Corwin*, 6,) b. Sept. 22d, 1803.
     M. Phineas Padden, Nov. 7th, 1820.
     Ch. Milton, Phineas, Sarah, William A., Sibyl T.

30 Anna Penney,[7] (*Annie Corwin*, 3,) b. July 27th, 1784.
     M. Stephen Sergeant.

31 Ann Elizabeth Hansen,[8] (*Abbiette Corwin*, 1.)

32 Anna Jane Weed,[8] (*Charity Corwin*, 1,) b. ——, 1820-30.
     M. Andrew Marvin.

33 Anna Mary Thompson,[8] (*Mary Corwin*, 25,) b. Jan. 27th, 1840.
     M. Samuel P. Clark
34 Anna Willinghouse,[8] (*Hannah J. Corwin*, 9.)

35 Ann Elizabeth Drake,[8] (*Arvilla Corwin*, 1.) b. ——, 1835-45.

36 Ann Augusta Bowers,[8] (*Elmina Corwin*, 1,) b. May 10th, 1838 ; d. Oct.
     24th, 1868.

37 Anna Girard,[8] (*Jerusha Corwin*, 2.)

38 Anna J. Van Wagener,[8] (*Maria E. Upright*, 13,) b. Feb. 26th, 1857.

39 Annie M. Meeker,[8] (*Harriet F. Corwin*, 4,) b. April 16th, 1863.

40 Ann Osborn,[8] (*Amelia Corwin*, 3,) b. Oct. 13th, 1848, d. Oct. 27th,
     1867.

41 Anna L. Byram,[9] (*Mary A. Corwin*, 34,) b. Feb. 7th, 1851.

42 Anna Paris,[9] (*Julia A. Blinn*, 18,) b. ——, 1856 ?

Anna Mary Thomson 33 (Mary Corwin 25)
born Jany 27. 1840.
Married Sept 24. 1873,
  Samuel T. Clark. (Sing-Sing N.Y,)
were.
176 Mary Corwin Clark 9? born June 29. 1874
117 William Archibald Thomson Clark
            born Feby 26. 1876.

Mary C. Clark married March 20. 1876
to Robert [illegible] [illegible]
me. [illegible]

       [illegible]

43 Ann M. Hackett,[9] (*Samuel B. Hackett*, 32,) b. June 24th, 1843.
  M. Albert Whitney, of Tunbridge, Vt., Sept. 14th, 1866.
  Ch. Minnie A., Orlando B.

44 Anne Thatcher,[8] (*Jesse Thatcher*, 10,) b. Nov. 29th, 1824.

45 Anna M. Lankton,[9] (*Phebe A. Padden*, 10.) b. May 25th, 1849.

46 Anna Lewis,[9] (*Emeline Corwin*, 2,) b. ——, 1850.

47 Ann Jennette Davis,[9] (*Polly Corwin*, 2,) b. ——, 1848.
  M. Orange J. Benson.
  Ch. Elmer E.

48 Anna Belle Van Wagener,[9] (*Sibyl T. Padden*, 4,) b. Mar. 31st, 1857.

49 Anna Wilson,[9] (*Julia E. Corwin*, 9.)

50 Annie E. Gray,[10] (*Sarah Corwin*, 56,) b. Dec. 11th, 1862.

51 Anna Colegrove,[9] (*Rev. Clinton Colegrove*, 1,) b. about 1855.

(1) Anna,[2] [*Jonathan*, (1),] b. Aug. 1st, 1687; d. March 20th, 1706.

(2) *Ann,[6] [*George*, (6),] b. about 1750?
  M. Joseph Hayes.

(3) *Anna Elizabeth,[8] [*Joab*, (1),] b. Jan. 12th, 1833.
  M. Henry W. Andress.

(4) *Anna F.,[9] [*Samuel L.*, (12),] b. Oct. 18th, 1855; d. Aug. 4th, 1858.

(A) ANNA MARIA, [Malcolm, (A),] b. in England about 1840.

1 ANNESTA MARIA,[9] (*Silas G.*, 12,) b. Feb. 23d, 1853; d. Sept. 27th, 1854.

1 ANSON,[7] (*William*, 4,) b. Feb. 14th, 1792; d. Oct. 4th, 1861.
  M. Elizabeth Halleck, Aug. 23d, 1817.
  Ch. Deborah A., George W., Daniel A., Anson L., Grotius S., Lucretia,
  Lucretia R. J.   (*Riverhead, L. I.*)

2 ANSON LEANDER,[8] (*Anson*, 1,) b. May 16th, 1833.
  M. Sarepta Edwards.
  Ch. Edward, Henrietta, Jesse.   (*Rockville Centre, L. I.*)

Anthony, (Latin,) priceless, praiseworthy.

1 ANTHONY,[8] ? (.) Bought land in Warren Co., N. J., 1855.

2 Anthony Moffat,[8] (*Sarah Corwin*, 11,) b. ——, 1810–20.

Antoinette, (Latin,) inestimable.

1 ANTOINETTE,[8] (*Phineas*, 1¾.)
        (Perhaps this name should be Angenette.)

Arabella, (Latin,) a fair altar.

1 ARABELLA,[8] (*Harrison*, 1,) b. May 12th, 1858, d. Jan. 3d, 1860.

2 ARABELLA,[9] (*William II.*, 38.)

Archibald, (German,) extremely bold, or a holy prince.

1 ARCHIBALD,[7] (*John*, 7,) b.
  M. Abigail Farman.
                    * Corwine.

Ch. Harriet, Theodore, Mary.

2 ARCHIBALD LITTLE,[8] (*Isaac N.*, 5,) b. Nov. 19th, 1840.

3 Archibald Tinney,[9] (*Lurelia Corwin*, 1,) b. ——.

1 ARDON,[7] (*Nathaniel*, 5,) b. ——, 1770–90.
 M. —— Philips.   (*Southampton, L. I.*)

1 ARMINDA,[8] (*Henry W.*, 8,) b. Nov. 27th, 1820, d. July 22d, 1825.

2 Arminda Brown,[7] (*Mary Corwin*, 8,) b. ——, 1790 ?
 M. Moses Meyers.
 Ch. Arminda, Rosetta, Henrietta, Sarah, Elizabeth, Jonathan, Mary.
                                                 (*Port Jervis, N. Y.*)

3 Arminda Meyers,[8] (*Arminda Brown*, 2,) b. ——, 1810–20.

4 Arminda Brown,[8] (*Daniel C. Brown*, 36.)

5 Arminda Brown Goble,[8] (*Elizabeth H. Corwin*, 19.)
 M. William Parsons.   (*Middletown, N. Y.*)

Arthur, (Celtic,) high, noble.

1 ARTHUR MILLS,[8] (*Rev. Eli*, 4,) b. March 24th, 1864, at Honolulu, Sandwich Islands.

2 Arthur Harlow,[8] (*Susan C. Corwin*, 2,) b. ——, 1840–50.

3 Arthur Reeve,[9] (*Charlotte M. Corwin*, 7,) b. ——, 1855.

(1) Arthur Barkalow,[8] [*Rachel Corwine*, (2).]
 M. Miss Dervey.

1 ARVILLA,[7] (*George*, 2,) b. April 22d, 1811, d. January 6th, 1851.
 M. James Drake.
 Ch. Eleazer, Mary C., Ann E., Laurin B., Cornelia.

2 ARVILLA,[8] (*Eleazer*, 1,) b. ——, 1835–45.

Asa, (Hebrew,) physician.

1 ASA,[5] (*Samuel*, 2,) b. May 29th, 1745, d. April 29th, 1761.   •

2 ASA,[6] (*Jonathan*, 1,) b. ——, 1750–60.
 M. Martha Raynor, December 23d, 1784. (*Aquebogue Records.*)
 Ch. Jonathan, William, Amelia, Esther, Nancy.   (*Aquebogue, L. I.*)

3 ASA,[6] (*David*, 3,) b. September 16th, (baptized October 26th, at Mattituck,) 1766. (*Aquebogue Records.*)
 M. Anna Chase, April 6th, 1788. She was b. September 16th, 1768.
 Ch. Philip, Rev. David, Richard W., Joshua C., George, Asa, Lucinda, Lydia A., Huldah.
 His wife and daughter Huldah united with the church of Aquebogue, August 1st, 1790.

4 ASA,[4] (*Asa*, 3,) b. ——, 1807.
 M. Jennette Brink.
 Ch. None.   (*Near Honesdale, Pa.*)

Asahel, (Hebrew,) whom God made.

1 ASAHEL,[7] (*Barnabas*, 2,) b. January 19th, 1819.   (*Knox Co., O.*)

1 Asenath Smiley,[7] (*Stephen*, 4,) b.

2 Asenath A. Campbell,[8] (*Daniel Campbell*, 34,) b. April 28th, 1836, d. April 24th, 1855.

1 Attison Everett,[7] (*Deborah Corwin*, 1,) b. ——, 1810–20.

1 Augusta,[9] (*Henry*, 20,) b. October 3d, 1849.

2 Augusta ——,[8] (*Hannah Corwin*, 6.)

1 Augustine,[8] (*Richard S.*, 7,) b. ——, 1820–30.

2 Augustine Reed,[8] (*Sophia Corwin*, 1,) b. July 4th, 1818.

1 Augustus Woodruff,[8] (*Erastus Woodruff*, 1,) b. ——, 1825–35.

2 Augustus Ball Shepherd,[8] (*Charlotte Corwin*, 6,) b. April 3d, 1856, d. September 12th, 1857.

1 Aura J. Austin,[9] (*Mahala Corwin*, 3,) b. December 5th, 1863.
(*Vermont.*)

1 Avanda,[8] (*Phineas*, 1¾.)

1 Azubah,[6] (*Silas*, 1,) baptized at Aquebogue, March 31st, 1765, d. April 26th, 1849.
M. Josiah Tuthill, of Middletown, N. Y., July 25th, 1803.
Ch. Irena, David W. (*Aquebogue Records.*)
Moved from Long Island to Orange Co., N. Y., about 1800; about 1815, to Cortlandt Co., N. Y.

2 Azubah,[7] (*Jacob*, 2,) b. ——, 1790–1800.
M. —— Beebe.
Ch. John E.

3 Azubah,[7] (*Peter*, 2,) b. August 19th, 1796.
M. David Carr.
Ch. Israel Y., William, Elizabeth, Corwin, Clarissa J., Azubah, David, Hiram. (*McGrawville, N. Y.*)

4 Azubah Everett,[7] (*Deborah Corwin*, 1,) b. ——, 1810–20.

5 Azubah Carr,[8] (*Azubah Corwin*, 3.) (*McGrawville, N. Y.*)

1 Barnabas,[4 or 5] (*George*, 1 ?) b. ——, 1760–70, d. ——, 1827, in Cayuga Co., N. Y.
M. Elenor ——.
Ch. Abraham, (or Abner.)
In 1795, bought land in Cayuga Co., N. Y. He was then of New-Wind-

2

sor, Orange Co., N. Y. Sold land in Cayuga, 1801. His executor was Daniel Corwin. In 1824 and 1827, he was of Lansing, Tompkins Co., N. Y.

2 BARNABAS,[6] (*Thomas*, 1,) b. December 3d, 1791, d. January 7th, 1870.
   M. (1) Harmony Aldrich, who was b. June 2d, 1793. (2) Elizabeth Coleman, who was b. December 3d, 1790.
   Ch. Asahel, Barnabas F., Timothy, Mary, Martha.
                                                                      (*Franklinville, L. I.*)

3 BARNABAS FRANKLIN,[7] (*Barnabas*, 2,) b. April 10th, 1822.
                                                                      (*Riverhead, L. I.*)

4 BARNABAS,[7] ? (*Henry.*)   (*Riverhead, L. I.*)

— 5 BARNABAS,[8] (*William*, 20,) b. September 6th, 1833.
   M. Jane Scott, July 9th, 1857.
   Ch. Jennie S., Thomas S., William H., Elizabeth S.

Entered service, July 9th, 1861, at Fort Hamilton, as First Lieutenant, 48th Regiment, New-York Volunteers. Promoted to be Major, and re-mustered, May 22d, 1863, 34th United States Colored Troops. Discharged by special order 325, of Adjutant-General's Department, September 30th, 1864, on account of sickness. Was in the following engagements : Port Royal, Fort Pulaski, Morris Island, Charleston, James Island, Pocataligo, etc. Resides in Brooklyn, N. Y., and is manager of Union Mutual Life Insurance Company, 151 Broadway, N. Y.

6 BARNABUS REEVES,[8] (*Daniel*, 19,) b. September 5th, 1803.
   M. (1) Melinda Homan. She b. March 1st, 1809. (2) Hannah Raynor. She b. March 19th, 1824.
   Ch. James M., Huldah E., John E., Mary C.

Bartholomew, (Hebrew,) son of Tolmai, or warlike son.

1 BARTHOLOMEW II.,[9] (*Abel*, 5 ?) b. April 29th, 1838.
   M. Eliza A. Dayton, December 11th, 1858. She b. February 14th, 1837.
   Ch. Mary H., Bessie, Jennie.   (*New-Haven Ct.*)

(1) Bartholomew,[4] [*George*, (3),] b. June 21st, 1693, d. May 9th, 1747.
   M. Esther, daughter of John Burt, of England.
   Ch. George, Richard, William, John, Joseph, Samuel.

He removed to Amwell, Hunterdon Co., N. J., in 1717, probably be-cause of his father's official relation to the Salem trials. His grandson Samuel (4) says, (in a letter yet preserved,) that Bartholomew visited England in 1731, but the fact is very doubtful. The books and papers which he left were destroyed, according to the same authority, in the re-volutionary war. The letter above alluded to, in possession of George R. Curwen, of Salem, Mass., and dated (if memory serves) 1751, is sealed with the Curwen arms. Trenton records show that he paid tax on three head of cattle and 100 acres of land in New-Jersey as early as 1721.

1 BARTLETT,[8] (*Richard S.*, 7,) b. ——, 1815-25.

1 Bela H. Colegrove,[7] (*Nancy Corwin*, 1½,) b. at Coventry, R. I., April
20th, 1797.

   M. (1) Eliza Ives, October 26th, 1825. She d. in 1852. (2) The
widow of Rev. Oliver J. Sprague, of Penn Yan, N. Y. Her father
was Abram Bennett, an early settler in that place.

   Ch. Rev. Clinton, James B.

He grew up on his father's farm in Rhode Island, and though with limited
means of education, early became a teacher in the district school there.
With these earnings he sought instruction at an academy. At the age of nineteen, he chose the profession of medicine and surgery, and entered the office
of Dr. Thomas Hubbard. His father dying soon, gave him a small patri-
mony, and he attended a course of medical lectures in New-Haven, Ct.,
and another course at Philadelphia, and graduated at the College of Physicians and Surgeons, in New-York City, in 1818. . In 1820, he emigrated
to Sardinia, Erie Co., N. Y., where he has since resided, practicing his profession, of which he has ever stood a most honored member. He was
elected to the State Legislature in 1841, and held the position of county
judge for six years.

2 Bela Horton Colegrove,[9] (*Rev. Clinton Colegrove*, 1,) b. July 27th, 1858.

3 Bela H. Colegrove,[9] (*James B. Colegrove*, 45½,) b. about 1867.

4 Bela H. Jewett,[9] (*Mary F. Colegrove*, 141½,) b. ——, 1855 ?

Belle, from Isabella.

1 Belle Stryker,[9] (*Mary Corwin*, 48,) b. December 1st, 1855.

(1) Belle Miller,[9] [*Sarah H. Corwine*, (6),] b. January 21st, 1860, d. July
26th, 1860.

Benjamin, (Hebrew,) son of the right hand.

1 BENJAMIN,[4] (*John*, 2,) b. ——, 1685–1700, d. ——, 1721.

   M.

2 BENJAMIN,[5] (*Samuel*, 2,) b. ——, 1733, d. April 18th, 1787.

   M. Mary ——.

   Ch. Abner, Experience R., Susannah, Sarah, are mentioned in his will.
Probably also Benjamin and Joseph, who lived in same locality.

His will in Secretary of State's office, at Trenton, N. J. Of Roxbury,
(Rocksbury,) Morris Co., N. J. Tombstone at Chester, N. J.

4 BENJAMIN,[6] (*Benjamin*, 2 ?) b. ——, 1750, d. 1821.

   M. Hannah ——.

   Ch. Joseph, Sarah, Mary, Elizabeth, Susan G., Rubama, Hannah,
Benjamin, William, John. In 1800 and 1801, Benjamin and Hannah
Corwin sold and bought land. (*Roxbury, Morris Co., N. J.*)

5 BENJAMIN,[9] (*Joshua*, 1,) b. ——, d. ——, 1848. (Will.)

   M.

Ch. William, Elizabeth, James, Gabriel L., Charity E.

(Mount Hope, N. Y.)

6 BENJAMIN,[6] *(James,* 1,) b. ——, 1780, d. June 24th, 1804, of yellow fever; was buried by the Brick Church in Beekman Street, New-York, where he had lived a few years. Unmarried.

7 BENJAMIN,[6] *(Benjamin,* 4,) b. ——, 1793 ? d. ——, 1848 ?
   M. (1) Susan Dickerson.   (2) Lois Young.   (3) Mary Hicks.
   Ch. Sarah E., John, Eliza E., Lewis D., Ann A.   *(Morris Co., N. J.)*

8 BENJAMIN,[6] *(Jeremiah,* 1,) b. ——, 1770 ?
   M.
   Ch. Abbiette, Joseph, John, Bethia, Benjamin, Mehetable.
                                          (Aquebogue, L. I.)

9 BENJAMIN V. R.,[7] *(Elisha,* 1,) b. ——, 1823, d. January 5th, 1863.
   M. Annie Jackson.
   Ch. William Jackson, Emily C., Ida V. R.   *(New-York City.)*

10 BENJAMIN HARVEY,[7] *(John,* 7,) b. ——, 1811.
   M. Esther Vail, who died April 19th, 1859.
   Ch. Henry, John, Ann.   *(Mount Hope, N. Y.)*

11 BENJAMIN WICKHAM,[7] *(John,* 10,) b. September 17th, 1806, d. April 12th, 1869.
   M. Sophia Roberts, September 17th, 1829.   She b. May 31st, 1808.
   Ch. Francis E., Helen Mary, Sophronia, Adaline, Henry, John E., Spencer W.

From 1829–39, at New-Vernon, Sullivan Co., N.Y.; 1839–50, Mount Hope, Orange Co., N. Y.; 1850–61, New-Vernon again; 1861–9, Middletown, N. Y.; often Justice of the Peace.

12 BENJAMIN,[7] *(Benjamin,* 8,) b. ——, 1790–1800.   Unmarried.

13 BENJAMIN FRANKLIN,[7] *(John,* 13,) b. ——, 1800–05.

14 BENJAMIN,[7] *(Henry,* 6,) b. May 7th, 1781, d. September 1st, 1832.
   M. Sarah Vail.   She b. September 2d, 1783, d. August 25th, 1857.
   Ch. Hewlett C., Henry W., Charles L., Mary B., Sarah M., Nancy A., Julia E., Harriet J.   *(Will at Riverhead, L. I.)*

15 BENJAMIN,[7] *(Matthias,* 5,) b. October 28th, 1785, d. January 23d, 1823.
   M. Rebecca Howell, May 14th, 1811.
   Ch. Francis, Clara, Thomas, Ann, Mary.

16 BENJAMIN,[7] *(Joseph,* 10.)

17 BENJAMIN,[7] *(Thomas,* 2,) b. ——, 1815–25.

18 BENJAMIN,[7] *(Edward,* 3,) b. April 27th, 1807.
   M. Betsy Chapin.
   Six Ch.

There was a Benjamin H. Corwin dealing in land in Tompkins Co., N. Y., in 1834.

18½ BENJAMIN LA FAYETTE,[8] *(Joshua,* 6,) b. January 4th, 1826, d. November 20th, 1827.

19 BENJAMIN,[8] (*Alsop*, 1,) b. ——,1810–20.

(*Troy, Susquehanna Co., Pa.*)

20 BENJAMIN,[8] (*William*, 11,) b. March 6th, 1838.   (*Pleasantville, Pa.*)

21 BENJAMIN O.,[8] (*James*, 18,) b. July 29th, 1862.

22 BENJAMIN FRANKLIN,[8] (*Epenetus II.*, 1,) b. ———, 1833, d. August 11th, '1834.

23 BENJAMIN HIRAM,[8] (*Shubal*, 1,) b. October 25th, 1822.
   M. Betsy Horton.
   Ch. Hiram B.   (*Lee Summit, Missouri.*)

24 BENJAMIN,[8] (*James Y.*, 11,) b. ——, 1810–20.

25 BENJAMIN W.,[8] (——,) b. ——, d., February, 1861.
   He was of Randolph Township, Morris Co., N. J. Josiah Meeker was executor of his will. He left a little property to his niece, Rhoda Hall, to his nephew, Joseph Hall, and the residue to his friend, Sarah C. Allen.

26 Benjamin Hutchinson,[3] (*Martha Corwin*, 1.)
   M.
   Ch. Elizabeth.   (*See Moore's Index of Southold*, p. 24.)

27 Benjamin Case,[4] (*Henry Case*, 1,) b. ——, 1692, d. November 14th, 1774.
   M. (1) perhaps Esther ——. (2) Mehitable Homan, November 11th, 1741, and who died October, 1751, aged 57. (3) Mary Overton, December 8th, 1755.
   Ch. Benjamin, Theodosia, Mary, Jemima, Abigail. (Will, New-York Lib. 29, p. 263.)   (*See Moore's Index of Southold*, p. 65.)

28 Benjamin Hutchinson,[4] (*Samuel Hutchinson*, 27,) b. —— 1704, d. ——, 1774.   (*Moore's Index*, p. 95.)

29 Benjamin Case,[6] (*Benjamin Case*, 27.)
   (*See Moore's Index of Southold*, p. 65.)

30 Benjamin Davis,[7] (*Elizabeth Corwin*, 4,) b. ——, 1780–90.

31 Benjamin F. Millspaugh,[7] (*Cynthia Corwin*, 1,) b. September 6th, 1820.
   M. Sarah Jane Albert, (widow of Washington Albert, and daughter of Vicesimus Knox Farr.)
   Ch. Mary C., Helen G., Alvirda, Emma, Laura K.
   (*Booneville Ind., now at Mount Pleasant, Iowa.*)

32 Benjamin F. Millspaugh,[8] (*William H. Millspaugh*, 71,) b. July 7th, 1845.   (*Columbia, O.*)

33 Benjamin McBride,[8] (*Elizabeth Corwin*, 12,) b. ——, 1830–40.

34 Benjamin Rinehart Honnell,[8] (*Susan Corwin*, 1,) b. ——, 1818.

35 Benjamin Gordon,[9] (*Sarah E. Corwin*, 43.)

(1) *BENJAMIN,[8] (*George*, 7¾,) b. 1810–20.   (*Indiana.*)

(2) *BENJAMIN F.,[8] [*Amos*, (1½).]   (*Broadwell, Ill.*)

* Corwine.

Benoni, (Hebrew,) son of grief.

1 Benoni Bradner Little,[7] (*Abigail Corwin*, 3,) b.
M. Mary Sellick.
Ch. Isaac, Fanny.   (*New-Milford, Pa.*)

Bertha, (Old German,) bright, beautiful.

1 BERTHA,[9] (*Charles N.*, 15½,) b. July 23d, 1869.

2 BERTHA,[9] (*Egbert*, 1,) b. November 18th, 1862.

1 Berthemia Fairchild,[9] (*Eliza E. Corwin*, 10.)

Bessie, see Elizabeth.

1 BESSIE,[10] (*Bartholomew*, 1,) b. December 5th, ~~1854~~. *1864*.

1 BESTIR,[8] (*Edward*, 3,) b. January 6th, 1793, d. February 11th, 1871.
M.
Ch. Esther, Fanny, Mercy, Electa.

Bethia, (Hebrew,) daughter of Jehovah.

1 BETHIA,[2] (*Theophilus*, 1 ?) b. ——, 1660-70 ?
This name occurs on census list, 1698, standing alone, her father being
probably dead.   (*Doc. Hist. N. Y.* i. 455.)

2 BETHIA,[6] (*Jeremiah*, 1,) b. ——, 1760-70.
M. Christopher Youngs. No children.

3 BETHIA,[6] (*Phineas*, 1,) b. May 27th, 1782, d. ——.
M. Jonah Chappel, October 4th, 1806.

3½ BETHIA,[7] (*Henry*, 6,) b. November, 1778, d. young.

4 BETHIA,[7] (*Benjamin*, 8,) b. ——, 1790-1800.
M. —— Odell.
Ch. James B., John F., William H., Mary A., Caroline A.

5 BETHIA,[7] (*Henry*, 6,) b. April 9th, 1786.
M. —— Chase.
Ch. Elizabeth A., Nathaniel, George, Albert, Oliver.

6 BETHIA MELISSA,[8] (*Gabriel*, 3,) b. March 20th, 1843, d. September 7th,
1843.

7 BETHIA JANE,[8] (*John*, 22,) b. August 12th, 1833, at Aquebogue.
M. Hiram H. Green.
Ch. Emma A., John H.   (*Mattituck, L. I.*)

8 BETHIA,[7] (*Rev. Joseph*, 5,) b. about 1798, d. December 13th, 1864.
M. Jacob Call, February, 1821.
Ch. William E., Ilia A., George, Mary.   (*Smithtown, L. I.*)

9 BETHIA,[7] (*John*, 9,) b. February 28th, 1790, d. March 22d, 1790.

9½ BETHIA ELEANOR,[8] (*David*, 12,) b. January 19, 1836.
M. Ezra Luckey Boisseau, September 8th, 1856.
Ch. Mary E., Julia B., Jessie F., Nellie C., Ezra E.   (*Southold, L. I.*)

10 Bethia Case,[5] (*Samuel Case,* 29.)
   M. Joel Bowditch, June 5th, 1735.   •

11 Bethia Hobart,[7] (*Mehetable Corwin,* 3,) b. July 24th, 1809.
   M. Henry Swina, July ——, 1827.
   Ch. Ira, William, Albert, Maria, Jennie, (or Jerome,) Lyman H., John,
   Emily, Myrin, Mary.   (*Wisconsin.*)

12 Bethia Little,[7] (*Desire Corwin,* 1,) b. April, 1816, d. October, 1834.

13 Bethia Taylor,[7] (*Fanny Corwin,* 1,) b. May, 1823.
   M. Etheridge G. Fowler, July, 1838.   (*Elbridge, N. Y.*)

14 Bethia Davis,[7] (*Elizabeth Corwin,* 4,) b. ——, 1780–90.

15 Bethia Squires,[8] (*Patty Corwin,* 1.)

Bethuel, (Hebrew,) man of God.

1 Bethuel E. Halleck,[5] (*Mary Corwin,* 29,) b. 1814, d. May 6th, 1841.
                                      (*Buried at Jamesport, L. I.*)

Betsy, see Elizabeth.

(1) * BETSY HOLMAN,[9] [*Samuel R.,* (7),] b. June 2d, 1856.

Beulah, (Hebrew,) married.

1 BEULAH,[6] (*James,* 2,) baptized May 24th, 1772, at Southold.

1 Bittridge Vail Little,[8] (*Henry Little,* 28,) b. ——, 1825–30.
   M. Albert S. Bennet, October 14th, 1851.

1 BREWSTER,[8] (*Daniel,* 16,) b. ——, d. young.

1 BUEL,[7] (*Abel,* 1,) b. ——, 1803, d. August 10th, 1820.

1 Burt Halsey,[8] (*Jefferson Halsey,* 1.)

1 BURTON DENNISON,[9] (*William H.,* 44,) b. April 25th, 1867.

1 Byron Swina,[9] (*Albert Swina,* 10,) b. ——, 1857.

Calcina, (Latin,) limy.

1 CALCINA,[8] (*Jeremiah,* 3½,) b.——, 1826, d. ——, 1837, from a lizard sting.

Caleb, (Hebrew,) a dog.

1 CALEB H.,[8]? (-   ,) b.
Was of the town of Lansing, Tompkins Co., N. Y., 1856, when he
bought land in Genoa, of John Corwin, and Melissa, his wife.   In 1867, he
appears as in town of Genoa, Cayuga Co., N. Y. ;

Calista, (Greek,) most beautiful.

1 CALISTA JANE,[7] (*Phineas,* 1¾.)
   M. Mr. Rouse, of Cazenovia, N. Y.

2 Calista Taylor,[8] (*John C. Taylor,* 70,) b. April, 1838.

3 Calista Fisher,[9] (*Helen M. Corwin,* 5½.)

                    * Curwen.

Carl, Carlos, Caroline, (Old German,) strong, noble-spirited.

1 CARL HENRY,[9] (*Marcus*, 2,) b. ———, 1866.

1 Carlos Smith,[9] (*Mary Corwin*, 53,) b. December 5th, 1858.

1 CAROLINE W.,[7] (*Elisha*, 1,) b. ———, 1810–20.

2 CAROLINE,[?] (*Phineas*, 3,) b. December 11th, 1817, d. March 5th, 1864.
M. Hezekiah Connor, November 13th, 1839.
Ch. Olivia, Leartus, Wellington, Sarah A., Milton.

3 CAROLINE,[7] (*Daniel*, 5,) b. ———, d. young.

4 CAROLINE L.,[5] (*John*, 32,) b. March 1st, 1829.
M. Harvey P. Hood, May 5th, 1850.
Ch. Laura C., Clara R., Nellie F., Charles H., Edward J., Gilbert H.
(*Vermont.*)

5 CAROLINE CONNOR,[8] (*Herrey*, 1,) b. December 27th, 1860.

6 CAROLINE,[8] (*David M.*, 13.)

6½ CAROLINE,[8] (*David*, *18*,) b. June 3, 1824.
M. Dr. Alfred Bennett, October 26, 1839.
Ch. Francis, Eliza, Caroline, Ella, Harriet, Alfred, Alice, Charles, William Totten, Adolph.   ·   (*Dunshore, Pa.*)

7 CAROLINE,[8] (*Gabriel*, 2.)

8 CAROLINE R.,[?] (*Thomas*, 4,) b. August 18th, 1836.
M. Dr. Charles Cropper, of Cincinnati, O., December 5th, 1865.
Ch. Thomas C., Maria L., Gertrude.   (*Lebanon, O.*)

9 CAROLINE,[8] (*Noah*, 2,) b. July 13th, 1833.
M. Charles Atwood.   He d. March 11th, 1863.
Ch. Willie.

10 CAROLINE,[?] (*Joshua C.*, 8⅜,) b. ———.

11 CAROLINE,[9] (*Samuel W.*, 25,) b. 1854.

12 CAROLINE GERTRUDE,[9] (*Silas G.*, 12,) b. April 3d, 1859.

12½ CARRIE MAY,[9] (*Hubbard*, 2,) b. May 13th, 1857.

13 Caroline Horton,[8] (*Eliza Corwin*, 2,) b. ———, 1830–40.
M. George Huff.
Ch. George.

14 Caroline Lippincott,[?] (*Elizabeth Corwin*, 18.)

15 Caroline Brown,[6] (*Elizabeth Woodhull*, 45,) b. May 21st, 1840.
M. Peter Van Arsdale.   (*Chester, N. J.*)

16 Caroline Amelia Odell,[8] (*Bethia Corwin*, 4.)

17 Caroline Eliza Wetherby,[?] (*Eliza Corwin*, 3,) b. April 14th, 1841.

18 Caroline Halsey,[8] (*Miller Halsey*, 2.)

19 Caroline Halsey,[8] (*Jefferson Halsey*, 1.)

20 Caroline Corwin Sage,[9] (*Emeline Corwin*, 1,) b. June 23d, 1856.

21 Carrie A. Whitney,[9] (*Helen M. Corwin*, 4,) b. September 16th, 1862.
(*Vermont.*)

22 CARRIE,[9] (*John*, 53,) b. ——, 18..

23 CARRIE MELINDA,[10] (*James*, 39,) b. March 12th, 1866.

(1) * CAROLINE ENDICOTT,[9] [*James B.*, (1),] b. January 7th, 1852.
(*Salem, Mass.*)

(2) * CAROLINE REA,[9] [*Samuel R.*, (7),] b. September 16th, 1851, d.
· January 6th, 1852. (*Salem, Mass.*)

(3) † CARRIE AUGUSTA,[9] [*Amos*, (2),] b. at New-Rochelle, 1866.

Catharine, (German,) pure.
Catharine. see Kate.

1 CATHARINE ANN,[7] (*Silas*, 5,) b. July 31st, 1824, d. April 27th, 1850.
M. James Mosher, 1849.

2 CATHARINE FITZ-RANDOLPH,[8] (*Thomas*, 4,) b. May 14th, 1827.

3 CATHARINE AMANDA,[9] (*John D.*, 47,) b.

4 Catharine M. Little,[7] (*Desire Corwin*, 1,) b. September, 1818, d. February, 1835.

5 Catharine Gerard,[8] (*Jerusha Corwin*, 2.)

6 Catharine Electa Millspaugh,[8] (*G. C. Millspaugh*,) b. September 13th, 1844.

7 Catharine C. Reed,[8] (*Sophia Corwin*, 1,) b. July 1st, 1821.

8 Catharine ——,[8] (*Hannah Corwin*, 6.)

9 Catharine Brown,[8] (*Elizabeth S. Woodhull*, 45,) b. July 7th, 1816.
M. (1) Edward Rose. (2) George Woodworth. (*Brooklyn, N. Y.*)

10 Catharine Willinghouse,[8] (*Hannah Corwin*, 9.)

11 Catharine A. Noyes,[9] (*Louisa Corwin*, 3,) b. April 17th, 1853. (*Vermont*)

12 Catharine Clark,[9] (*Helen A. Colgrove*, 10,) b. about 1867.

Cecil, Cecilia, (Latin,) dim-sighted.

1 CECIL SHERMAN,[8] (*Rev. Eli*, 4,) b. February 26th, 1860, at Honolulu, Sandwich Islands.

1 CECILIA ANN,[8] (*Nathaniel*, 7,) b. November 9th, 1819.
M. George Clark, November 10th, 1840.
Ch. Georgiana.

Celeste and Celia, (Latin,) heavenly.

1 Celeste Clark,[8] (*Sarah Jane Millspaugh*, 66,) b. February 3d, 1841.

1 CELIA,[7] (*Samuel*, 9,) b. April 13th, 1816. Unmarried.
· (*Middle Hope, N. Y.*)

1 Celinda Harding,[9] (*Polly M. Corwin*, 4.)

Charity, (English, from Latin, caritas,) dearness, high regard, love.
Charity, see Charry.

1 CHARITY,[7] (*Daniel*, 4,) b. ——, 1780–90.
M. Linas Weed.

* Corwen. † Corwine.

Ch. Daniel C., Harriet E., Ann J., Lucretia, Martha.

(*Montgomery, N. Y.*)

2 CHARITY ESTHER,[7] (*Benjamin*, 5,) b. ——.

M. ——, Reed ?

3 CHARITY,[7] (*Phineas*, 3,) b. July 10th, 1812.

4 CHARITY A.,[8] (*Daniel*, 11,) b. ——, 1824.

M. Silas G. Corwin, 8.

Charles, (Old German,) noble, high-spirited.

1 CHARLES S.,[7] (*Joseph*, 4,) b. May 18th, 1809, d. Aug. 4th, 1843.

2 CHARLES EMMOT,[7] (*Moses*, 1,) b. October 5th, 1816, d. June 13th, 1837.

3 CHARLES H.,[7] (*Matthias*, 6,) b. ——, 1790–1800.

M. Mary E. Downs, March 29th, 1847.   She was daughter of William Downs and Sarah Furman.

Ch. Matthias, Ida, Margaret, Helen, Isabel, Sophia.

4 CHARLES,[7] (*John*, 13,) b. ——, 1780–90.

5 CHARLES,[7] (*Thomas*, 2,) b. ——, 1815–25.

6 CHARLES B.,[7] (*Richard*, 3,) b. April 8th, 1855.

7 CHARLES,[8] (*George S.*, 12,) b. February 12th, 1835.
M. (1) Mary E. Force, May 23d, 1860 ; (2) Sarah Williams. *Sept. 22 1869*
Ch. Charles S., Ida, Eliz a etti .   (*Catasauqua, Pa.*)

8 CHARLES ABEL,[8] (*Rev. Eli*, 4,) b. January 6th, 1857, at Newburg, N. Y.

9 CHARLES E.,[8] (*William*, 11,) b. December 22d, 1839.

10 CHARLES E.,[8] (*John*, 22,) b. September 21st, 1826.
M. —— Rogers.   (*Riverhead, L. I.*)

11 CHARLES G.,[8] (*Manasseh R.*, 1,) b. ——, 1805–15.

12 CHARLES,[8] (*Moses H.*, 2,) b. ——, 1852 ?

12½ CHARLES B. NOTT,[8] (*Rev. Ira*, 4,) b. at Fairfield, Ohio, December 23d, 1853, d. August 3d, 1862, at same place.

13 CHARLES FREDERICK,[8] (*Joseph*, 16,) b. July 25th, 1845. (*Newark, N. J.*)

14 CHARLES,[8] (——.)   (*See Mary Frances*, 36.)

15 CHARLES MATTHIAS,[8] (*Matthias*, 8.)

15½ CHARLES NICHOLSON,[8] (*David*, 12,) b. January 5th, 1834.
M. Electa Boisseau, November 29th, 1855.   She b. 1836.
Ch. Eugene, John, Annie, Bertha.   (*Southold, L. I.*)

16 CHARLES,[8] (*Jeremiah*, 4.)

16½ CHARLES LODEWICK,[8] (*George W.*, 18,) b. March 13th, 1840 ; lost at sea, December 13th, 1859.

17 CHARLES T.,[8] (*Daniel*, 16,) b. ——, 1821, d.
M. Corinna ——.
Ch. Isabel.

18 CHARLES,[8] (*Hudson*, 1,) b. March 6th, 1829, d. August 10th, 1850.

uildren of Charles Corwin ? and Sarah Williams,
dith Frances Corwin, b. Nov 1. 1873.
Ella Williams Corwin b. July 17. 1875
           died Feby 1881.
Henry Arthur Corwin b. Jany 29. 1877
           died Feby 1887.
Thomas Sherwood Corwin b. Oct 3. 1882.
Helen Avis Corwin, b. Feby 28. 1884.
Russell Wood, Corwin b Dec 30, 1889.

19 CHARLES LANKTON,[5] (*Benjamin*, 14,) b. February 9th, 1819.
   M. Ann M. Terry, March 27th, 1845. She b. July 25th, 1826.
   She was daughter of Isaiah Terry and Mary Woodhull, of Wading River,
L. I., who were married December 16th, 1803, by Rev. Jacob Corwin.
     Ch. Mary E., Lois A., Hannah M., Charles H., Isaac T., Huldey T.,
     Susan M.   (*Greenport, L. I.*)

20 CHARLES WARREN,[8] (*Joshua C.*, 8½.)

20½ CHARLES,[8] (*Alonzo*, 1,) b. June 6th, 1840, d. June 6th, 1863.
   M. Nettie Soles, March 28th, 1862.
   Ch. Cora, b. March 20th, 1863.

•21 CHARLES B.,[8] (*Parker*, 1,) b. September 2d, 1833.
   M. Matilda Ann Korah, March 5th, 1856.
   Ch. Ira W., Gertrude E., Charles P., Grace L.

   He and his wife both studied in the Wilson Collegiate Institute, in
Niagara Co., N. Y.   (*Ransomville, Niagara Co., N. Y.*)

22 CHARLES DANA,[8] (*Jabin*, 1,) b. January 21st, 1831, at Tunbridge, Vt.
   M. Margaret Sweeney, June 4th, 1860.
   Ch. Charles J., Walter F.   (*Boston, Mass.*)

22¼ CHARLES,[8] (*Henry*, 11,) b. April 18th, 1837, d. March 18th, 1862.

23 CHARLES ELLET,[8] (*Edward P.*, 6,) b. July 4th, 1861. (*New-York City.*)

23½ CHARLES LYMAN,[8] (*Rev. Jason*, 4,) b. at Augusta, N. Y., August 5th, 1840.

   A graduate of Michigan University. Printer, lawyer, and more recently
editor of the *Real Estate Journal*, Chicago, Ill.; now associate-editor with
General Pierce of the Fort Smith (Ark.) *Patriot.*

24 CHARLES JABIN,[9] (*Charles D.*, 22.) b. January 11th, 1864.

25 CHARLES,[9] (*Amos* 1.)

26 CHARLES H.,[9] (*Charles L.*, 19,) b. May 27th, 1855.

27 CHARLES,[9] (*Henry*, 21,) b. ——, 1866.

28 CHARLES,[9] (*John D.*, 47.)

29 CHARLES S.,[9] (*Frederic*, 6,) b. October 9th, 1857.

30 CHARLES H.,[9] (*Samuel B.*, 23,) b. March 20th, 1852. (*Greenport, L. I.*)

31 CHARLES EDWARD,[9] (*Rev. Edward T.*, 7,) b. at Millstone, N. J., September 7th, 1868.

32 CHARLES WINFIELD,[9] (*Silas G.*, 12,) b. March 26th, 1857.

33 CHARLES STANLEY,[9] (*Charles*, 7,) b. March 2d, 1861.

33½ CHARLES,[9] (*Isaac*, 6,) b. December, 1831, d. at the age of two weeks.

34 CHARLES,[9] (*Isaac*, 11.)

35 CHARLES,[9] (*Samuel W.*, 25,) b. ——, 1852.

36 CHARLES P.,[9] (*Charles B.*, 21,) b. May 12th, 1863.

37 CHARLES M.,[9] (*Daniel A.*, 23,) b. ——, 1853-60.

38 CHARLES RUSSELL,[9] (*John A. R.*, 60,) b. ——, 1851-60.

33 *Charles Stanley [illegible handwriting] b. March [illegible]*
M. Jan'y [illegible]
Ch. [illegible]

41½

38½ CHARLES A.,[9] (*William A.,*) b. November 10th, 1864.

39 CHARLES,[9] (*Isaac,* 6.)

39½ CHARLES HENRY,[13] (*John,* 67,) b. January 17th, 1857.

40 Charles Hunter,[7] (*Mary A. Corwin,* 27,) b. May 13th, 1864.

41 Charles Petit,[8] (*Hannah Corwin,* 12,) b. ——, 1849.

42 Charles Emmot Flannery,[8] (*Mary Corwin,* 33,) b. July 7th, 1843.

 M. Elizabeth L. Cleveland, October 17th, 1869. She was born at Buffalo, N. Y., May 9, 1849.   (*Oil City, Pa.*

43 Charles Woodruff,[5] (*Silas Woodruff,* 15,) b. ——, 1825–35.
 M. Kate Sly.   (*New-Windsor, N. Y.*) •

44 Charles Horton,[8] (*Eliza Corwin,* 2,) b. ——, 1830–40.

45 Charles Augustus Miller,[9] (*Sarah J. Corwin,* 32,) b. ——, 1825–35.
 M. Rosa Cooper.

46 Charles M. Lankton,[9] (*Phebe A. Padden,* 10,) b. April 2d, 1847.

47 Charles Joseph Howell,[8] (*Sophia Woodruff,* 5,) b. June 4th, 1837.

48 Charles E. Little,[8] (*John A. Little,* 69,) b. ——, 1852.

49 Charles Taylor,[9] (*Hendrick Taylor,* 2,) b. May, 1856.

50 Charles Edwin Wells,[8] (*Puah Corwin,* 1.)
 M. Elizabeth ——.

51 Charles Marvel Howell,[8] (*Harriet N. Corwin,* 3,) b. November, 1845.

52 Charles Waters,[8] (*Eliza Corwin,* 8.)

53 Charles Henry Howell,[8] (*Elizabeth Corwin,* 20.)
 M. Susan Aldrich.

54 Charles Henry Ketcham,[9] (*Nancy A. Corwin,* 3,) b. February, 1841.
               (*Warsaw, Ind.*)

54½ Charles M. Byram,[8] (*Mary A. Corwin,* 34,) b. May 13th, 1864.

55 Charles Murray,[9] (*Lucinda Corwin,* 1.)

56 Charles Tinney,[9] (*Lucilia Corwin,* 1.)

57 Charles Omer Howes,[9] (*Theresa Corwin,* 2.)

58 Charles Wheeler,[9] (*Sarah Corwin,* 45.)

59 Charles Wesley Concklin,[9] (*Sarah M. Corwin,* 34,) b. May 11th, 1836.

60 Charles Jacob Davaw,[9] (*Helen Hunsike,* 7.)

61 Charles Lewis Gordon,[9] (*Sarah E. Corwin,* 43.)

62 Charles Derrick,[9] (*Mary M. Blinn,* 137,) b. ——, 1866 ?

63 Charles Ira Concklin,[9] (*Lucretia R. J. Corwin,* 3,) b. ——, 1846.
 M. Anna Sinclair.
 Ch. Leveret.

64 Charles W. Noyes,[9] (*Louisa Corwin,* 3,) b. March 29th, 1850.
              (*Vermont.*)

70 Charles W. Call.[9] (Wm E. Call 90½) b. Sept 19, 1846.
m. Mildrew Edwards Nov 16, 1872.

65 Charles H. Hood,[9] (*Caroline L. Corwin*, 4,) b. February 26th, 1860.

66 Charles F. Dunham,[10] (*Alida M. Durkee*, 1,) b. November 4th, 1861, d. June 10th, 1863. (*Vermont.*)

67 Charles F. R. Baker,[10] (*Edna Kyte*, 3,) b. March 16th, 1869.

68 Charles C. Sears,[9] (*Eliza N. Colgrove*, 26¼,) b. about 1850.

69 Charles A. Clark,[9] (*Helen A. Colgrove*, 10,) b. about 1865.

(In 1848, Charles S. Corwin, of Deer Park, died. His wife's name was Clarissa. They left no children, but had an adopted daughter. Joseph Corwin, of Lumberland, was a witness to will, which is at Goshen, N. Y.)

(1) *CHARLES FREDERIC,[9] [*Samuel R.*, (7),] b. July 18th, 1853.

(*Salem, Mass.*)

(2) †CHARLES,[9] [*George*, (11¼),] b. ——, 1818. (*Carthage, Mo.*)

Charlotte, feminine of Charles.

1 CHARLOTTE,[6] (*Eli*, 1,) b. in Scotchtown, N. Y., March 30th, 1800.
M. Moses Young, October 13th, 1818. He was b. in same place, June 11th, 1797, and d. there December 21st, 1838. His parents came from Germany. In 1849, Mrs. Young removed from Scotchtown to Plymouth, Ill.
Ch. Harriet O., Daniel, Eli H.

2 CHARLOTTE,[7] (*James*, 8,) b. ——, 1800-10. (*Camillus, N. Y.*)

3 CHARLOTTE,[7] (*Jemuel*, 1,) b. April 26th, 1844.

4 CHARLOTTE,[7] (*James*, 4,) b. ——, 1790-1800.
M. David Hubbard. (*Cutchogue, L. I.*)

5 CHARLOTTE,[7] (*Martin Luther*, 1,) b. ——, 1794, d. ——, 1867.
M. Thomas P. Terry.
Ch. Nancy, Laura, Silas, Mary, William, Oliver, Thomas, Adaline.
(*New-Vernon, N. Y.*)

6 CHARLOTTE,[7] (*James*, 6,) b. January 10th, 1815.
M. Lawrence Vanderveer Shepherd, June 5th, 1842.
Ch. Elisha, George, Emma, Henry, Augustus. (*Bound Brook, N. J.*)

7 CHARLOTTE MARIA,[8] (*Shubal*, 1,) b. May 31st, 1834.
M. Edward T. Reeve.
Ch. Arthur, Fanny. (*Near Aquebogue, L. I.*)

8 CHARLOTTE,[8] (*Jeremiah*, 4.)

9 Charlotte Hobart,[7] (*Mehetable Corwin*, 3,) b. March 16th, 1817.
M. Daniel Chappel, October —, 1844.
Ch. Howard, Amanda, Jonah C., Cynthia, Perry F. (*Elbridge N. Y.*)

10 Charlotte Millspaugh,[7] (*Cynthia Corwin*, 1,) b. May 26th, 1826.
M. N. E. Armstrong.
Ch. Alice V., Marcella, Sarah B. (*Booneville, Ind.*)

11 Charlotte De Boyce,[8] (*Sarah Corwin*, 20,) b. ——, 1820-30.

* Curwen.   † Corwine.

12 Charlotte Petit,[3] (*Hannah Corwin*, 12.)

13 Charlotte Hurtin,[8] (*Sarah A. Corwin*, 27.)

1 CHARLTON JASON,[8] (*Rev. Jason*, 4,) b. at' Deposit, N. Y., October 20th, 1824.

At first a printer, now editor of the *Jefferson City*, (Mo.) *State Journal.*

Charry, see Charity.

1 CHARRY,[7] (*Nathan*, 2,) baptized at Aquebogue, May 7th, 1809.
M.
Ch. Nelson G.

2 CHARRY AUGUSTA,[8] (*Gilbert*, 4,) b. May 2d, 1836, d. February 25th, 1839.

1 CHATHAM,[8] (*Nathaniel*, 7,) b. August 19th, 1839, d. in army, 1861–5.

2 Chatham Davis,[7] (*Elizabeth Corwin*, 4,) b. ——, 1780–90.

1 CHAUNCEY,[7] (*Joshua*, 4,) b. ——, 1810.   (*Southampton, L. I.*)

2 CHAUNCEY,[8] (*Daniel*, 19,) b. September 28th, 1829, d. February 2d, 1840. Unmarried.

3 CHAUNCEY FRANKLIN,[9] (*Daniel*, 24,) b. September 27th, 1839, d. February 2d, 1850.

Chloe, (Greek,) blooming.

1 Chloe North,[8] (*Mehetable Campbell*, 10,) b. September 17th, 1829.
M. Alonzo Streeter.

Christiana, (Greek,) Christian.

1 CHRISTIANA,[7] (*George*, 2,) b. ——, 1813, d. young.

2 CHRISTIANA,[8] (*Franklin*, 8,) b. ——, 1870.

3 Christiana Sinclair,[8] (*Frances Corwin*, 1,) b. August 4th, 1855.

4 Christiana May Cooper,[8] (*Martha C. Corwin*, 4½,) b. ——, 1861.

5 CHRISTIANA CASE,[8] (*Ezra*, 2½.)

Christopher, (Greek,) Christ-bearer.

1 CHRISTOPHER,[8] (*James Y.*, 11,) b. ——, 1810–20.

Clara, (Latin,) bright, illustrious.

1 CLARA,[8] (*Benjamin*, 15,) b.   , d.   .   (*Ohio.*)

2 CLARA,[8] (*William S.*, 9,) b. October —, 1851.

3 CLARA MAY,[9] (*Cornelius G.*, 2,) b. June 12th, 1864.

4 Clara Thompson,[8] (*Mary Corwin*, 25,) b. January 1st, 184[8]. *d. March*

5 Clara Millspaugh,[8] (*William E. Millspaugh*, 72,) b. September 1st, 1847.

6 Clara Belle Gordon,[9] (*Sarah E. Corwin*, 43.)

7 Clara Miller,[9] (*Sylvanus K. Miller*, 1.)

10 Clara Estella Thomson [9] *(Edward Thomson 16) b. Oct 9. 1880.*

8 Clara R. Hood,[9] (*Caroline L. Corwin*, 4,)b. October 22d, 1854.

(*Vermont*.)

9 Clara M. Boyd,[9] (*Emeline Corwin*, 7,) b. January 16th, 1867, d. September 27th, 1868. (*Vermont*.)

(1) *CLARA,[9] [*George*, (11½,)] b. (*Carthage, Mo.*)

Clarence, (Latin,) bright, illustrious.

1 Clarence Sinclair,[x] (*Emeline Corwin*, 1,) b. April 14th, 1853.

1 CLARINDA,[8] (*James II.*, 13,) b. February 14th, 1856.

2 CLARINDA,[6] (*Parker*, 1,) b. November 27th, 1828.
   M. Francis Kyte, April 2d, 1849.
   Ch. Edna E., George L., Sarah E., Francis E., Marion L., Clarinda L., Frederic W.

Clarinda studied in Wilson Collegiate Institute, Niagara Co., N. Y. Francis Kyte was the son of William Kyte, who was a soldier of 1812–14. William K. emigrated from Eastern New-York to Niagara Co. in 1830. His wife was Larinda Culver. (*Ransomville, Niagara Co., N. Y.*)

3 Clarinda L. Kyte,[9] (*Clarinda Corwin*, 2,) b. July 29th, 1862.

Clarissa, same as Clara.

1 CLARISSA,[8] (*William*, 11,) b. February 10th, 1844.

2 CLARISSA REBECCA,[8] (*Hervey*, 1,) b. ——, 1856, d. ——, 1859.

3 CLARISSA T.,[9] (*William II.*, 38,) b. July 17th, 1845, d. December 19th, 1859.

4 Clarissa Taylor,[7] (*Fanny Corwin*, 1,) b. June —, 1835.
   M. Albert Crego, September —, 1851.
   Ch. Edgar C., George P., Horatio A., Elmer A., Orlando L., Duane C., Theresa.

5 Clarissa Amanda Wetherby,[8] (*Eliza*, 3,) b. July 8th, 1839.

6 Clarissa Taylor,[8] (*David J. Taylor*, 30,) b. July, 1835.
   M. Henry Monroe.
   Ch. Dora, Elmore, Clarissa.

7 Clarissa Jane Carr,[9] (*Azubah Corwin*, 3.)
   M. Ransom McIlhenny. (*McGrawville, N. Y.*)

8 Clarissa Monroe,[9] (*Clarissa Taylor*, 6,) b. May, 1862.

(¹) *CLARISSA,[7] [*Amos*, (1),] b. ——, 1790–1800 ?
   M. —— Ficklin. (*Paris, Bourbon Co., Ky.*)

(2) *CLARISSA,[8] [*Samuel*, (6½,)] b. December 25th, 1829.
   M. Mason Jones, March 8th, 1849.
   Ch. Eight, all residing at Omega, Ohio.

1 Clayton Drake,[8] (*Mary Corwin*, 20,) b.

* Corwine.

1 Rev. Clinton Colegrove,[8] (*Bela II. Colegrove*, 1,) b. in Sardinia, N. Y.,
   June 9th, 1827.
M. Elarcia M., daughter of George and Martha Bigelow, October 28, 1850.
Ch. Francis I., Adelaide M., Edith S., Belah II., George B., Alice E.,
   Mary B.

He entered Hamilton College (now Madison University) in 1841, and
remained two years. Studied medicine with his father, and also attended
lectures, and graduated from the Medical College of Buffalo, in 1850.
Practiced medicine in Sardinia, N. Y., for seven years. He then entered
the ministry of the Baptist denomination, and was pastor at Holland, Erie
Co., N. Y., 1858–9; Springville, N. Y., 1859–60, nearly two years; his
health then required an intermission of labor; pastor again, at Sardinia,
1862–5; Yorkshire Centre, 1867; Yorkshire, N. Y., 1868, to present time,
1871. He holds Adventist views. He has written and published considerably in the denominational papers of his faith. He was a member of the
Sanitary and Christian Commissions during the rebellion.

1 Coe,[7] ? (———.)
M. Sarah Ann ———.
Sold land in Southport, Chemung Co., N. Y. Vol. 36, p. 268 of Deeds.

2 Coe Hawkins,[8] (*James II.*, 13,) b. May 24th, 1854.

3 Coe Mullock, (*Mary Corwin*, 22.)

(1) * Conde May,[9] [*Rev. Jesse D. II.*, (1),] b. in Kentucky, May 2d, 1862.

Cora, (Greek,) maiden.

1 Cora Ellen,[9] (*Hubbard*, 2,) b. September 29th, 1851.

2 Cora Estell Hackett,[9] (*Samuel B. Hackett*, 32,) b. April 22d, 1858.
   (*Vermont.*)

3 Cora Whitney,[10] (*Emma E. Hackett*, 19,) b. January 11th, 1870. (*Vt.*)

Cordelia, (Latin,) warm-hearted.

1 Cordelia,[8] (*Phineas*, 1¾.)

2 Cordelia Chase,[9] (*Phebe Corwin*, 8¼,) b.

(1) * Corliss Hitt,[7] [*Rev. Jesse D. II.*, (1),] b. in Kentucky, November
   16th, 1855.

Cornelia, feminine for Cornelius.

1 Cornelia,[7] (*Elisha*, 1,) b. ———, 1825–35.

2 Cornelia,[8] (*Rev. David*, 15.)

3 Cornelia Brown,[8] (*Parmenas Brown*, 2.)

4 Cornelia Drake,[8] (*Arvilla Corwin*, 1,) b. ———, 1835–45.

5 Cornelia R. Easton,[9] (*Hannah E. Corwin*, 16,) b. November 21st, 1849.

6 Cornelia Lydia Newman,[9] (*Moses B. Newman*, 9,) b. at Wyandotte,
   Kansas, March 21st, 1859, d. July 19th, 1859.

* Corwine.

Cornelius, (Latin,) perhaps a little horn.

1 CORNELIUS,[7] (———,) b. ——.   (*Tyre, N. Y.*)

2 CORNELIUS G.,[8] (*George W.*, 18.) b. March 12th, 1842.
  M. Mary L. King, July 25th, 1863. She b. June 22d, 1846.
  Ch. Clara M., Ernest N.   (*Greenport, L. I)*

3 Cornelius Wood,[8] (*Mary Corwin*, 16.)

4 Cornelius Hackett,[9] (*James Hackett*, 51.)   (*Boston, Mass.*)

(1) * CORNELIUS T.,[8] [*Gideon R.*, (1,)] b. Nov. 6th, 1829.
  M. Mary, daughter of John Hart, November, 1859, of Harborton, N. J.
  No ch.   (*Pennington, N. J.*)

1 CORTLANDT ELIAS,[7] (*Ezra*, 2,) b. April 18th, 1811.
  M. Julia, daughter of Joseph Eggleston, of New-Windsor, Ct., and
    Hannah Patrick, of Colebrook, Ct., who were married about 1804,
    and moved to Chenango, N. Y.
  Ch. Hannah E., Melissa, Adelia.

2 Cortlandt De Boyce,[8] (*Sarah Corwin*, 20,) b. ——, 1820–30.

3 Cortlandt W. Milnor,[8] (*Harriet A. Millspaugh*, 19,) b. March 6th, ~~1860.~~ *1840*

4 Cortlandt W. Clark,[8] (*Sarah J. Millspaugh*, 66,) b. April 17, 1836.

Corwin, (Latin.) a raven.

1 Corwin Carr,[8] (*Azubah Corwin*, 3,) b. ——.
  M. Ardilla Inman.   (*McGrawville, N. Y.)*

2 Corwin Sage,[9] (*Evelina Corwin*, 1.) b. May 31st, 1859.

1 Crissa P. North,[8] (*Sibyl Campbell*, 3,) b. May 17th, 1848.

Cynthia, (Greek,) belonging to Mount Cynthus.

1 CYNTHIA,[6] (*Eli*, 1,) b. October 3d, 1782, d. May 1st, 1835.
  M. James Millspaugh.
  Ch. Gilbert C., Gideon H., Mary A., Sarah J., Harriet A., John W.,
    Benjamin F., William H., Charlotte, Cynthia.
                              (*Ohio, afterward Mt. Pleasant, Iowa.*)

2 CYNTHIA,[6] (*Nathaniel*, 5,) baptized August 20th, 1769, at Southold, L. I.

3 CYNTHIA,[5 or 6] (———,) b. ——, 1774, d. October 13th, 1824.
  M. Nathaniel Wells. He was b. February 9th, 1775 ; (was the son of
    Nathaniel Wells, b. 1740, and Jerusha Wickham, who was b. 1743,
    and d. 1817.) Nathaniel, after Cynthia Corwin's death, married, suc-
    cessively, Abigail Wickham and Frances Belknap.
  Ch. Nathaniel C.   (*Middletown, N. Y.*)
  Perhaps Cynthia, 3, was a grand-daughter of David, 2.

4 CYNTHIA,[7] (*Jemuel*, 1,) b. January 3d, 1824, d. January 27th, 1824.

5 CYNTHIA,[7] (*Daniel*, 9,) b. ——, 1800–10.

6 CYNTHIA,[7] (*David*, 8.)

                        * Corwine.

3

7 CYNTHIA SOPHIA,[8] (*Rev. Eli*, 4,) b. October 13th, 1854, at San José,
  Cal.

8 CYNTHIA LOTTY,[9] (*John D.*, 47.)

9 Cynthia Little,[7] (*Abigail Corwin*, 3,) b. April 4th, 1811.[•]
  M. Adam Crans, January, 1841, who was b. August 10th, 1817.
  Ch. Adam, Irene, Henrietta.   (*Goshen, N. Y.*

10 Cynthia Hobart,[7] (*Mehetable Corwin*, 3,) b. December 26th, 1799, d.
  May, 1841.
  M. Jacob Halstead, December 16th, 1819.
  Ch. Mary, Eliza, Amanda, Margaret, Cynthia.

11 Cynthia Millspaugh,[7] (*Cynthia Corwin*, 1,) b. ——, 1828 ?

12 Cynthia E. Millspaugh,[7] (*Dorothy Corwin*, 2,) b. ——, 1828 ? d. ——,
  1845 ?

13 Cynthia Millspaugh,[8] (*William E. Millspaugh*, 72,) b. October 16th,
  1845.

14 Cynthia Chappel,[8] (*Charlotte Hobart*, 9,) b. February ——, 1851.

15 Cynthia Mullock,[8] (*Elizabeth Corwin*, 11,) b. ——, 1820-30.
  M. Peter Jackson.   (*Mongaup, N. Y.*)

16 Cynthia Halstead,[8] (*Cynthia Hobart*, 10,) b. April 8th, 1841.

17 Cynthia Jones,[9] (*Amanda Halstead*, 6.)

Daniel, (Hebrew,) a divine judge.

1 DANIEL,[2] (*Theophilus*, 1,) b. ——, 1660-70, d. before 1719.
  M. Mary, daughter of Simon and Mary Ramsay, after 1698.
  Ch. Daniel ? Henry ? Simeon ?
  She was a widow in 1719. In 1703, he received a deed from Theophi-
lus, his brother, for land at Aquebogue. This might have been either
Theophilus, 1, or his son Theophilus, 2. The latter is no doubt correct, as
Matthias, the first settler, mentions only John, Martha, and Theophilus, in
his will, which see. (Matthias, 1.) In 1698, Daniel's name stands alone in
the census list, intimating that his father was probably dead. (*Doc.
Hist.* i. 454.) In the same list, p. 455, occurs the name of Mary, daughter
of Simon Ramsay. Hence Theophilus was not yet married at that date,
at least to her.   (*Southold, L. I*)

2 DANIEL [JEDEDIAH,][1] (*Daniel*, 1, ?) b. about 1690, d. September 7th, 1747,
  (or 1744.)
  M. (1) ——. (2) Elizabeth Cleaves, January, 1722-3.
  Ch. Nathan, Pelatiah, Mary, Michal,* Lucas, Jedediah, Silas, John
  Daniel, Edward ? and another who married an Armstrong. His will,
  in Lib. 16, N. Y. C., p. 261. His name occurs as a freeholder of
  Southold, L. I., in 1737. (*Doc. Hist. N. Y.* iv. 132.)
  An Elizabeth Corwin, widow of Daniel, died March 30th, 1774. (*Aque-*

* This name is feminine.

logue Records.) He mentions in his will, besides his children, two grand-sons, Edward and Separate, and a granddaughter, Mehetable Armstrong.

3 DANIEL,⁵ (*Daniel*, 2,) b. ——, 1725–30, d. about 1800.
 M. Temperance Baily, about 1750.
 Ch. Betsy, Phebe, Mahala, Elsie, Temperance, Polly, Lucinda, Dilly, Ruth, Daniel, Henry, Nathaniel.   (*Southold, L. I.*)

4 DANIEL,⁶ (*David*, 4,) b. ——, 1759–64, d. ——, 1829.
 M. Anna Hulse.
 Ch. Daniel, David, Mary, Martha, Charity, Anna, Abigail. Bought land in Orange Co., N. Y., 1798.   (*Wallkill, N. Y.*)

5 DANIEL,⁶ (*William*, 2,) b. April 13th, 1788, in Morris Co., N. J. Yet living in 1870.
 M. (1) Mary Hamill, (2) Elizabeth Hamill, (3) Elizabeth Spinning, (4) Elizabeth Brace, b. 1796, and died November 13th, 1866.
 Ch. Daniel W., Caroline, Mary J., George, Marselus, Sarah.
             (*Moved to Oxford, O.,* 1808.)

6 *DANIEL⁶? (*George*, 1,) b.——, 1770–80.

7 DANIEL,⁶ (*Jedediah*, 2, ?) b. ——, 1760 ? d. ——.
 M.
 Ch. Dilly, Henry. In 1828, a Daniel Corwin of Southampton died. Perhaps this one.

8 DANIEL,⁶ (*Silas*, 1,) b. February 28th, baptized April 13th, 1773, at Aque-bogue, d. April 5th, 1837.
 M. Mary Little, who was b. October 25th, 1777.
 Ch. Rosetta.   (*A physician at Minisink, N. Y.*)

9 DANIEL,⁶ (*John*, 6,) b. February 8th, 1773, d. ——, 1818.
 M. Mary Tuthill, October 12th, 1800. (*Aquebogue Records.*)
 Ch. Hendrick, George W., Daniel T., Lucinda, Elvira, Cynthia.

10 DANIEL,⁶ (*Daniel*, 3,) b. ——, 1750–60.
 In 1795, Daniel Corwin, and Ruth his wife, of New-Cornwall, Orange Co., N. Y., sold land in Cayuga Co. to William Corwin. In 1801 and 1802, sold land to Barnabas Corwin. Sold land again in 1803 and 1808. In 1808, they were of Milton, Cayuga Co., N. Y.
 In 1827–35, we find a Daniel Corwin having a son Daniel dealing in land in Cayuga Co. He is said to be of the town of Genoa, formerly Milton. In 1830, Daniel, Jr., buys land of his father.

11 DANIEL,⁷ (*Daniel*, 4,) b. ——, 1779, yet living.
 M. (1) Irene Smith, (2) Mary Reynolds.
 Ch. Rev. Gabriel S.; Anna M., Eliza J., George W., Alma Irene, Har-riet N., Joanna R., Charity A.   (*Howells, N. Y.*)

12 DANIEL YOUNG,⁷ (*Jemuel*, 1,) b. September 6th, 1836.
          (*Hopewell, Orange Co., N. Y.*)

* Spelled his name *Curran*, though of the Corwin family.

1151617

13 Daniel Wells,[7] (*John II.*, 8,) b. October 10th, 1822, d. October 11th, 1826.

14 Daniel,[7] (*Moses*, 1,) b. November 22d, 1831.
   M. Sophia B. Barlow, March 25th, 1862.   She was born November 9th, 1837.
   [7] *Ch.* Florilla M., Ella.   (*Middletown, N. Y.*)

15 Daniel William,[7] (*Daniel*, 5,) b. July 5th, 1811.   (*In the West.*)

16 Daniel,[7] (*Jeremiah*, 2,) b. December 29th, 1788, d. ——, 1859.
   M. Irene Tyler.
   Ch. Hannah J., George W., Brewster, John, Charles T., James, Ann Kate.

17 Daniel,[7] (*Jabez*, 1,) b. October 29th, 1808, d. ——, 1825 ?
   Unmarried. A former son Daniel, b. July 6th, 1802, lived only a few days.

18 Daniel Tuthill,[7] (*Daniel*, 9,) b. ——, 1795–1805.
   (A Daniel T. Corwin and Sarah his wife were of Newfield, Tompkins Co., N. Y., 1855.)

19 Daniel,[7] (*Henry*, 6,) b. September 13th, 1776, d. November 20th, 1862.
   M. Charity Reeve, January 15th, 1803.   She b. April 25th, 1785, d. February 3d, 1871.
   Ch. Barnabas R., Joshua, Daniel, Henry, George, La Fayette, Chauncey, Hampton, Sarah C.   (*Riverhead, L. I.*)

20 Daniel,[7] (*Nathaniel*, 5,) b. ——, 1770–90.
   M. Sarah ——.
   Ch. Daniel.
   (A Daniel Corwin married Sarah Corwin, December 7th, 1826.)
                                              (*Mattituck Records.*)

21 Daniel A.,[8] (*David*, 17,) b. June 19th, 1815.
   M. Mary Land.
   Ch. Edward M., Frederick.   .   (*Eddyville, Iowa.*)

22 Daniel M.,[8] (*Gilbert*, 4,) b. October 16th, 1847.

23 Daniel Ardent,[8] (*Anson*, 1,) b. May 20th, 1822.
   M. (1) Mary Frances Corwin, (2) Maria M. Hallock.
   Ch. Charles M., Emma F., William, Edward, Daniel, Franklin.

24 Daniel,[8] (*Daniel*, 19,) b. April 23d, 1808.
   M. Huldey Goodale.   She b. May 29th, 1815.
   Ch. Rosetta, Chauncey.

25 Daniel,[8] (*Daniel*, 20.)

26 Daniel W.,[8] (*De Witt C.*, 2,) b. May 22d, 1844.

27 Daniel ?[8] (*James T.*, 21,) b. ——, 1840–50.

28 Daniel Terry,[8] (*George*, 10,) b. ——.
   M. Elizabeth ——.   (*Orange Co., N. Y. ?*)
   (There was a Daniel Corwin, and Mary his wife, (late Mary Emmons,)

who sold land in Essex Co., N. J., in 1811. He was a grocer in New-York City, 1810–13.)

28½ DANIEL,[8] (*Henry,* 11,) b. March 10th, 1825.
 M. (1) Mary Smith, November 23d, 1848 ; (2) Diantha Rundell.
 Ch. William D., Oriet, Lewis, Marietta.   (*Lake City, Minn.*)

29 DANIEL,[9] (*Oliver,* 5.)

30 DANIEL JOSHUA,[9] (*Silas G.,* 12,) b. February 3d, 1861.

31 DANIEL,[9] (*Daniel A.,* 23,) b. ——, 1860–5.

32 Daniel Jennings,[6] (*Sarah Corwin,* 4.)

33 Daniel Reeve,[7] (*Mary Corwin,* 12,) b. ——, 1810–20

34 Daniel Campbell,[7] (*Sarah Corwin,* 6,) b. November 16th, 1807.
 M. Mary Rand, April 30th, 1834.
 Ch. Asenath, Theodore C., Daniel E., Albert B., Mary P., Florence D., Theodore, Daniel E.

35 Daniel Young,[7] (*Charlotte Corwin,* 1,) b. March 21st, 1825.
 M. Jane F. Cozzens, January 13th, 1847.
 Ch. Harriet C., Daniel C.   (*Morning Sun, Iowa.*)

36 Daniel Corwin Brown,[7] (*Mary Corwin,* 8,) b. ——, 1800–5.
 M. Jane Van Nest.
 Ch. Martha, Silas Y., Emeline, Leah A., Mary Arminda.
       (*Millsburg, Orange Co., N. Y.*)

37 Daniel Weed,[8] (*Charity Corwin,* 1,) b. ——, 1820–30.

38 Daniel Squires,[8] (*Elizabeth Corwin,* 21.)

39 Daniel E. Campbell,[8] (*Daniel Campbell,* 34,) b. August 11th, 1840.
 M. Mary J. Fleming, December 12th, 1860.
 Ch. Mary D.

40 Daniel Brown,[8] (*Parmenas Brown,* 2.)

41 Daniel F. Flannery,[8] (*Mary Corwin,* 33,) b. at New-London, O., January 18th, 1855.   (*Cleveland, O.*)

42 Daniel Cozzens Young,[8] (*Daniel Young,* 35.)

43 Daniel Corwin Bishop,[8] (*Edward Bishop,* 15,) b. about 1820.
       (*Hamburg, Mich.*)

44 Daniel Halsey Howell,[8] (*Elizabeth Corwin,* 20,) d. when seven years of age.

45 Daniel Halsey Howell,[8] (*Elizabeth Corwin,* 20,) b. ——.

46 Daniel Irvin Cox,[8] (*Mary S. Corwin,* 44.)

(1) Daniel Corwine Stevenson,[9] [*Sarah A. Corwine,* (5,)] b. December 18th, 1851, at Aberdeen, Ohio.
 Now in Transylvania University, Lexington, Ky.

David, (Hebrew,) beloved.

1 DAVID,[3] (*Theophilus*, 1, ?) b. about 1675? d. December 25th, 1739.
  M. Jerusha Terrill ?   (*Mattituck*, L. I.)
  "(In 1735, a David Corwin bound himself to learn the hatter's trade.)

2 DAVID,[4] (*John*, 2,) b. about 1705–10, d. before 1782.
  M. Deborah Wells, 1732.   She was b. 1717, d. November 24th, 1798.
  Ch. David, Joshua, Joseph, Phineas, Eli, Annie, and perhaps other
  daughters.

In 1776, on census list, 403, having in his family one male over fifty,
and one female over sixteen.   He moved to Orange Co., N. Y., probably at
the opening of the Revolutionary War.   He was buried at Middletown, in
that county, but no tombstone has been found.   This is, no doubt, the
David whose name occurs as a freeholder of Southold, in 1737. (*Doc. Hist.
N. Y.* iv. 132.)   His name is often written Curwin.   His descendants are
very many.

3 DAVID,[5] (*Samuel*, 2,) b. October 12th, 1739, d. April 18th, 1801.
  M. Anna Tuthill. She died February 10th, 1801. (*Aquebogue Records.*)
  Ch. Henry B., Asa, Samuel.
  (A David Corwin married Anna Terrill, 1762, according to *Mattituck Re-
  cords.*)

4 DAVID,[5] (*David*, 2,) b. ——, 1730? d. ——, 1794.
  M. (1) Mary Wells, 1750.   (She was one of the eleven children of Capt.
  Daniel Wells, who was born 1702, and died 1761.) (2) Abigail Davis.
  Ch. Daniel, David, Mary, Jesse, Eli, Abigail, William, Joseph, Mche-
  table, Elisha, Nebat, Phineas, Deborah, Naboth D.

In 1775, he signed an engagement to support Congress; in 1776, on
census list, 370, having in his family one male, two females over sixteen,
and four children.   In 1770, in company with George McNish, Ab. Gale,
and Jonathan Smith, he bought 787 acres of land at Minisink, Ulster
Co., N. Y.   In this patent, now in possession of Mrs. Phebe Corwin,
(widow of Nathan H. Corwin,) of Newark, N. J., his name is spelled *Curwin.*

5 DAVID,[5] (*Theophilus*, 4, or *Timothy*, 3,) b. in Little Britain, Orange Co.,
  N. Y., June 9th, 1777, d. Jan. 25th, 1843.
  M. Sarah Christy, 1803.   She was daughter of John and Anna Christy.
  Ch. Anna, Sarah, Maria, Richard T., William, David, John C., George
  H., Jane A.

Was a blacksmith.   He lived successively in Little Britain, Orange Co.,
in Dutchess Co., and in Ulster Co., N. Y.

6 DAVID,[6] (*David*, 4,) b. ——, 1750–60, d. ——, 1802 ?
  M. (1) —— Corey, (2) —— McNish, (3) —— Burchon.
  Ch. Phineas ? Thomas, and a daughter, who m. —— Bard.   (*Ohio.*)

7 DAVID W.,[6] (*Eli*, 1,) b. July 17th, 1793, d. Feb. 11th, 1853.
  M. Catharine Miller.   (She now lives in Matteawan, N. Y.)

Ch. Sarah, Jane, Olivia, Mary II.

His descendants live in Liberty, Sullivan Co., N. Y. (*Middletown, N. Y.*)

8 DAVID,[6] (*Jesse,* 2,) b. Aug. 5th, 1776, in Fayette Co., Pa. Yet living, 1871.

M. Hannah Roberts.

Ch. Maria, Elias, Elvira, Harvey, John, Cynthia, Minerva, Harvey, Amanda, Thomas.

In 1785, he was taken by his father to Kentucky. In 1798, he removed to Lebanon, Ohio, and became a merchant. Had been a farmer. While engaged as a merchant, his son sat on a keg of powder with a lighted cigar, and a spark caused an explosion, killing the boy, and blowing up the store. He then engaged in speculation in lower Kentucky, for a number of years, and subsequently farmed in different localities in Ohio. He once kept a hotel at Urbana, Ohio. On his ninety-fourth birthday, 1870, a picnic of the Corwins of Warren Co., Ohio, was held in his honor. He has black eyes, dark skin, hair once dark, now snowy white, and is noted for his proverbial kindness of heart.

9 DAVID,[6] (*Samuel,* 6,) b. ——, 1771? d. ——.

M. Lydia Hart.

Ch. Uriah, and a daughter who m. a Mr. Fisher.

He went to sea when about eight years of age. He was afterward impressed in the British army, and for years was supposed by his friends to be dead. When he subsequently returned, his only brother, Samuel, cheerfully divided his inherited property with him, though he had long supposed it was all his own.

10 DAVID,[6] (*David,* 5,) b. Oct. 20th, 1815.

M. Charlotte Anna, daughter of Thomas and Philena Mulleminx, Feb. 20th, 1845.

Ch. Josephine M., Sarah E., Thomas J., Mary E., George D., David.

(*New-Paltz, N. Y.*)

He writes that his grandfather fell in battle at Fort Montgomery, and thinks that his name was Timothy, (the records being lost ;) but more probably it was Theophilus. His grandmother Corwin's maiden name was Anna Jaynes.

11 DAVID,[6] (*Joshua,* 1,) b. Oct. 2d, 1762, d. ——, 1824?

M. Naomi Corwin, (No. 1,) Dec. 5th, 1782. (*Aquebogue Records.*)

No Ch. (*A tailor, Calhoun, N. Y. ?*)

12 DAVID,[7] (*Rev. Joseph,* 5,) b. July 22d, 1791, at Aquebogue, d. July 28th, 1835.

M. Nancy Reeve, March 4th, 1817.

Ch. David H., Nancy P., Eliza A., Joseph W., Frances M., George H., Mary J., Charles N., Bethia E.

(*Southampton, Aquebogue, Cutchogue, L. I.*)

13 DAVID M.,[7] (*Henry,* 5,) b. ——, 1825?

M. —— ——.

     Ch. James, Carrie.   (*Policeman, Brooklyn, N. Y.*)

14 DAVID,[7] (*Henry,* 6,) b. Nov. 30th, 1789, d. Jan. 14th, 1861.
     Unmarried.

15 REV. DAVID,[7] (*Asa,* 3,) b. Dec. 12th, ~~1796.~~ - *1798*
     M. —— ——.

     Ch. Warren, Cornelia, Emily.

A Baptist clergyman, who lived for a while in the eastern part of New-York State, perhaps not far from Albany.

     (↓ David, son of Asa Corwin, was baptized at Aquebogue, June 1st, 1804.)

16 DAVID,[7] (*David,* 10.) b. June 15th, 1858.

17 DAVID,[7] (*Daniel,* 4,) b. Sept. 6th, 1793, d. May 23d, 1857.
     M. Catharine Houston.

     Ch. Daniel A., Sarah J., Mary, Emeline, Abigail, Thomas.   (*Iowa.*)

18 DAVID,[7] (*Abner,* 1,) b. at Mt. Hope, Orange Co., N. Y., July 18th, 1790, d. Sept. 26th, 1839.

     M. Hester Totten, of Mt. Hope, Oct. 4th, 1817. She b. Dec. 10th, 1799. [She was daughter of John Totten, (b. Aug. 5th, 1770, d. Oct. 13th, 1864,) and Lydia Jacks, (b. Oct. 2d, 1771, d. March 1st, 1848,) of Morristown, N. J.]

     Ch. John T., Stephen O., Benjamin F., (b. May 26th, 1822, drowned Feb. 5th, 1843,) Caroline, Henry W., Andrew Jackson, (b. Jan. 29th, d. Feb. 2d, 1829,) Jemima J., Lydia A., Sarah E., David R. P.

He was a soldier in the war of 1812, in the coast guard of Long Island Sound. His widow received bounty land, but no pension, on account of not being married at the time of his enlistment. He lived for many years at New-Vernon, Orange Co., N. Y. In 1836, he removed to Buttermilk Falls, on the Susquehannah, about twenty miles above Wilkesbarre, Pa.

19 DAVID,[7] (*John,* 7,) b. ——, d. ——, young.

20 DAVID HENRY,[8] (*Hector,* 1,) b. ——, 1830–40.

20½ DAVID RITTENHOUSE PORTER,[8] (*David,* 18,) b. July 18th, 1838.
     M. Susannah Irwin, Oct. 18th, 1860.

     No ch.

He served in the R. R. Corps of U. S. Army during the Rebellion; was stationed at Chattanooga, Georgia, in charge of the forwarding department. Honorably discharged on account of sickness, July 4th, 1864. Now Secretary and Treasurer of the Pittsburg, Virginia, and Charleston R. R. Co.   (*Pittsburg, Pa.*)

21 DAVID SWEEZEY,[8] (*Joshua,* 6,) b. April 15th, 1813, d. Oct. or Nov., 1870.
     M. Sarah Mills.

     Ch. Eliza.

22 DAVID,[8] (*John,* 30,) b. ——.

23 DAVID,[8] (*Alsop*, 1.) b. ——, 1810–20.   (*Near Montrose, Pa.*)

24 DAVID,[8] (*David*, 12,) b. Dec. 15th, 1817.
  M. Ann Amelia Fairbanks.
  Ch. Mary D., Amelia A., Ella V., Omer R., Juliette, Emma, Addie.
                                                        (*Sag Harbor, L. I.*)

25 DAVID B.,[8] (*Robert G.*, 2,) b. Nov. 27th, 1839.
  Lieut.-Col. 2d Ind. Reg. ; afterward Col. 5th Ind. Reg.   Now City Soli-
    citor for Dayton, Ohio.

26 DAVID JAMIESON,[8] (*Isaac*, 8,) b. July 19th, 1855.

27 David Everett,[7] (*Deborah Corwin*, 1,) b. ——, 1815–25.

28 David Penney,[7] (*Annie Corwin*, 3,) b. April 25th, 1786.
  M. Fanny Smith, Nov. 14th, 1805.

29 David Thatcher,[7] (Jemima Corwin, 1,) b. ——, 1803.   (*Missouri.*)

30 David J. Taylor,[7] (*Fanny Corwin*, 1,) b. May —, 1813.
  M. Sarah Wilson, Jan. —, 1833.
  Ch. Clarissa, Eliza, John, Sarah, Elnora, Fanny, Wilmer, Abbie.
                                                        (*Savannah, N. Y.*)

31 David W. Tuthill,[7] (*Azubah Corwin*, 1,) b. ——, 1805.
  M. Sarah Jane Mills.

32 David Reeve,[7] (*Mary Corwir*, 12,) b. ——, 1810–20.
                                                        (*Middletown, N. Y.*)

33 David Davis,[7] (*Elizabeth Corwin*, 4,) b. ——, 1780–90.

33½ David Bishop,[7] (*Mary Corwin*, 11,) b. ——, 1786 ; living.
                                                (*Barrington, Yates Co., N. Y.*)

34 David Carr,[8] (*Azubah Corwin*, 3.)
  M. Jerusha Underwood.   (*East-Homer, N. Y.*)

35 David Taylor,[8] (*John C. Taylor*, 70,) b. Sept. —, 1852.

36 David T. Van Wagener,[8] (*Maria E. Upright*, 13,) b. Sept. 26th, 1852.

(1) [8]DAVID MORRISON,[8] [*Joab*, (1,)] b. December, 1824.
  He is the resident Agent of the Pacific Mail Steamship Co., at Panama,
New-Granada, which position he has held since 1858.   Previously he was
Receiving Teller of the Commercial Bank of Cincinnati, O.

Deborah, (Hebrew,) a bee.

1 DEBORAH,[6] (*David*, 4,) b. ——, 1786 ?
- *M.* Ephraim Everett.
  Ch. Gabriel, Abigail, Lansing, Olive, Attison, Ezra, Azubah, Lewis,
    Ephraim, George, David, Deborah.

2 DEBORAH,[6] (*Joseph*, 1,) b. ——, 1774 ?
  M.   (*Ohio.*)

* Corwine.

3 DEBORAH,[6] (*Thomas*, 1,) b. ——, 1790–1800.
  M. Chester Gregg.

4 DEBORAH,[7] (*Isaac*, 1,) b. June 6th, 1780.
  M. Amos Leek.
  Ch. Isaac H., John, and about 10 others, all in the West.

5 DEBORAH,[7] (*Silas*, 5,) b. ——, 1810–20.

6 DEBORAH,[7] (*Jacob*, 2,) b. ——, 1790–1800.

7 DEBORAH ANN,[8] (*Anson*, 1,) b. May 4th, 1820.
  M. James Downes.
  Ch. Elizabeth M.

8 Deborah Osman,[4] (*Sarah Corwin*, 1,) b. 1700–1710.
  (Name not in census-list, 1698. (*Doc. Hist.* i. 450.) Hence born after 1698.)

9 Deborah Ann Everett,[7] (*Deborah Corwin*, 1,) b. ——, 1815–25.

10 Deborah Penney,[7] (*Annie Corwin*, 3,) b. July 4th, 1793.
   M. Ephraim Clark, December 7th, 1816.

11 Deborah A. Aldrich,[8] (*Deborah Corwin*, 6.)

1 DE FOREST,[6] (*Pollydore B.*, 1,) b. October 12th, 1830.
  M. Abigail A. Barton, November 26th, 1854. She b. July 9th, 1838.
  Ch. Alonzo B., Ezra P., Mary F.

Delilah, (Hebrew,) feeble.

1 DELILAH,[9] (*Ezra*, 3,) b. ——, 18⁣48, d. 1850.

2 DELILAH,[9] (*Noah*, 3,) b.

Deliverance, from deliberare, (Latin,) a setting free.

1 DELIVERANCE,[6] (*John*, 6,) b. May 2d, 1764, d. November 1st, 1821.
  M. William Horton, November 11th, 1787. (*Aquebogue Records.*)
  Four *Ch.*

Delphina, (Greek?) a dolphin?

1 Delphina Murray,[9] (*Lucinda Corwin*, 1.)

2 Delphina Davaw,[9] (*Helen Hunsike*, —.)

1 DENCY,[7] (*John*, 9,) b. January 25th, 1792, at Aquebogue. Yet living.
  M. Jonathan B. Concklin.

1 DENNISTON,[7] (*Samuel*, 11,) b. ——, ——, d.
  Unmarried.   (*Peconic, L. I.*)

2 DENNISTON,[8] (*Henry*, 11,) b. December 22d, 1829.
  M. Sarah Smith, January 29th, 1857.
  No Ch.   (*Peconic, L. I.*)

Desire, from desiderare, (Latin,) a longing.

1 DESIRE,[6] (*Phineas*, 1,) b. April 6th, 1788, d. June 15th, 1834.

M. William Little, October 8th, 1807.

Ch. Lucretia, Mehetable, William C., George C., Bethia, Catharine, Desire, John A., Oscar L., Theodore W.   (*Elbridge, N. Y.*)

2 Desire Hobart,[7] (*Mehetable Corwin,* 3,) b. April 20th, 1822.

3 Desire Little,[7] (*Desire Corwin,* 1,) b. January —, 1821.

1 De Witt C.,[7] (*Nebat,* 1,) b. ——, 1810–20.

2 De Witt Clinton,[7] (*Alexander,* 1,) b. ——, 1813.

M. Harriet Amelia Chilson.

Ch. George W., Daniel W., William H., Louisa J., Sarah F.   (*Newburg, N. Y.*)

Diana, (Latin.) a goddess.

1 Diana,[7] (*Edward,* 3,) b. November 3d, 1812.

M. W. Edson.

Three *Ch.*

Diantha, (Greek,) flower of Jove; a pink.

1 Diantha Smith,[8] (*Philenia Corwin,* 1,) b. April 18th, 1830, d. —— 1861.

M. Franklin C. Thorne, 1850.

Ch. Carrie, Freddie.   (*Vermont.*)

1 Dicey Thatcher,[8] (*Jesse Thatcher,* 10,) b. April 21st, 1827.

1 Dilly,[6] (*Daniel,* 3,) b. ——, 1750–60.

M. Zadok Eldridge.

2 Dilly,[7] (*Daniel,* 7,) b. ——, 1790–1800.

Dora, (Greek,) gifts, or from Dorothea.

1 Dora Monroe,[9] (*Clarissa Taylor,* 6,) b. August —, 1857.

1 Dorastus Brown,[7] (*Mary Corwin,* 8,) b. ——, 1800?

M. Esther Brink.

Ch. Esther, James W., Theodore, Mary E., Elsie A.   (*Greenville, Orange Co., N. Y.*)

2 Dorastus Brown,[8] (*Silas C. Brown,* 16.)

Dorothy, or Dorothea, (Greek,) a gift of God.

1 Dorothy, (Dolly,) Jane,[7] (*Jason,* 1,) b. February 28th, 1811.

Unmarried.   (*Scotchtown, N. Y.*)

2 Dorothy,[6] (*Eli,* 1,) b. March 26th, 1790.

M. Samuel Millspaugh, March 26th, 1814. (His first wife was Dorothy Brown, by whom he had two children, Gilbert Brown and Adam White.) He d. 186–.

Ch. Mary, John, William E., Isaac L., Cynthia E.

3 Dorothy,[7] (*Eli II.,* 2,) b. May 12th, 1809, d. March 10th, 1846.

4 Dorothy Jane,[7] (*Jemuel,* 1,) b. December 30th, 1840.

M. Ezra Smith.   (*Hopewell, Orange Co., N. Y.*)

5 Dorothy T.,[7] (*Ezra,* 2,) b. April 2d, 1803.

M. Jesse Pike.

Ch. Orsamus, Mary A., Olivia, Le-Roy, Fitz-Gerald, Rollin.

G Dorothy Little,[8] (*Henry Little*, 28,) b ——, 1825-30.

   M. —— King.

1 Douglass M. Van Wagener,[9] (*Sybil T. Padden*, 4,) b. October 31st, 1859.

1 DRAKE,[8] (*Nathaniel*, 3.)

   His wife was burned to death at Hope, N. J.   (*Buttsville, Pa.*)

1 Duane C. Crego,[8] (*Clarissa Taylor*, 4,) b. February, 1862.

1 DUDLEY,[8] (*Pollydore B.*, 1,) b. November 1st, 1849.

1 DWIGHT,[10] (George, 48,) b. ——, 1863.   *1 1 4 1 ᒣᒧ᚛ ᑑ ᓇᑭ*

2 DWIGHT TOWNSEND,[10] (*Henry II.*, 26,) b. ——, 1867.

Eben, (Hebrew.) a stone.

1 EBEN,[8] (——,) b.

   Lived in Newark, N. J., 1868-9. (See *Directory for* 1868-9.)

1 E. BALDWIN,[7] (*Ezra*, 2,) b. December 27th, 1794. A mute.

             (*Cortlandt, N. Y. ; now Wis.*)

Ebenezer, (Hebrew,) stone of help.

1 EBENEZER,[6] (*William*, 2,) b. October 13th, 1790, d. April 8th, 1851.

   M. (1) Elizabeth Skellinger, (2) —— Hatch.

   Three *Ch.*   (*Prattsville, O.*)

2 EBENEZER,[6] (*Silas*, 1,) baptized October 9th, 1777, at Aquebogue, d. May 13th, 1780. (*Aquebogue Records.*)

2 EBENEZER,[6] (*Amaziah*, 1,) b. ——, 1790-1800.   (*Blodgett's Mills.*)

3 EBENEZER,[7] (*Ezra*, 2,) b. May 14th, 1796, baptized at Aquebogue, L. I., July 13th.

   M. (1) Nancy Knox, January 24th, 1821, (2) Nancy Ames, April 5th, 1843.

   Ch. Sarah J., Maria, Rosetta, Silas D., Marietta, James A., Harriet.

4 EBENEZER,[9] (*John*, 58.)

Edgar, (Anglo-Saxon.) a javelin, for defense of property.

1 EDGAR,[8] (*William H. II.*, 18,) b. March 27th, 1850.

2 Edgar Crego,[8] (*Clarissa Taylor*, 4,) b. May, 1852.

3 Edgar Cox,[8] (*Mary E. Corwin*, 44.)

4 Edgar E. Davis,[9] (*Rachel Corwin*, 4,) b. ——, 1839. (*Yates Co., N. Y.*)

Edith, or Ada, (Old German,) happiness.

1 EDITH GERTRUDE,[8] (*Albert*, 2,) b. January 28th, 1855.

2 EDITH RECTOR,[8] (*James*, 20,) b. June 2d, 1845.

   M. ————, September 21st, 1864.   (*Cuba, Fulton Co., Ill.*)

3 Edith May Burton,[9] (*Abbey Charlotte Corwin*, 2,) b. ——, 1869.

4 Edith Sela Colegrove,⁹ (*Rev. Clinton Colgrove*, 1,) b. December 24th, 1855.

Edmund, (Anglo-Saxon,) defender of property.

1 EDMUND D.,⁸ (*William*, 19,) b. February 28th, 1865, d. March 16th, 1866.

2 Edmund Philips Newman,⁹ (*Moses B. Newman*, 9,) b. at Wyandotte, Kansas, September 5th, 1860.

3 Edmund E. Smith,⁹ (*George Smith*, 74,) b. ——, 1854–60.
(*St. Lawrence Co., N. Y.*)

Edna, (Hebrew,) pleasure.

1 EDNA,⁹ (*John*, 56,) b. February 17th, 1868.

2 Edna F. Kyte,⁹ (*Clarinda Corwin*, 2,) b. May 7th, 1850.
M. Stephen C., son of Stephen C. Baker and Hannah Durand, March 3d, 1868.
Ch. Charles F. R., Parker C. (*Ransomville, N. Y.*)

Edward, (Anglo-Saxon,) guardian of property.

1 EDWARD,⁵ (*Daniel*, 2 ?) b. about 1710, d. March 16th, 1732–3.
M.
Ch. Edward ? Separate ?
Edward and Separate are mentioned in will of Daniel, 2, as his grandchildren; hence probably their father was dead at time of Daniel's death in 1747. This Edward, No. 1, may be their father. The dates permit, and the sameness of name favors. Edward was a family name in this branch.

2 EDWARD,⁶ (*Edward*, 1 ?) b. about 1731–2, d. about 1761.
M. Hannah Horton, September 27th, 1753. Both of Mattituck when married.
Ch. Phineas, Elizabeth, Nancy H., Edward, Sarah. (*Aquebogue Records.*)
They had their children, Phineas and Elizabeth, baptized February 22d, 1756. On July 29th, 1764, the widow, Hannah Corwin, had three children, Hannah, Edward, and Sarah, baptized. In 1765, the widow, Hannah Corwin, joined the church. Now, a Nancy is well known to have been a sister of this Phineas and Edward; and her son Bela says that she wrote her name Nancy H. Hence either the baptismal record is imperfect, or she may have afterwards prefixed the name of Nancy to Hannah, for reasons of her own. Edward was taken from his family in the night by a press-gang, in the time of the French war, (1755,) put on a ship, where he died in about three or four years, without ever getting back to his family. He was grandson of Daniel, 2, though his father's name is not positively ascertained. There is a will of an Edward Corwin, dated May 6th, 1760, in Liber 23, p. 62, in New-York City; but it contains no names except that of Daniel Wells, executor. It was probated February 9th, 1761. It may be the will of Edward, 2, though it is remarkable that no names are mentioned.

3 EDWARD,⁷ (*Edward*, 2,) b. February 13th, 1759, d. September 15th, 1849.

M. (1) Esther, daughter of Yet-once Barstow, of Franklin, Ct., November 4th, 1784. She was born, May 9th, 1766, d. August 20th,
1797. (2) Olive Colgrove, of Rhode Island, March 4th, 1798. She
was niece of the wife of his brother Phineas, 1½. She was b. May
1st, 1775, d. December 31st, 1859.

Ch. Bestir, Esther, Ghordis, Olive, Edward, Benjamin, Eliza, Nancy,
Diana, Philetus, John.

He removed from Long Island to Connecticut about the opening of the
Revolution. His son Ghordis writes, that his father was six years a Revolutionary soldier, enlisting at the age of seventeen; that he was born on
Long Island, married in Connecticut; that he removed to Cazenovia, Madison Co., N. Y., about 1804, and to Norwich, McKean Co., Pa., about
1822; was a farmer, and both he and his wife were members of the Baptist
Church for more than sixty years.

When he enlisted in the Revolutionary army, his name was recorded as
Currin, (according to pronunciation,) and he received his pension from the
government, under that name, up till his death in 1849, and his mother,
after her husband's death, till her death in 1859. After the close of the
Revolution, Edward and his brother Phineas were merchants for a time in
New-York, under the name of Currin, but they then corrected it to the
proper spelling, of which they were well aware, namely, Corwin. Edward
was taken prisoner by the Indians, and they delivered him to the British
authorities, and he was not exchanged for about twenty-two months.
Hiram Payne, of Waverly, N. Y., undertook to write a sketch of this
Edward; but, it is believed, he never completed it.

The name of Edward Corwin, (Curwain, Curwine, Curvin,) late of Ulster
Co., N. Y., appears on the Revolutionary Bounty-Book, (copies of which
the author consulted in the county clerk's office at Auburn and Syracuse,)
as a private of 2d Regiment New-York, Company 6th, (Fowler's Company,)
of Revolutionary troops. He received 500 acres of bounty land, being
lot 66, in town 10, Cayuga Co., N. Y., on July 8th, 1790. His land was
located by Connolly's Survey. (His deed was filed in the county office
before recording was practiced; these filed deeds have since been recorded
in separate volumes for easy reference, and are known as *"The Filed
Deeds."*) On May 20th, 1792, he assigned to Eliphalet Kellogg, an attorney,
of Saratoga, all such bounty lands as the Secretary of War might award.
This attorney sold for £25 a plot coming to him, which was in the (then)
county of Herkimer, (late Montgomery Co.,) being lot 66, in the town of
Pompey, embracing 600 acres. On December 19th, 1793, Edward sold or
assigned land to Daniel Delavan, but my memoranda do not give the particulars. Edward's name occurs both in the Revolutionary Bounty-Book
and on the Cayuga County records. It is said that a man forged a title to the
whole, or a part, of this bounty land, and Edward gave an attorney $50 to
prosecute his claim, and when the attorney wanted more money, Edward
assigned the land to the attorney, (probably in trust,) upon receiving back
his $50. But the land was never recovered, and it was believed that there was
a great swindle somewhere. But while such are the orthographies of the

name, as taken from the Bounty-Books, the family assert that he received his pension under the orthography of Currin. He was in the battle of Monmouth, at the surrender of Burgoyne, saw André hung, was with Sullivan in his expedition up the Susquehannah, after the massacre of Wyoming. He had many thrilling incidents to relate. He lived to the age of ninety.

3 . EDWARD,[7] (*Edward*, 3,) b. January 11th, 1803, d. November 23d, 1844. Unmarried. A school teacher.

    4 EDWARD BRUSH,[7] (*George*, 5,) b. November 10th, 1806, in New-York City, d. April 19th, 1856, in New-York City.
      M. Hannah Close Knapp, October 5th, 1847. She was born in Greenwich, Ct., June 4th, 1823.   (*New-York City*.)

He entered the Tradesmen's Bank in 1835, and the Chemical Bank in 1840, as first Book-keeper. Subsequently became first Teller. Resigned September, 1847.

He possessed a splendid library of rare, curious, and valuable books, tracts, autographs, manuscripts, engravings, paintings, etc. It embraced almost every department of literature, but was especially rich in bibliography, poetry, almanacs, penmanship, short-hand, illustrated books, Bibles and biblical literature, old theology and sermons, history and biography; it contained many valuable books and tracts relating to the early history of America, and its literature. Duyckinck, in his *Cyclopedia of American Literature*, (Vol. i. p. ix.,) refers to this private collection as one which had been freely opened to him, and from which he had received assistance in the preparation of his great work.

He possessed a rare taste, tact, and knowledge in collecting books. Many of the books had neatly penciled memoranda, indicative of extensive reading, while his large acquaintance with bibliography, in which department he possessed many treatises of great value, was employed in the notice of many rare books, but especially of tracts on America. His illustrated books were especially distinguished for the superiority of the engravings. The library was especially rich in versions of the Bible, the gem of the collection being *Tyndale's Pentateuch*, the only other perfect copy known to exist belonging to Lord Granville. This book was "emprinted at Marlborow, in the land of Hesse, by me Hans Luft, in the year of our Lord, 1530." It was 18mo in size, and in 𝔅𝔩𝔞𝔠𝔨 𝔏𝔢𝔱𝔱𝔢𝔯, with numerous curious wood-cuts. It was bound in antique morocco, by Hayday, with fac-similes by Harris. This volume brought at the sale $500.

The library also contained copies of John Eliot's Indian Bible, and Psalms, and Testament, printed at Cambridge, Mass., in 1661–63, 2 vols. 4to, *calf*. They belonged to the first edition. The type was set up by an Indian. This was also the first edition of the Bible printed in America. The Psalms are in Indian verse. It is in the Natick, or Nipmuck language. These books brought $200 at the sale.

The collection contained 457 books on Miscellaneous Literature, 455 on

Bibliography, 57 classified as Polygraphic Authors, 76 volumes of Addresses, Orations, or Discourses, 45 Novels and Romances, 532 volumes of Poetry and the Drama, 81 on Natural and Philosophical Sciences, 193 on Mathematics, Astronomy, Engineering, Military Works, etc., 209 on Calligraphy and Stenography, 96 on Games and Sports, 91 volumes on the Arts 248 Illustrated Books, Scrap-Books, Engravings, etc., 100 Bibles, Prayer-Books, Psalters, etc., 133 volumes on Biblical Literature, 417 on Theology, 21 on Jurisprudence, 74 on General History, 472 works relating to America, (of this number, not a few were pamphlets tied in lots, and counted as lots,) 164 on Biography and Correspondence, 35 on Voyages and Travels, besides many autographs, manuscripts, etc. To these must be added 1292 engravings. It was a most splendid collection of books, and the catalogue itself is a curiosity.

5 * EDWARD CALLWELL,[7] (*James*, 5,) b. near Middletown, Orange Co., N. Y., December 30th, 1807, d. in Jersey City, August 22d, 1856, of apoplexy.

M. Mary Ann, daughter of Christian Shuart, in New-York City, June 16th, 1829.†

Ch. James Horton, Leah Margaret, Edward Tanjore, George Brainard.

He removed to New-York City with his father in 1824 ; became ferry-master at the Canal street (Hoboken) ferry in 1826, which position he retained for twelve years; then became a contractor, especially of wharves and bridges, building some of the largest docks, and most of the ferries about New-York City, prior to 1850. He also engaged in the steamboat business, being one third owner of the steamboat Napoleon, plying between New-York and Albany, in 1839, and several years thereafter. In the spring of 1847, he removed to Jersey City, having contracted to build the Cunard Docks, in that place, for the British steamers.

* His mother's sister married a Callwell,

† Christian Shuart was a descendant of Olferts Suert, who is found living on Broadway, (New-York City,) below Trinity Church, in the year 1697. In 1698, his name again occurs in Valentine's *Manual of New-York City*, as a freeman of the city. The records of the Dutch Church give the baptisms of the following children : Ytje, (Grietje,) baptized July 3d, 1698 ; Johannes, April 6th, 1701 ; Luykus, May 24th, 1704 ; Cornelis, December 25th, 1705, who died soon after; and another Cornelius, August 17th, 1707. One or more of these sons moved to Bergen Co., N. J., with the early settlers of that county, about 1725–30. In the baptismal records of the Reformed Church of Paramus, in that county, we find the names of Joost Schyourt and Elizabeth, his wife, having a son Lewis baptized in 1754; of Adolph Scoort, and Margrietje, his wife ; of Hannes Syourt, and Elizabeth, his wife ; and of William Syourt, and Catrina, his wife ; the latter couple had the following children baptized at Paramus, namely, Grietje, 1753 ; Isaac, 1756, whose wife was Grietje, and who had sons, William, 1781, and John, 1786 ; William, 1758; Adolph, 1764, whose wife was Altje Bush, and who left the following children: William ; Margaret, born March 3d, 1792; David, July 1st, 1794; Catrina, September 19th, 1796 ; Jacobus, (James,) October 17th, 1798 ; Daniel, February 1st, 1801 ; Margrietje, July, 1803; Christian, December 23d, 1806 ; Abram, December 23d, 1806 ; Elizabeth, October 15th, 1808; John, 1810, and Henry, 1812; the last son of William and Catrina was Christian, born July, 1771. When William died, his wife Catrina married a Mr. Zabriskie, at Paramus, and lived on the place subsequently owned by Abram Carloch, in what is called the Point Neighborhood. Christian Shurte first married Margaret Demarest, who died April, 1808, aged 37 years and 3 months. Their children were William, b. 1796 ;

6 EDWARD PAYSON,[7] (*John II.*, 8,) b. June 27th, 1833.
   M. Mary F. Frisbie, January 1st, 1860.
   Ch. Charles E.   (*New-York City.*)

7 REV. EDWARD TANJORE,[*8] (*Edward C.*, 5,) b. in New-York City, July
   12th, 1834.
   M. Mary Esther Kipp, (daughter of Nicholas Kipp[†] and Mary Freshour,)

David, b. 1799, d. 1860, who married successively Hannah Van Houten, b. 1791, d. 1811, Margaret Vanderhoof, and Jane Hopper, in 1830; Isaac, b. 1800, mar. Margaret Shurte; Peter, b. 1802; Hester, b. 1804, mar. John Clark; Catharine, b. 1806, d. 1836. After the death of his first wife, Christian Shuart married Leah Kipp, widow of John Van Riper, having already the following children; Garret, b. October 6th, 1793, d. December 30th, 1850, who mar. Sarah Vanderhoof; Nicholas, b. February 9th, 1796, d. April 4th, 1803; Cornelius, b. December 19th, 1799, mar. Ann Winans; Sophia, b. March 26th, 1801, mar. Harry Masker; Catharine, b. October 4th, 1803, d. January 22d, 1851, mar. Robert Clark; John, b. October 24th, 1806, d. May 21st, 1838; the children of Leah Kipp and Christian Shuart were Mary Ann, who married Edward C. Corwin, No. 3, as above: Margaret, b. July 28th, 1811, mar. John Ludlum, May 14, 1834; Elizabeth Jones, b. July 10th, 1813, mar. Isaac Vanderhoof, February 20th, 1831; William, b. June 7th, 1815, d. March 23d, 1850, mar. (1) Victoria Francisco, (2) Elizabeth Travers; and Leah, b. March 31st, 1819, d. November 16th, 1847, mar. (1) Samuel Haviland, (2) Thomas Walker. Christian Shuart was drowned in the Passaic River, at Paterson, N. J., March 22d, 1824. His wife Leah Kipp, (daughter of Nicholas Kipp and Leah Mandeville,) was born February 5th, 1777, and d. November 21st, 1851. She was mar. a third time, to Jasper Dodd, on ———. (See Note on Rev. Edward T., 7)

* TANJORE is the name of a large city in Southern Hindostan. The name was used by the celebrated authoress, Catharine G. Ward, as a given name to one of her characters in her work entitled *The Mysterious Marriage, or, The Will of my Father.* The first American edition was published in New-York in 1834, from the sixteenth London edition. The termination *jore*, or *yure*, means village. The meaning of *Tan* is unknown.

† Nicholas Kipp was a descendant of Roeloff de Kype, of Bretagne, France. Roeloff was born about 1510-20, and was a warm partisan of the Guises. He fled to Holland. Here he joined the party of the Earl of Anjou, and fell in battle near Jarnac. His son, Roeloff de Kype, lived at Amsterdam, Holland. In 1588, a company was formed for the purpose of exploring a north-east passage to the Indies, around the northern coasts of Europe and Asia. Hendrik Kype, born 1576, the son of Roeloff, was among the members. He left Amsterdam with his family, and came to New-Netherlands in 1635. In 1642, he received a tract of land on the north side of Bridge Street, in New-York City. He soon after returned to Holland, and died there. He left three sons in America, Henry, James, and Isaac, all of whom were born in Holland. Isaac married, in 1653, Catalina Hendricks de Suyens, and in 1675, Mrs. Maria Vermilyea de la Montaigne. Isaac's children were Henry, b. 1654; Tryntje, b. 1656; Abraham, b. 1659; Isaac, b. 1662; and James, b. 1663. Henry and James were the co-patentees of the manor of Kipsburgh, on the Hudson. From James, Bishop Kip, of California, and Rev. Francis M. Kip, for many years pastor at Fishkill, are descended. Isaac owned a part of what is now the City Hall Park, and Nassau Street was first called, after him, Kip Street.

James, another son of the first Hendrik, was born in 1631. He married Maria de la Montaigne in 1654. His children were John, b. 1655; James, b. 1656; Abraham, b. 1658; Jesse, b. 1660; Rebecca, b. 1664; Rachel, b. 1661; Maria, b. 1666; Hendrik, b. 1669; Catharine, b. 1671; Petrus, b. 1674; Benjamin, b. 1678; Samuel, b. 1682; Solomon, b. 1682.

Henry, the eldest son of the original Hendrik, was born about 1628. He married Anna de Sille, (daughter of Nicasius de Sille and Maria de la Montaigne,) on February 29th, 1660. He probably removed to Pollifly, near Hackensack, N. J. His son Nicasie, (Nicholas.) named after his maternal grandfather, married Antje Bryant, and reared a large family in New-Jersey, whose descendants are now very numerous. Nicasie's children were Henry; Isaac, b. 1697, who married Williamyntje Bordan in 1723; James, who moved to Schraalenbergh; Peter, b. 1703, who married Elsie Vanderbeek; Jacob, b. 1702; Catharine; perhaps

4

at Geneva, N. Y., July 25th, 1861. She was b. at Geneva, August 21st, 1840.

Ch. Euphemia Kipp, (a son, unnamed, b. June 23th, d. June 26th, 1866,) and Charles Edward.

He was graduated from the College of the City of New-York, in the first class which that institution sent forth, in 1853, having entered at its opening in January, 1849. In the fall of 1853, he entered the Theological Seminary of the Reformed (Dutch) Church at New-Brunswick, N. J., and was graduated thence in 1856. The following year was spent in New-Brunswick as a resident licentiate. In July, 1857, received a call to the Reformed Church of Paramus, N. J., and was installed September 22d of same year. Farewell sermon was preached November 29th, 1863, having accepted a call to the Reformed Church of Millstone, N. J., in which he was installed December 29th, 1863. In 1859, published *The Manual and Record of the Church of Paramus ;* in the same year, *The Manual of the Reformed Protestant Dutch Church in North-America,* giving the names and settlements, with dates, of all the ministers of that denomination, with succinct accounts of its educational and benevolent operations; in 1866, published *The Millstone Centennial;* and in 1869, *The Manual of the Reformed Church in America,* being a revised and enlarged edition of the former Manual, with about one hundred sketches or characterizations of deceased ministers; with an introductory chapter on the history of the denomination, and appendices, containing the history of Rutgers College, of the New-Brunswick Theological Seminary, and of the several Benevolent Boards of the church.

8 EDWARD TANJORE,[8] (*George S.,* 8,) b. January 15th, 1855, d. January 25th, 1856.

9 EDWARD,[8] (*William,* 16,) b. ——, 1825–30.

10 EDWARD M.,[8] (*George S.,* 12,) b. February 18th, 1843.

M. Frances Couzens, December 2d, 1869.   (*Newark, N. J.*)

11 EDWARD MANNING,[9] (*Daniel A.,* 21,) b. ——.

Elizabeth, who married Jan Hopper ; and Cornelius, b. December, 1699, bapt. January 1st, 1700, at Hackensack, N. J. Cornelius married Eva Berdan. In connection with George Doremus, he bought six hundred acres of land at Preakness, N. J., in 1723. His son Nicholas, b. September 15th, 1726, d. December 3d, 1808, married Leah Mandeville, of Pompton Plains, N. J., February 14th, 1757. Their children were Eva ; Henry, who married such cessively Catrina Doremus and Sarah Doremus ; Cornelius, b. 1762, d. 1840, and married Christina Demarest in 1783 ; Nicholas. b. July 25th, 1780, d. January 2d, 1856, married Hester Johnson, (she was daughter of John Johnson, b. 1752, d. 1833, and Mary Cooper, b. 1751, d. January 31st, 1840;) Annie, who married John Vader and David Hennion ; Catharine, married Tunis Hennion ; Elizabeth, married Edward Jones ; Mary, married Garret Haulenbeck ; and Leah, b. February 5th, 1777, d. November 21st, 1851. (See Edward C., No. 5.)

The above Nicholas, who married Hester Johnson, after living in Newark, N. J., moved to Ontario Co., N. Y., about the year 1814. Their children were Nicholas, of Geneva, b. 1809, married Mary Freshour ; Mary, married James Jones ; Leah, married Frederick Backenstose ; John, married Jane Burt. The children of Nicholas Kipp and Mary Freshour are Charles, b. August 18th, 1834, married Delia Woodward ; Catharine E., b. August 15th, 1836, married Edward Huntington ; and Mary Esther, who married Edward T. Corwin, No. 7.

Edward Thomson 16. (Mary Corwin 25)
  born Jany 11. 1845
married first Oct 5. 1870,
  Sarah Tapscott Shanks,
  she died July 10. 1872.
Married secondly, Sept 29. 1874
    Emma Talbot. daughter of the
  Hon Chas C. Talbot of Brooklyn. N.Y.
  issue —
5 Lillian Genevieve Thomson ?
        born Augt 30. 1875
10 Clara Estella Thomson
        born Oct 9. 1880.

10 - Edward M. Corwin b. Dey 16 1848
m. Dec 2. 1869. Fannie Conyers.
Chos - Annie Busby b. Dec 19, 1871,
    Susie Knight, b. July 9, 1872,
    George Wood, b. March 14, 1874,
    Frank Edward, b. March 17 1876.

Annie Busby Corwin, m. Sept 20. 1887
George W. Cole, issue,
William Edward b. May 18. 1888,
Helen Lucile, b. Oct 9. 1890.
May Gertrude, b. July 16, 1893 d. mar b 20 /95
Preston Corwin. b. June 22, 1895.
        (Jacksonville, Florida

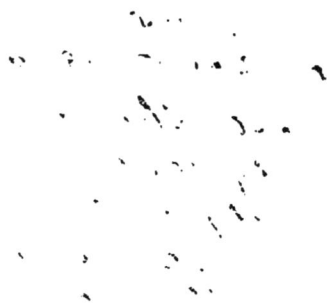

11½ EDWARD,[9] (*Warren*, 1,) b. ——.

12 EDWARD LEMMA,[9] (*John*, 53,) b. December 5th, 1848.

13 EDWARD,[9] (*Daniel A.*, 23,) b. ——, 1856–65.

14 EDWARD,[9] (*Anson L.*, 2,) b. ——, 1855–65.

15 Edward Bishop,[7] (*Mary Corwin*, 11,) b. August 23d, 1795 ; living.
    M. ——.
    Ch. Daniel C.
He left L. I. in 1811, served in the war of 1812, afterward settled at Minisink, N. Y., and subsequently at Hamburg, Mich.

16 Edward Thompson,[8] (*Mary Corwin*, 25,) b. January 11th, 1845.

17 Edward Watkins,[8] (*Marietta Woodruff*, 2,) b. June 11th, 1828.
    M. Mary Jane Marsh.   (*Williamsport, Pa.*)

18 Edward Lewis Seybolt,[9] (*Adaline Corwin*, 2,) b. July 6th, 1857.

19 Edward E. Noyes,[9] (*Maria L. Smith*, 17,) b. January 1st, 1869.   (*Vt*)

20 Edward J. Hood,[9] (*Caroline L. Corwin*, 4,) b. October 19th, 1863. (*Vt.*)

21 Edward Harvey,[10] (*Elizabeth Corwin*, 40,) b. ——, 1861.

Edwin, (Anglo Saxon,) gainer of property.

1 EDWIN HATFIELD,[8] (*Silas*, 7,) b. ——.   (*Drowned.*)

2 EDWIN,[9] (*Egbert*, 1,) b. March 4th, 1867.

2½ Edwin Little,[7] (*Abigail Corwin*, 3,) b. May 12th, 1821, d. October 13th, 1822.

3 Edwin Little,[8] (*Henry Little*, 28,) b. 1825–30.

4 Edwin Stevens Howell,[8] (*Harriet N. Corwin*, 3,) b. January 6th, 1842.

5 Edwin Concklin,[9] (*Lucretia R. J. Corwin*, 3,) b. ——.

Effie, for Eva, (Hebrew,) life.

Effie, see Euphemia.

1 EFFIE LEE,[9] (*James*, 36½,) b. ——, d. about 8 years old.

Egbert, (Old German,) the sword's brightness.

1 EGBERT CHAUNCY,[8] (*Richard S.*, 7,) b. March 30th, 1835.
    M. Huldah E. Corwin.
    Ch. Addison L., Justina, Bertie, Edwin.   (*Riverhead, L. I.*)

2 Egbert Little,[8] (*William C. Little*, 65,) b. 1842.

3 Egbert Brown,[8] (*Parmenas Brown*, 2.)

4 Egbert Tompkins,[9] (*Naomi A. Corwin*, 3,) b. 185–.

Elbert, same as Albert.

1 ELBERT OLIN,[9] (*George H.*, 24,) b. May 16th, 1851.

2 Elbert H. Gray,[10] (*Sarah Corwin*, 56,) b. September 14th, 1858.

Eleanor, same as Helen.

1 ELEANOR,[7] (*Rev. Joseph*, 5,) b. 1800–10.
    M. (1) Luther Tapping. (2) Thomas Smith.

Eleazer, (Hebrew,) whom God helpeth.

1 ELEAZAR,[7] (*George,* 2,) b. August 22d, 1800.
   M. Mary Whiteford.
   Ch. George, William, Arvilla.

2 Eleazar Drake,[8] (*Arrilla Corwin,* 1,) b. 1835–45.

3 Eleazar Peters,[8] (*Mary Corwin,* 30,) b. 1830–40.
   M. Matilda Lenhart.
   Ch. Mary E., Horace S.

Electa, (Latin,) chosen.

1 ELECTA,[8] (*Bestir,* 1.)
   M. J. Grimes.
   Six Ch.
      2. Electa M. Millspaugh,[6] (*Gideon H. Millspaugh,* 1,) b. August 30th,
      1842.

Electra, (Greek,) amber, or beauty.

1 Electra M. Hobart,[8] (*Peter Hobart,* 9,) b. May 8, 1833.

Eli, (Hebrew,) height.

1 ELI,[6] (*David,* 2,) b. April 1st, 1757, d. March 16th, 1833.
   M. Dorothy Horton, May 1st, 1777. (*Aquebogue Records.*) She was
   born April 3, 1756, d. January 5, 1840.
   Ch. Eli H., Abigail, Cynthia, Jason, Silas, Jemuel, Dorothy, Joseph,
   John H., David W., Olivia, Charlotte.   (*Near Middletown, N. Y.*)
   In 1775, he signed an engagement to support Congress. Bought land in
Orange Co., N. Y., in 1809. Will at Goshen.

2 ELI H.,[6] (*Eli,* 1,) b. April 3, 1779, at Riverhead, L. I., d. October 25th,
   1864. Buried at Newburgh, N. Y.
   M. Mary Wells, January 15th, 1803. She was b. October 8th, 1778, d.
   December 5th, 1847.
   Ch. Nathaniel, Abel, Dorothy, John W.

3 ELI,[6] (*David,* 4,) b. January 27th, 1765, d. June 1st, 1826.
   M. (1) Mary Wickham, December 10th, 1783. She was b. July 31st,
   1764, and d. August 12th, 1842. (Her father, Samuel Wickham,
   was b. August 13th, 1713, d. November 24th, 1778.)
   Ch. Samuel W., Sarah, Nathan H., William O., Orthenal, Wickham.
   Bought land near Minisink, N. Y., 1789. Will at Goshen, N. Y.

4 REV. ELI,[7] (*John H.,* 8,) b. October 30th, 1824, at Wallkill, Orange Co.,
   N. Y.
   M. Henrietta S. Howell, July 16th, 1851. She was b. February 4th,
   1830, at Newburgh, N. Y.
   Ch. John H., Cynthia S., Charles A., Cecil S., Arthur M., Walter B.
   He was graduated from Williams College, Mass., 1848, and from Union
Theological Seminary, New-York City, 1851 ; appointed chaplain of Sea-

men's Friend Society for San Francisco, California, and ordained June 22d, 1851. He sailed October 1st of same year in ship Corinth, *via* Cape Horn. Was pastor of the Presbyterian Church, San José, California, 1852–8; also Secretary of State Agricultural Society 1856. He then removed to the Sandwich Islands, and became pastor of the Fort street church, Honolulu, October, 1858, to October, 1868, when he returned to California, and became the pastor of the Second Congregational church, Oakland, Cal., October, 1868, to January, 1870; of Green Street Congregational church, San Francisco, January, 1870, to September, 1871. Now connected with Mills Seminary, Brooklyn, Cal.

5 Eli Egbert,[8] (*Isaac S.,* 5,) b. May 18th, 1834.

6 Eli Horton Young,[7] (*Charlotte Corwin,* 1,) b. November 16th, 1828.
   M. Angeline Brewster, daughter of John F. Brewster, of Scotchtown, N. Y., July 21, 1853. She was b. February 1st, 1828.
   Ch. Mary C., Moses E., Lewis W.   (*Plymouth, Ill.*)

7 Eli Squires,[10] (*Mary M. Concklin,* 168,) b. 1870 ?

Elias or Elijah, (Hebrew,) my God is Jehovah.

1 Elias,[7] (*John C.,* 11,) b. 1790–1800.
   M. Mary ——.
   Ch. John, Joseph, Frederick.

2 Elias,[7] (*David,* 8.)

3 Elias Woodruff,[7] (*Sibyl Corwin,* 1,) b. 1800–10.   (*A Physician.*)
   M. Annie ——.

4 Elias Cortlandt Sperry,[8] (*Mary Corwin,* 40,) b. September 11th, 1828.
   M. Mary Padon.

5 Elias Woodruff,[8] (*Silas Woodruff,* 15,) b. ——, 1825–35.

6 Elias W. North,[8] (*Sibyl Campbell,* 3,) b. February 10th, 1833.

1 Elijah Reeve,[7] (*Mary Corwin,* 12,) b. ——, 1810–20. (*Middletown, N. Y.*)

2 Elijah Thatcher,[7] (*Jemima Corwin,* 1,) b. ——, 1792, (?) in Ohio, d. in Upper Sandusky. Was a soldier in war of 1812.

Elisha, (Hebrew,) God is his salvation.

1 Elisha,[6] (*David,* 4,) b. September 14th, 1782, d. May 20th, 1868. Buried at Hamptonburg, N. Y.
   M. Christiana Smith, May 24th, 1808. She was b. ——, 1786, d. October 10th, 1856.
   Ch. Sarah J., Caroline W., Harriet N., William S., Benjamin V. R., Emeline, Cornelia, Frances.   (*Coal merchant, New-York City.*)

2 Elisha Shepherd,[8] (*Charlotte Corwin,* 6,) b. December 30th, 1843.
   M. Louisa Harris.   (*Bound Brook, N. J.*)

Eliza, (Hebrew,) worshiper of God ; contracted from Elizabeth.

1 ELIZA,[2] (*John*, 2,) b. about 1680, d. April, 1713. On census list, 1698, directly under John Corwin and Sarah, his wife.

(*Doc. Hist. N. Y.* i. 450.)

2 ELIZA,[7] (*Joseph*, 9,) b. ——, 1808 ?
   M. Gabriel Horton.
   Ch. Julia, Henry, Abigail, Sarah, Caroline, Mary, Adelia, Charles, Harriet.

3 ELIZA,[7] (*James*, 5,) b. May 7th, 1804, d. November 30th, 1842
   M. Henry Corwin Wetherby, July 13th, 1834. He was b. February, 1812, d. August 21st, 1869. He was son of John Wetherby, who removed from New-Hampshire to Luzerne County, Pa., about 1804. His mother was Sarah Simrall, sister of Mary Simrall, who married James Corwin, No. 5. Sarah Simrall was born July, 1790, and lived with her sister from 1794–1807, when she removed to Greenfield, Pa. In 1810, she married John Wetherby. She died October 7th, 1869. After Eliza Corwin's mother's death, in 1825, Eliza lived with her aunt, Sarah Wetherby, and afterward married her own cousin.
   Ch. George C., James H., Andrew B., Clarissa A., Caroline E.

(*Scott, Luzerne Co., Pa.*)

4 ELIZA,[7] (*John C.*, 11,) b. ——, 1805.
   M. —— Brown.
   Ch. John.

5 ELIZA,[7] (*William*, 6,) b. November 28th, 1804.
   M. Henry C. Beach, July 13th, 1825.
   Ch. Martha, Emeline, William C., Sarah V. (*Sparta, Sussex Co., N. J.*)

6 ELIZA,[7] (*Ichabod*, 1,) b. August 25th, 1797, d. July 22d, 1822.
   M. William Newman, April, 1815. He was from London, Eng.
   Ch. Moses B.

   A few months after this marriage, he started on business for Georgetown, Ky., and was never heard of again.

7 ELIZA J.,[7] (*Richard*, 3,) b. about 1836.
   M. Byron, son of Joseph and Susan Moore, October 29th, 1869.

8 ELIZA,[7] (*Richard*, 2,) b. ——, 1810–20.
   M. John H. Waters.
   Ch. Elizabeth H., Charles.

8½ ELIZA,[7] (*Edward*, 8,) b. October 13th, 1809.
   M. A. Stull.
   Five Ch.

9 ELIZA JANE,[8] (*Daniel*, 11,) b. ——, 1810.
   M. William Finn.

10 ELIZA E.[8] (*Benjamin*, 7,) b. ——, 1825–30.

M. Ephraim Fairchilds.

Ch. Berthemia, Louisa, William M., Andrew T., Elnora.

(*Wyoming, Iowa.*)

11 Eliza,[8] (*Robert G.*, ½,) b. March 12th, 1858.

12 Eliza Terry,[8] (*Richard S.*, 7.) b. ——, 1820–30.

13 Eliza Ann,[8] (*David*, 12,) b. June 5th, 1821, d. April 12th, 1860.

14 Eliza Jane,[8] (*Thomas*, 5,) b. ——, d. ——.

15 Eliza J.,[9] (*Jabin*, 2,) b. February 12th, 1863, at Chelsea, Vt.

16 Eliza,[9] (*William H.*, 38,) b. December 6th, 1850, d. September 1st, 1851. (*Vermont.*)

17 Eliza,[9] (*David S.*, 21.)

18 Eliza M.,[8] (*Alva*, 1,) b. January 3d, 1824.

M. William Baker, October, 1848.

19 Eliza Osman,[4] (*Sarah Corwin*, 1,) b. ——, 1690–1700.

20 Eliza Philips,[7] (*Hannah Corwin*, 2,) b. ——.

21 Eliza Ann Halsey,[7] (*Prudence Corwin*, 1,) b. ——, 1790–1800. *Aug 3*

22 Eliza A. Aldrich,[8] (*Deborah Corwin*, 6,) b. ——.

23 Eliza Crassons,[8] (*Sarah J. Corwin*, 13,) b. ——, 1830–40.

24 Eliza Smith,[8] (*Sarah Corwin*, 24,) b. ——.

25 Eliza Skidmore,[8] (*Mary Corwin*, 28,) b. ——, 1810–20.

26 Eliza Dunlevy,[8] (*Lucinda Corwin*, 4,) b. September 15th, 1826.

M. Rev. William Ashmon, Baptist missionary at Swatow, China.

26½ Eliza N. Colgrove,[8] (*Bela H. Colgrove*, 1,) b. May 11th, 1829.

M. Thomas P. Sears, May 9th, 1849.

Ch. Charles B., Mary, John B. (*Buffalo, N. Y.*)

27 Eliza Jane Drake,[8] (*Mary Corwin*, 20)

M. —— Simonton. (*Cincinnati, O.*)

28 Eliza Elston,[8] (*Esther Corwin*, 3,) b. ——, 1825–35.

M. —— ——. Missionary to the Indians.

29 Eliza Halstead,[8] (*Cynthia Hobart*, 9,) b. May 7th, 1823.

M. —— ——, November 13th, 1851. (*Elbridge, N. Y.*)

30 Eliza North,[8] (*Sarah Hobart*, 65,) b. December, 1848.

31 Eliza Taylor,[8] (*David J. Taylor*, 30,) b. July, 1838.

32 Eliza Corwin Newman,[9] (*Moses B. Newman*, 9,) b. at Sidney, Ohio, November 23d, 1853.

33 Eliza Lockwood,[10] (*Lucretia Corwin*, 4.)

Elizabeth, (Hebrew,) God is her oath; that is, worshiper of God.

1 Elizabeth,[4] (*John*, 2,) b. about 1700.

1½ Elizabeth,[5] (*John*, 3,) b. about 1730–40.

1¾ Elizabeth,[5] (*Edward*, 2,) baptized February 22d, 1756, at Mattituck.

(An Elizabeth Corwin, of Aquebogue, married Nathaniel Patty, of
Wading River, August 18th, 1774. Probably this one. *Aquebogue
Records.*)

2 ELIZABETH,[6] (*William*, 2,) b. December 6th, 1785, d. December 7,
1860.
M. Henry Halsey.
Ch. Pithom, Lucinda, Mary, William, Stephen, Jefferson H.
(*Morris Co., N. J.*)

3 ELIZABETH,[6] (*Benjamin*, 4,) b. ——, 1780–90.
M. Jabez Coleman.
Ch. ——.   (*Morris Co., N. J.*)

4 ELIZABETH,[6] (*Jeremiah*, 1,) b. ——, 1705?
M. —— Davis.
Ch. Chatham, Davis? Benjamin, Mary, Elizabeth, Bethia.

5 ELIZABETH,[6] (*Matthias*, 4,) b. ——, 1760?
M. ——.
No Ch.

6 ELIZABETH,[6] (*Samuel*, 6,) b. ——, d. ——.
M. Hedges Osborne.

7 ELIZABETH ANN,[6] (*Amaziah*, 1,) b. ——, 1800?

8 ELIZABETH,[6] (*Timothy*, 2,) b. ——, 1790–1800, d. ——, 1848?
M. William Whitefield.
No Ch.   (*Fishkill, N. Y.*)

9 ELIZABETH,[6] (*Thomas*, 1,) b. ——, 1790–1800.
M. Albert Porter, February 13th, 1826, at Mattituck.
Ch. Albert B., Jane E., Levisa, Ruth A.

9½ ELIZABETH,[6] (*Silas*, 1,) bapt. May 14th, 1769, d. November 26th, 1769.
(*Aquebogue Records.*)

9¾ ELIZABETH,[6] (*Silas*, 1,) bapt. June 19th, 1776, d. June 24th, 1776.
(*Aquebogue Records.*)

10 ELIZABETH,[6] (*James*, 1,) b. ——, 1781? d. 1820?
M. Eliphalet Warner.
Ch. Mehetable.

10½ ELIZABETH,[6] (*Daniel*, 3,) b. ——, 1750–60.
M. ——.

10¾ ELIZABETH,[6] (*Jonathan*, 1,) b. ——, d. March 10th, 1783.
(*Aquebogue Records.*)

11 ELIZABETH,[7] (*John*, 7,) b. ——, 1790–1800.
M. William Mullock.
Ch. Cynthia.

12 ELIZABETH,[7] (*Benjamin*, 5,) b. ——, 1813?
M. John McBride.
Ch. Benjamin.

13 ELIZABETH WELLS,[7] (*John II.*, 8,) b. February 28th, 1818.
M. Merville Saunders, January 21st, 1847.
Ch. William M. S.   (*Newbury, N. Y., and Elizabeth, N. J.*)

14 ELIZABETH,[7] (*John*, 9,) b. August 27th, 1779, d. September 22d, 1841.
M. John Clark.

15 ELIZABETH,[7] (*Isaac*, 1,) b. ——, d. young.

16 ELIZABETH,[7] (*William*, 5,) b. ——, 1785–90.
M. John Wright.
Ch. John, Sarah A.   (*Brooklyn, N. Y.*)

17 ELIZABETH,[7] (*Moses*, 1,) b. at Walkill, N. Y., November 25th, 1818.
M. Peter Taylor, at Penn Yann, N. Y., October 16th, 1843. He was
born at Kinderhook, N. Y., May 25th, 1818.
Ch. Martha A., George L., James A., Eva E.
                    (*Moved to Rochester, Ohio,* 1844)

18 ELIZABETH,[7] (*Nathaniel*, 2,) b. June, 1811.
M. Rev. Caleb Lippincott.
Ch. William, Sarah, Caroline, Hekshedah.

19 ELIZABETH HALLOCK,[7] (*Peter*, 2,) b. at Florida, Orange Co., N. Y.,
June 5th, 1788, d. July 19th, 1819.
M. George W. Goble, February 8th, 1806. He was son of George Goble
and Julia Wisner.
Ch. Elizabeth, Milton, Uriah H., Fanny, Arminda B.
                    (*Cortland, N. Y.,* 1814.)

20 ELIZABETH,[7] (*Jabez*, 1,) b. September 16th, 1810.
M. Charles Ransom Howell.
Ch. Daniel H., Charles H., Mary E., Sarah B., Daniel H.
                    (*Riverhead, L. I.*)

21 ELIZABETH,[7] (*Henry*, 6,) b. December 21st, 1783.
M. —— Squires.
Ch. Ellis, Daniel, Rachel.

22 ELIZABETH,[7] (*Martin*, 2,) b. ——, 1805–15.
M. Archibald Daly.   (*Balmville, N. Y.*)

23 ELIZABETH,[7] (*Samuel*, 9,) b. May 18th, 1813 ; unm. (*Middle Hope, N. Y.*)

24 ELIZABETH,[7] (*Thomas*, 2,) b. ——, 1815–25.

24½ ELIZABETH,[7] (*Phineas*, 1¾,) b. ——.

25 ELIZABETH,[7] (*Selah*, 1,) b. ——, 1790–1800.
M. Alfred Goble.

26 ELIZABETH,[7] (*Matthias*, 5,) b. January 27th, 1783, d. April 29th, 1814.
M. Alexander Van Pelt, July 4th, 1800.

27 ELIZABETH BAKER,[7] (*Stephen*, 4.)

28 ELIZABETH,[7] (*Joseph*, 4,) b. April 30th, 1794, d. September 12th, 1819.
M. —— Vail.
Ch. Hiram H.

29 Elizabeth,[8] (*Matthias*, 8.)

30 Elizabeth,[8] (*Jeremiah*, 4.)

31 Elizabeth,[8] (*Tuthill*, 1.)
    M. —— Graves.

32 Elizabeth,[8] (*Abel*, 3.)

33 Elizabeth Stevenson,[8] (*Nathaniel*, 7,) b. March 19th, 1825.
    M. George Lyon, May 6th, 1844.
    Six Ch.   (*Green Point, L. I.*)

34 Elizabeth,[8] (*Isaac*, 4,) b. July 5th, 1818.
    M. John S. Wheat, November 30th, 1848.
    No Ch.

35 Elizabeth,[6] (*Richard W.*, 5.)
    M. Edward H. Badger.

36 Elizabeth,[8] (*Peter*, 4,) b. ——, 1835–40.

37 Elizabeth,[5] (*John E.*, 19,) b. ——, 1845–50.

37½ Elizabeth Alice,[6] (*James*, 16,) b. June 23d, 1866.

38 Elizabeth,[9] (*James H.*, 13,) b. July 10th, 1852.

39 Elizabeth,[8] (*Francis F.*, 1.)

40 Elizabeth Ann,[9] (*Isaac*, 11,) b. February 2d, 1834.
    M. Joseph P. Harvey, September 5, 1853.
    Ch. Sylvester, Sarah, Edward.   (*Hackettstown, N. J.*)

41 Elizabeth Mary,[9] (*Frederic*, 6,) b. September 12th, 1855.

42 Elizabeth,[9] (*John*, 58.)

43 Elizabeth Scott,[9] (*Barnabas*, 5,) b. October 29th, 1865, d. October 27th, 1869.

43½ Elizabeth,[9] (*Ezra*, 3,) b. 1839.
    M. Thomas Grinolds, 1855.
    Ch. Selah E.

43¾ Elizabeth Priscilla,[10] (*James*, 39,) b. August 12th, 1862, d. September, 1862.

44 Elizabeth Osman,[4] (*Sarah Corwin*, 1,) b. ——, 1700–1710.

44½ Elizabeth Case,[8] (*Samuel Case*, 29.)
    M. Peter Hobart.   (*See Moore's Index of Southold*, p. 64.)

44¾ Elizabeth Brockway,[6] (*Mary Corwin*, 5,) b. about 1800, d. soon after marriage.
    M. —— Smith.

45 Elizabeth Smith Woodhull,[7] (*Hannah Corwin*, 2,) b. December 16th, 1795.
    M. Peter Brown, April 27th, 1815.   He b. May 8th, 1794, d. October, 1866.
    Ch. Catharine, William, Elizabeth, Alfred, Hannah, Mary S., Peter, Sarah, Priscilla, Caroline.   (*Chester, N. J.*)

46 Elizabeth Thatcher,[7] (*Jemima Corwin*, 1,) b. ——, 1787? in Warren
     Co., O., d. in same county.

47 Elizabeth Davis,[7] (*Elizabeth Corwin*, 4,) b. ——, 1780–90.

48 Elizabeth Bertholf,[8] (*Mahala Corwin*, 2,) b. September 2d, 1851.
     M. Josiah Otis, October, 1870.

49 Elizabeth De Boyce,[8] (*Sarah Corwin*, 20,) b. ——, 1820–30.

50 Elizabeth Reed,[8] (*Sophia Corwin*, 1,) b. February 27th, 1801.
     M. Elias J. Skellinger, ——,1819?   (*Chester, N. J.*)

51 Elizabeth Goble,[8] (*Elizabeth H. Corwin*, 19,) b. ——.
                                                  (*Cortland, N. Y.*)

52 Elizabeth Carr,[8] (*Azubah Corwin*, 3.)
     M. Adolph Barker.   (*McGrawville, N. Y.*)

53 Elizabeth Howell,[8] (*Elizabeth Corwin*, 20.)

54 Elizabeth Ann Chase,[8] (*Bethia Corwin*, 5.)

55 Elizabeth Hannah Waters,[8] (*Eliza Corwin*, 8.)

56 Elizabeth Meyers,[8] (*Arminda Brown*, 2.)

57 Elizabeth Pellet,[8] (*Mary Corwin*, 24,) b. ——, 1820–30.
     M. —— Stout.   (*Orange, N. J.*)

58 Elizabeth Thompson,[8] (*Mary Corwin*, 25,) b. *July 15*1843, d. *Feby 19, 1844*

59 Elizabeth Millspaugh,[8] (*Gideon H. Millspaugh*, 1,) b. February 24th,
     1848.

60 Elizabeth Brown,[8] (*Elizabeth Woodhull*, 45,) b. April 15th, 1821.
     M. Isaac Nesbit.   (*Farmingdale, Morris Co., N. J.*)

61 Elizabeth Jane Concklin,[9] (*Lucretia R. J. Corwin*, 3,) b. ——, 1843.
     M. James Deel.
     Ch. William, Elizabeth, Margaret.

62 Elizabeth Ketcham,[9] (*Rebecca J. Corwin*, 3,) b. May 23d, 1860.

63 Elizabeth Stryker,[9] (*Mary Corwin*, 48,) b. February 13th, 1849.

64 Elizabeth Paris,[9] (*Julia A. Blinn*, 18,) b. March, 1854.

65 Elizabeth Mehetable Downs,[9] (*Deborah A. Corwin*, 7,) b. February 20th,
     1845.

66 Elizabeth Deel,[10] (*Elizabeth J. Concklin*, 61,) b. ——, 1865–70.

67 Elizabeth Robinson,[10] (*Hannah Corwin*, 22.)

68 Elizabeth Tuthill,[9] (*Ann Corwin*, 22.)

   (1) *ELIZABETH,[2] [*George*, (1),] bapt. July 2d, 1648.

   (2) *ELIZABETH,[3] [*John*, (1),] b. April 28th, 1668.

   (3) ELIZABETH,[3] [*Jonathan*, (1),] b. May 5th, 1678, d. May 19th, 1706.
        M. James, eldest son of Timothy and Mary (Veven) Lindall, of Salem,
        Mass., December 15th, 1702. He was born February 1st, 1675–6.

                              * Curwen.

69. *Elizabeth Call*[9] (*Wm E. Call*[8] 90½) *b. Feby 20. 1852, d. Dec 31. 1853*

70 *Elizabeth M. Call*[9] (*Wm E. Call*[8] 90½) *b. March 10. 1854, m. June 17.
     1883. Whitford J. Lee.*

Ch. Elizabeth, Mary.

Her name is spelled Corwin, Corwine, and Curwen.

(3½) †Elizabeth,[6] [*George* (6),] b. ——, d. young.

(3¾) †Elizabeth,[7] [*Richard* (2),] b. ——, 1780–90, d. young.

(4) †Elizabeth,[8] [*Samuel*, (6½),] b. April 23d, 1820, d. September 22d, 1865.

M. George T. Saxon, September 25th, 1858.

No Ch.   (*Flora, Clay Co., Ill.*)

(4½) †Elizabeth,[9] [*Amos B.*, (2),] b. at New-Rochelle, September 29th, 1862.

(5) †Elizabeth,[8] [*Amos*, (1½),] b. ——.

M. R. T. Wilson.   (*Kinmundy, Ill.*)

(6) Elizabeth Wolcott,[3] [*Penelope Curwen*, (1),] b. March 30th, 1688.

(7) Elizabeth Lindall,[4] [*Elizabeth Curwen*, (3),] b. September 29th, 1703.

(8) Elizabeth Bellos,[7] [*Alice Corwine*, (1).]

Ella, contraction of Eleanor.

1 ELLA,[8] (*Daniel*, 14,) b. December 25th, 1870.

1½ ELLA,[8] (*James*, 16,) b. May 27th, 1852.

2 ELLA V.,[9] (*David*, 24,) b. ——.

3 ELLA,[9] (*Melvin*, 1,) b. ——, 1868–9.

4 ELLA GERTRUDE,[9] (*John E.*, 45,) July 15th, 1868, d. August 24th, 1869.

5 ELLA,[10] (*William H.*, 59.)

6 Ella Gordon,[9] (*Sarah E. Corwin*, 43.)

7 Ella Tinney,[9] (*Lucilia Corwin*, 1.)

8 Ella Jane Crater,[10] (*Mary Corwin*, 82.)

Ellen, diminutive of Eleanor.

1 ELLEN,[8] (*Peter*, 3,) b. ——, 1810–20.

2 ELLEN C.,[8] (*James*, 18,) b. October 6th, 1860.

3 ELLEN MARY,[9] (*John A. R.*, 60,) b. ——, 1855–60.

4 Ellen North,[8] (*Sarah Hobart*, 65,) b. April, 1840.

5 Ellen ——,[8] (*Hannah Corwin*, 6.)

6 Ellen Garlock,[9] (*Julia Horton*, 13.)

7 Ellen Blinn,[9] (*Adam Blinn*, 3,) b. ——, 1865.

8 Ellen Leach,[9] (*Mary Corwin*, 126,) b. ——, 1843–50.

                                        (*St. Lawrence Co., N. Y.*)

Ellis, same as Elisha.

1 Ellis Squires,[8] (*Elizabeth Corwin*, 21.)

† Corwine.

Elma, probably the same as Alma.

1 ELMA,[7] (*William*, 4,) b. July 5th, 1810.
M. Patrick Mongown, March 6th, 1835.
Ch. John, Margaret A.   (*Aquebogue, L. I.*)

Elmer, (Anglo-Saxon,) noble, excellent.

1 ELMER ELLSWORTH,[8] (*Howard*, 1,) b. July 7th, 1861, d. January 12th, 1862.

2 Elmer Woodruff,[8] (*Silas Woodruff*, 15,) b. ——, 1825–35.
M. Julia Brown.   (*Scotchtown, N. Y.*)

3 Elmer A. Crego,[8] (*Clarissa Taylor*, 4,) b. April, 1858.

4 Elmer F. Whitney,[9] (*Helen M. Corwin*, 4,) b. July 10th, 1865.
                    (*Vermont.*)

5 Elmer E. Benson,[9] (*Ann J. Davis*, 47,) b. ——, 1868–70.

6 Elmer Baker,[9] (*Rosalinda Corwin*, 1,) b. ——.   (*Niagara Co., N. Y.*)

1 ELMINA,[7] (*John*, 14,) b. November 20th, 1808.
M. Daniel Bower, November 12th, 1829. He d. April, 1852.
Ch. John, Stephen, Rebecca, Ann A., Samuel.
                    (*Morristown, N. J., and Newburg, N. Y.*)

2 ELMINA,[8] (*William*, 23,) b. December 3d, 1852.
M. Hugh Douglass, of Lackerbie, Scotland, June 8th, 1869.

Elmira, or Almira, (Arabic,) a princess.

1 ELMIRA HARDING,[9] (*Polly M. Corwin*, 4.)

1 ELMORE H.,[8] (*James*, 20,) b. June 29th, 1847, near Mansfield, Ohio.
M. Bernice Coykendall, June 6th, 1869.
Ch. Ernest.
In 1855, removed to Canton, Ill. Was a member of Co. K, 7th Ill. Cavalry, and was wounded at the battle of Nashville, December 16th, 1864.

2 Elmore Monroe,[9] (*Clarissa Taylor*, 6,) b. August, 1859.

1 Elnora Taylor,[8] (*David J. Taylor*, 30,) b. and d. 1851.

2 Elnora Fairchild,[9] (*Eliza E. Corwin*, 10.)

1 ELSIE,[6] (*John*, 6,) b. ——, 1763 ? d. ——.
M. Silas Corwin, (6.)

2 ELSIE,[6] (*Daniel*, 3,) b. ——, 1755–65, d. ——.
M. Silas Corwin, (6.)

3 ELSIE M.,[9] (*Theodore*, ½.)

4 Elsie A. Brown,[8] (*Dorastus Brown*, 1.)

1 ELURY Y.,[7] (*Jedediah*, 3,) b. ——, 1807 ? d. May 31st, 1849.
M. David T. Warner, September 14th, 1825.   (*Aquebogue, L. I.*)

Elvira, (Latin,) white.

1 ELVIRA,[7] (*Daniel*, 9,) b. ——, 1800-10.

2 ELVIRA,[7] (*Ichabod*, 1,) b. April 23d, 1805.
M. Rev. Daniel Bryant, April 14th, 1824.

3 ELVIRA,[7] (*David*, 8.)

1 Elwood Wells,[8] (*Puah Corwin*, 1.)

Emeline, (Old German,) energetic, industrious.

1 EMELINE W.,[7] (*Elisha*, 1,) b. ——, 1820-30.
M. Rev. James Sinclair, ——, 1851.
Ch. Clarence.
He is a Presbyterian minister, now of Smithtown Branch, L. I.

2 EMELINE,[8] (*David*, 16,) b. March 2d, 1829.
M. Dr. —— Lewis.
Ch. Anna, Willard, James, Sarah. (*Hampton, N. Y.*)

3 Emeline,[8] (*John*, 16,) b. December 7th, 1832.

4 EMELINE,[8] (*James*, 18,) b. June 6th, 1848.

5 EMELINE,[8] (*Matthias*, 7,) b. ——.
M. —— Van Harlingen, M.D. (*Lebanon, O.*)

6 EMELINE EVANS,[8] (*George*, 10,) b. ——, d. aged 18 years.

7 EMELINE C.,[8] (*John*, 32,) b. June 10th, 1830.
M. Alfred Boyd, January 28th, 1858.
Ch. John A., Fannie E., Sarah, Clara M., Everet W. (*Vermont.*)

8 EMELINE,[8] (*William*, 19,) b. January 27th, 1858, d. February, 1859.

9 EMELINE MAUD,[9] (*Hubbard*, 2,) b. December 28th, 1859.

10 EMELINE,[10] (*Samuel*, 26,) b. March 10th, 1866.

11 Emeline Brown,[8] (*Daniel C. Brown*, 36.)

12 Emeline Beach,[8] (*Eliza A. Corwin*, 5,) b. February 10th, 1830.

1 Emeret Smith,[9] (*George Smith*, 74,) b. ——, 1855-65.
(*St. Lawrence Co., N. Y.*)

1 EMERSON,[8] (*Henry W.*, 8,) b. ——, 1825-30.

Emily, (Old German,) energetic.

1 EMILY C.,[8] (*Benjamin V. R.*, 9,) b. ——, 1845-50.

2 EMILY,[8] (*James*, 18,) b. November 20th, 1848.

3 EMILY ELMORE,[8] (*Jabez*, 3,) bapt. at Aquebogue, June 19th, 1836.
M. Leander Howell.

4 EMILY,[8] (*John*, 30,) b. ——.

5 EMILY,[8] (*Rev. David*, 15.)

6 Emily Hobart,[7] (*Mehetable Corwin*, 3,) b. January 24th, 1820.

7 Emily S. Brown,[8] (*Silas C. Brown*, 16.)

Emma Thomson 12, (Mary Corwin 25)
born Augt 29. 1841, died June 20. 186
Married Feby 15. 1883
Robert C. Shaw. of New York City —

8 Emily Borgy,[8] (*Harriet N. Corwin*, 1,) b. ——, 1825–35.
(*Missionary in China.*)

9 Emily M. Aldrich,[8] (*Deborah Corwin*, 6,) b. ——.

10 Emily Swina,[8] (*Bethia Hobart*, 11.)

11 Emily Smith,[9] (*John C. Smith*, 92,) b. ——, 1842, d. ——, 1857.
(*St. Lawrence Co.*, N. Y.)

Emma, (Old German,) industrious, energetic.

1 EMMA,[8] (*Nathan H.*, 3,) b. June 22d, 1850.

2 EMMA,[8] (*Gilbert*, 2,) b. ——, 1840–50.

3 EMMA,[8] (*George E.*, 13,) b. ——, 1830–40.

4 EMMA,[8] (*Samuel B.*, 16.)

5 EMMA L.,[8] (*William*, 23,) b. April 2d, 1841.
M. Henry F. Williams, of New-York City, September 16th, 1862.

6 EMMA,[9] (*John*, 14½,) b. ——.

7 EMMA,[9] (*Alva*, 1,) b. February 4th, 1837.
M. George W. Hallock, February, 1857.
Ch. Georgiana.   (*Upper Aquebogue, L. I.*)

8 EMMA,[9] (*John*, 56,) b. June 25th, 1860.

9 EMMA,[9] (*James Horton*, 34,) b. in Jersey City, N. J., December 2d, 1854.

9½ EMMA,[9] (*David*, 24,) b. ——.

10 EMMA F.,[9] (*Daniel A.*, 23,) b. ——, 1855–60.

10½ Emma Shepherd,[9] (*Charlotte Corwin*, 6,) b. December 2d, 1850, d.
June 6th, 1854.

11 Emma Finch,[8] (*John*, 21,) b. March 15th, 1854.

12 Emma Thompson,[8] (*Mary Corwin*, 14,) b. ——, 1847. - *m.* Robert C Shaw.

13 Emma Millspaugh,[8] (*Benjamin F. Millspaugh*, 31,) b. ——, 1850–55.

14 Emma Beach Petit,[8] (*Hannah Corwin*, 12.)

15 Emma Beyea,[9] (*Sophronia Corwin*, 1,) b. February 12th, 1857.

16 Emma A. Green,[9] (*Bethia Corwin*, ——.) b. ——, 1853, d. ——, 1854.

17 Emma L. Noyes,[9] (*Louisa Corwin*, 3,) b. March 21st, 1848. (*Vermont.*)

18 Emma L. Smith,[9] (*Loren Smith*, 1,) b. ——, 1848–55.
M. Hiram Snell, 1869.
Ch. Judson.   (*St. Lawrence Co.*, N. Y.)

19 Emma E. Hackett,[9] (*Samuel B. Hackett*, 32,) b. August 4th, 1845.
M. Azro B. Whitney, of Tunbridge, Vt., November 1st, 1866.
Ch. Myron E., Cora.

(1) *EMMA,[9] [*George*, (11¼),] b. ——.   (*Carthage, Mo.*)

* Corwine.

20. Emma J. Call[9] (*Wm E. Call.*[8] 90½) b. March 5. 1850
m. John M. Tyler. March 11. 1869.

Emory, (Anglo-Saxon,) powerful, rich.

1 EMORY,[8] (*William*, 39,) b. ——, 1855–60.

Enoch, (Hebrew,) initiated ; hence, consecrated. ·

1 ENOCH BOUTON,[7] (*John H.*, 8,) b. July 21st, 1820, d. November 16th, 1838.

1 EPENETUS HAVENS,[7] (*William*, 4,) b. July 8th, 1807, d. April 1st, 1843.
M. Mary B. Corwin.
Ch. Oliver H., William H., Benjamin F., Mary A.   (*Riverhead, L. I.*)

2 EPENETUS,[7] (*William*, 4,) b. May 7th, 1803, d. 1806.

3 EPENETUS LESTER,[8] (*Joseph W.*, 12,) b. November 2d, 1841, d. February 27th, 1842.

Ephraim, (Hebrew,) double land ; others, very fruitful.

1 EPHRAIM II.,[8] (*Jabin*, 1,) b. December 25th, 1818, d. ——, 1836.
(*Vermont.*)

2 Ephraim Everet,[7] (Deborah Corwin, 1,) b. ——, 1815–25.

Erastus, (Greek,) lively, amiable.

1 Erastus Woodruff,[7] (*Sybil Corwin*, 1,) b. ——, 1800 ?
M. (1) Fanny Belknap ; (2) Amelia West.
Ch. Augustus, Harriet N., Milton, Judson A., Mary.

2 Erastus North,[8] (*Mehetable Campbell*, 10,) b. October 17th, 1833.
M. Harriet Carpenter, July, 1858.

Ernest, (German,) earnest.

1 ERNEST NORMAN,[9] (*Cornelius*, 2,) b. February 27th, 1867.

2 ERNEST,[9] (*Elmore H.*, 1,) b. ——, 1870, d. ——, 1870.

3 ERNEST ALBERT,[9] (*John A. R.*, 60,) b. ——, 1855–60.

(1) *ERNESTINE,[9] [*Amos* (2),] b. in Panama, August 28th, 1858.

1 Ernestus II.,[7] (——,) b. ——, 1809, d. April 1st, 1843.   Tombstone at Upper Aquebogue, L. I.

1 ERI COLEMAN,[6] (*Joshua*, 6,) b. November 8th, 1823.
M. Cynthia Mapes.
No Ch.   (*Middletown.*)

1 ERIN,[8] (*Jesse*, 6,) b. ——.   (*Ohio.*)

Esther, (Persian,) a star, good fortune.

1 ESTHER,[6] (*Samuel*, 6,) bapt. May 2d, 1773, at Southold.

2 ESTHER,[7] (*Joseph*, 4,) b. February 12th, 1812, d. February 9th, 1852.
M. —— Drake.

* Corwine.

3 ESTHER,[7] (*John* 7,) b. ——, 1790–1800.
M. Jacob Elston.
Ch. Margaret, Lemuel, Eliza, Mary.

4 ESTHER,[7] (*Asa,* 2,) bapt. at Aquebogue, June 19th, 1794.

5 ESTHER,[7] (*Edward,* 3,) b. April 22d, 1800.
M. J. Southwick.
Seven Ch.

5½ ESTHER,[5] (*Bestir,* 1.)
M. Gilbert Webster. .
Five Ch.

5 Esther Vail Terry,[7] (*Keturah Corwin,* 1,) b. ——, 1780–90.

6 Esther Penney,[7] (*Annie Corwin,* 3,) b. July 24th, 1791, d. October 9th, 1793.

7 Esther Brown,[8] (*Dorastus Brown,* 1.)

(1) *ESTHER,[6] [*George,* (6),] b. October 31st, 1742.
M. Jacob Snyder, (or Sneder.)
Ch. Rebecca.   (*Moved to Kentucky about* 1783.)

1 Etta D. Byram,[8] (*Mary A. Corwin,* 34,) b. March 18th, 1854.

Eugene, (Greek,) well-born, noble.

1 EUGENE PHILO,[8] (*Rev. Jason,* 4,) b. at Penfield, N. Y., June 21st, 1827. Inventor and patentee of the famous buckle without a tongue, manufactured in St. Louis, Mo.   (*Washington, Ill.*)

2 EUGENE,[9] (*Charles N.,* 15½,) b. July 22d, 1857.

3 Eugene Pickard,[9] (*Margaret Halstead,* 10,) b. July 12th, 1851.

Eunice, (Greek,) happy victory.

1 EUNICE,[6] (*Samuel,* 6,) bapt. December 18th, 1783, at Southold, d. ——. Unm.

2 EUNICE OLIVE,[8] (*Isaac S.,* 5,) b. October 11th, 1843.

3 EUNICE COOK,[8] (*Stephen,* 4,) b. ——.   (*Mansfield, O.*)

4 EUNICE,[6] (*Salem,* 1,) b. ——.
M. John McPhail.   (*Green Point, L. I.*)

5 Eunice Shannon,[8] (*Mehetable Corwin,* 5,) b. ——, 1820–30.

6 Eunice Carts,[9] (*Maria Swina,* 15.)

Euphemia, (Greek,) of good report.

1 EUPHEMIA KIPP,[9] (*Rev. Edward T.,* 7,) b. at Paramus, N. J., June 26th, 1863.

1 Eva Ermina Taylor,[8] (*Elizabeth Corwin,* 17,) b. at Rochester, Ohio July 13th, 1859.

* Corwine.

5

Eveline, diminutive of Eva, (Hebrew,) life.

1 EVELINA AMELIA,[8] (*Thomas*, 4,) b. July 14th, 1831.
   M. George R. Sage, May 22d, 1855, of Cincinnati, O.
   Ch. Caroline, Corwin, and a daughter b. March 26th, 1858, d. April 5th, 1858.

1½ EVELINE,[6] (*Phineas*, 1¾.)
   M. Harry Lee.   (*Kansas.*)
   (Possibly this name should be Emeline.)

2 EVELINA,[9] (*George W.*, 23,) b. April 7th, 1862.

3 EVELINE BLANCH,[9] (*Hubbard*, 2,) b. April 29th, 1848.
   M. Waldo Wells, February 2d, 1870.

1 EVERETT,[9] (*George W.*, 23,) b. September 2d, 1859, d. January 19th, 1862.

2 Everett W. Boyd,[9] (*Emeline Corwin*, 7,) b. April 18th, 1869.

Experience, (Latin,) a trial of (a thing.)

½ EXPERIENCE,[3] see Samuel, 2. It seems that a Samuel married an Experience Corwin, or, perhaps, her surname was thus written by mistake.

1 EXPERIENCE,[7] (*Isaac*, 1,) b. January 12th, 1801, d. April 30th, 1856.
   M. Nathan C. Hunt. He b. ——, 1798, d. July 31st, 1860.
   Ch. George, Jerusha.   (*Succasunna, N. J.*)

2 EXPERIENCE ANN,[8] (*Stephen*, 3,) b. ——, 1810–20.
   M. Lewis B. Ayres, December 30th, 1851.
   Ch. Laura A., Sylvester E.   (*Oskaloosa, Iowa.*)

3 EXPERIENCE REEVES,[6] (*Benjamin*, 2,) b. ——.   (*Morris Co.*)

Ezra, (Hebrew,) help.

1 EZRA,[5] (*Samuel*, 3,) b. ——.

2 EZRA,[6] (*Silas*, 1,) b. September 27th, 1759, bapt. October 14th, 1759, at Mattituck, L. I., d. April 24th, 1840.
   M. (1) Dorothy Tuthill, who was b. September, 1760, d. December 15th, 1795, *no Ch.*; (2) Hannah Cook, who was b. January 30th, 1772, d. April 23d, 1841.
   Ch. E. Baldwin, Ebenezer, Orsamus, Polydore B., Dorothy T., Mary, Samuel, Cortlandt E., Julia N.
He united with the church of Aquebogue, November 16th, 1783.
(*Minisink or Weston, Orange Co., N. Y.*, 1800; *Cortland Co., N. Y.*, 1808.)

2½ EZRA,[7] (*Jabez*, 1,) b. October 22d, 1825.
   M. Mary S. Penney.
   Ch. Martha E., Christiana C.

3 EZRA,[8] (*John*, 30,) b. ——, 1809.
   M. (1) Jane Wyckoff; she d. 1852; (2) Jane Gordon, 1854.
   Ch. Almira, Lucretia, Delilah, Elizabeth, Theresa; John, James.
               (*White Lake, Oakland Co., Mich.*)

4 EZRA,[6] (*Polydore B.*, 1,) b. March 14th, 1833.
(*Washington, D. C., and New-York City.*)

5 EZRA P.,[9] (*De Forest*, 1,) b. August 11th, 1857.

5½ Ezra Everett,[7] (*Deborah Corwin*, 1,) b. ——, 1810-20.

6 Ezra Skult,[8] (*Fanny Hobart*, 9,) b. ——, 1820-30.
M. Melissa Dairs. (*Elbridge, N. Y.*)

7 Ezra Millspaugh,[8] (*Gilbert C. Millspaugh*, 6,) b. August 10th, 1829.

8 Ezra Ernest Boisseau,[9] (*Bethia Corwin*, 9½,) b. June 30th, 1871.

9 Ezra Lockwood,[10] (*Lucretia Corwin*, 4,) b. ——, 1855-60.

Fanny, a diminutive of Frances. ..1

FANNY, see also FRANCES.

1 FANNY,[6] (*Phineas*, 1,) b. May, 1791.
M. John Taylor, December 5th, 1810.
Ch. George P., David J., James P., John C., Fanny M., Bethia F.,
Mehetable, Hendrik, Clarissa. (*Elbridge, N. Y.*)

1½ FANNY,[6] (*James*, 23,) bapt. May 4th, 1806, d. May 7th, 1806.
(*Aquebogue Records.*)

2 FANNY,[7] (*George*, 2,) b. ——, 1802? d. ——, 1811?

3 FANNY JANE,[7] (*Moses*, 1,) b. January 13th, 1810, d. January 11th, 1811.

4 FANNY JANE,[8] (*Joshua*, 6,) b. October 5th, 1811.
M. A. T. Gordon.
Ch. Priscilla. (*Centreville, Orange Co., N. Y.*)

5 FANNY,[8] (*Richard*, 4.)

5½ FANNY,[8] (*Bestir*, 1.)
M. L. Chapin ; one daughter.

6 FANNY,[9] (*George W.*, 23,) b. August 6th, 1864.

6½ FANNY HEWLETT,[9] (*Hulbard*, 2,) b. April 12th, 1867.

7 Fanny M. Taylor,[7] (*Fanny Corwin*, 1,) b. June, 1820.

8 Fanny Little,[7] (*Abigail Corwin*, 3,) b. August 17th, 1807.
M. James Hortimer Reeves. (*Scotchtown, N. Y.*)

9 Fanny Hobart,[7] (*Mehetable Corwin*, 3,) b. February 17th, 1798.
M. William Skult, May 24th, 1827.
Ch. William, Ezra, Mehetable.

10 Fanny Shannon,[8] (*Mehetable Corwin*, 5,) b. ——, 1820-30.
(*Little Valley, N. Y.*)

11 Fanny Taylor,[8] (*David J. Taylor*, 30,) b. October, 1853.

12 Fanny M. Knox,[8] (*Julia Corwin*, 7,) b. December 27th, 1843

13 Fanny Fowler,[8] (*Mehetable Taylor*, 12,) b. May, 1863.

14 Fanny Cooper,[6] (*Martha Corwin*, 4.)

15 Fanny Goble,[8] (*Elizabeth H. Corwin*, 19.)
  ; *M.* Charles Kelly.   (*Baldwinsville, Onondaga Co., N. Y.*)
16 Fanny Little,[8] (*Benoni B. Little*, 1,) b. ——, 1830?
17 Fanny Rosetta Cooper,[8] (*Martha C. Corwin*, 4½,) b. ——, 1850, d. young.
18 Fanny E. Boyd,[9] (*Emeline Corwin*, 7,) b. November 22d, 1862.
                                                              (*Vermont.*)
19 Fanny Beyea,[9] (*Sophronia Corwin*, 1,) b. August 13th, 1852.
20 Fanny Harding,[9] (*Polly M. Corwin*, 4.)
21 Fanny Reeve,[9] (*Charlotte M. Corwin*, 7,) b. ——, 1857.
22 Fanny Robinson,[10] (*Hannah Corwin*, 22.)
 (1) Fanny Miller,[9] [*Sarah H. Corwine*, (6),] b. November 24th, 1850, in
    New-York City.

Ferdinand, (Old High-German,) brave, valiant.

1 Ferdinand Crassons,[8] (*Sarah J. Corwin*, 13,) b. ——, 1830–40.
                                                        (*Williamsport, Pa.*)

Fidelia, (Latin,) faithful.

1 Fidelia Chase,[9] (*Phebe Corwin*, 8½,) b. ——.
  FILE, see PHILE.

1 FINEY,[7] (*Nathan*, 2,) b. ——, 1780–90.
  M. Daniel Corwin.

Fitz-Alan, (Celtic,) son of harmony.

1 Fitz-Alan E. North,[8] (*Sibyl Campbell*, 3,) b. March 16th, 1835.
  M. Julia Gray, October 23d, 1855.
  Ch. Wellington, John.   (*Bloomingdale, Mich.*)

Flora, (Latin,) a flower.

1 FLORA E.,[8] (*James*, 16,) b. November 15th, 1858.
2 FLORA,[9] (*John*, 58·)
3 FLORA MARIA,[9] (*John A. R.*, 60,) b. ——, 1855–60.
4 Flora W. Dunham,[10] (*Alida Durkee*, 1,) September 29th, 1867.
                                                              (*Vermont.*)

Florence, (Latin,) blooming.

1 FLORENCE,[9] (*Walter S.*, 2,) b. ——, 1845–50.
2 Florence D. Campbell,[8] (*Daniel Campbell*, 34,) b. June 5th, 1859, d.
   November 18th, 1860.
3 Florence Taylor,[8] (*Hendrik Taylor*, 2,) b. March, 1858.
4 Florence Millspaugh,[8] (*William E. Millspaugh*, 22,) b. August 10th,
   1849.

1 FLORILLA M.,[8] (*Daniel*, 14,) b. February 13th, 1868.

1 Floy Lusetta Little,[8] (*George C. Little*, 53,) b. August, 1857.

Frances, (French.) free.

FRANCES, see also FANNY.

1 FRANCES,[7] (*Elisha*, 1,) b. ——, 1825–35.
   M. Rev. H. Sinclair, ——, 1853.
   C. Christiana.

2 FRANCES E.,[8] (*Benjamin W.*, 11,) b. July 16th, 1830, d. September 21st, 1830.

3 FRANCES,[8] (*Nathaniel*, 7,) b. ——, 1825–30, d. aged 13.

4 FRANCES,[8] (*George S.*, 12,) b. March 12th, 1830, d. April 17th, 1852.
   M. David L. Garrigues.
   Ch. George C.   (*Littleton and Newark, N. J.*)

5 FRANCES ALTHEA,[8] (*Joseph W.*, 12,) b. February 9th, 1850.

6 FRANCES,[8] (*Lewis W.*, —,) b. ——, 1825–35.

7 FRANCES C.,[8] (*John*, 22,) b. June 5th, 1818.
   M. George Howell, November 5th, 1835.

8 FRANCES,[8] (*Nathaniel*, 7,) b. September 1st, 1834, d. August 18th, 1859.

9 FRANCES MATILDA,[8] (*David*, 12,) b. August 17th, 1826.
   M. (1) Henry A. Loper ; (2) Azariah Anderson.
   Five Ch.

10 Frances J. Aldrich,[8] (*Deborah Corwin*, 6.)

11 Frances Murray,[9] (*Lucinda Corwin*, 1.)

12 Frances Allen,[9] (*Susannah Hobart*, 3,) b. June 22d, 1850.

12 Frances Marietta Howell,[8] (*Sophia Woodruff*, 5,) b. April 6th, 1832.
   M. William James Embler, November 26th, 1851 ; he was b. September 14th, 1825.
   Ch. Sophia E.   (*Walden, N. Y.*)

13 Frances H. Leach,[9] (*Mary Corwin*, 126,) b. ——, 1843, d. ——, 1859.

Francis, (French,) free.

FRANCIS, see also FRANK.

1 FRANCIS FOURNIER,[7] (*Silas*, 5,) b. October 7th, 1821.
   M. Sarah Elizabeth, daughter of Lewis B. Stiles, of Morristown, N. J.
   Ch. Lewis F., Silas R., Elizabeth, Silas R., William O., Lizzie.
                    (1852–9, *Baskingridge, N. J. ;* 1859, *Newburg, N. Y.*)
   A brick manufacturer by steam power. He lost his right hand by its being caught in machinery.

2 FRANCIS NICHOLAS WEST,[8] (*Joseph*, 16,) b. July 4th, 1840.
   M. Louisa Westervelt, October 16th, 1867.   She d. April 6th, 1869.
                                               (*Newark, N. J.*)

3 FRANCIS,[8] (*Benjamin*, 15,) b. ——.   (*Ohio.*)

4 FRANCIS P.,[8] (*Jabez*, 3,) b. ——.
   M. —— Tuthill.
   (Possibly this name should be *Frances*.)

5 Francis Osborn,[6] (*Amelia Corwin*, 2,) b. August 24th, 1838, d. September 5th, 1839.

6 Francis Dunlevy,[6] (*Lucinda Corwin*, 4,) b. April 1st, 1821, at Chicago, Illinois.

7. Francis Cox,[6] (*Mary E. Corwin*, 44.)

8 Francis Mapes,[9] (*Mary Corwin*, 49.)

Frank, same as Francis.

1 FRANK,[8] (*Ira C.*, 1.)

2 FRANK,[8] (*Samuel B.*, 16.)

3 FRANK,[8] (*John*, 30.)

4 FRANK,[9] (*John*, 62,) b. October 29th, 1858.   (*Vermont.*)

5 FRANK LIVINGSTON,[9] (*John*, 53,) b. June 15th, 1859.

6 FRANK,[9] (*George*, 40,) b. ——, 1869.   (*Michigan.*)

7 FRANK,[9]? (*Theodore*, —.)

8 FRANK JEFFREYS,[10] (James, 39,) b. July 24th, 1861.

9 Frank C. Norris,[8] (*Marcia Corwin*, 1,) b. August 8th, 1860.

10 Frank Young,[8] (Mary Corwin, 37.)

11 Frank Halsey,[8] (Jefferson Halsey, 1.)

12 Frank Galloway,[8] (Amanda Corwin, 2,) b. ——.
          (*Lawyer, Morristown, N. J.*)

13 Frank Wells,[8] (Laura Corwin, 4,) b. ——, 1810–20.

14 Frank Topping,[8] (Julia Corwin, 4,) b. July 6, 1842.

15 Frank Swina,[9] (Albert Swina, 10,) b. ——, 1855.

16 Frank A. Noyes,[9] (Louisa Corwin, 3,) b. February 9th, 1846.

17 Frank Ketchum,[9] (*Nancy A. Corwin*, 3,) b. December, 1851.
          (*Warsaw, Ind.*)

18 Frank Wilson,[9] (*Julia E. Corwin*, 9.)

19 Frank Blinn,[9] (*George D. Blinn*, 56,) b. November 7th, 1869.

20 Frank Perinchief Newman,[9] (Moses Newman, 9,) b. at Wyandotte, Kansas, June 18th, 1857.

21 Frank Kyte,[9] (*Clarinda Corwin*, 2,) b. January 16th, 1859.
          (*Niagara Co., N. Y.*)

22 Frank A. Hackett,[10] (*George H. Hackett*, 89,) b. October 14th, 1869.
          (*Vermont.*)

23 Frank Crater,[10] (*Mary Corwin*, 83.)

1 FRANKLIN EZRA,[7] (*Jabez*, 1,) b. ——, 1825.
M. Mary S. Penney.
· *Ch.* Martha A., Christiana.   (*Jamesport, L. I.*)

2 FRANKLIN,[8] (*George E.*, 13,) b. ——, 1830–40.

4 - Frank H. Call[9] (W. E. Call 90½) b. Sept 28. 1864. m Bertha Collins June 21. 1884

3 FRANKLIN,[8] (*Matthias*, 7,) b. ——.
    M.
    Ch.
He was frequently a member of the Ohio Legislature from Clinton Co. He afterward moved to La Salle Co., Ill., and has been more than once a member of the Illinois Legislature. He is now speaker of the House of Representatives in that State.   (*Ottawa, Ill.*)

4 FRANKLIN,[9] (*Daniel A.*, 23,) b. ——, 1860 ?
5 Franklin Girard,[8] (*Jerusha Corwin*, 2.)
6 Franklin Millspaugh,[8] (*Gilbert C. Millspaugh*, 6,) b. January 21st, 1838.
7 Franklin Smith,[9] (*Mehetable J. North*, 18,) b. September, 1863.
8 Franklin Drake,[9] (*Mary Terry*, 128.)
9 Franklin Newton Wetherby,[9] (*James H. Wetherby,*) b. October 28th, 1865.
10 Franklin Corwin Cooper,[8] (*Martha C. Corwin*, 4½.)
11 Franklin L. Lewis,[10] (*Imogene Davis*, 1,) b. ——, 1865–70.

    Frederic, (Old High-German,) abounding in peace, or peaceful ruler.

1 FREDERIC,[7] (*John*, 13,) b. ——, 1780–90.
2 FREDERIC,[7] (*Abel*, 1,) b. ——, 1790–5.
3 FREDERIC,[8] (*Elias*, 1,) b. ——, 1824–30.
4 FREDERIC,[8] (*William*, 17,) b. October 1st, 1840.
5 FREDERIC VICTOR,[8] (*John*, 25.)
6 FREDERIC H.,[8] (*Parker*, 1,) b. September 13th, 1818.
    M. Louisa J., (daughter of Barnum Treadwell and Mary Raze,) on November 13th, 1847. Barnum T. was then of Royalton, Niagara Co., N. Y., but originally of Bridgeport, Ct. He was among the earliest settlers in Western New-York, and d. 1861, aged 84. Mary Raze was of Vermont, b. 1780, living yet in Niagara Co., N. Y. Louisa J. Treadwell, b. May 15th, 1825.
    Ch. Orin E., Parker E., Libbie M., Charles S., Louisa G.
        (*Justice of Peace for many years, Ogden Centre, Mich.*)
7 FREDERIC M.,[9] (*George H.*, 38,) b. April 1st, 1856.
8 FREDERIC,[9] (*Daniel*, 21.)
9 FREDERIC,[9] (*John*, 56,) b. April 16th, 1870.
10 FREDERIC,[9] (*John*, 62,) b. October 29th, 1858.   (*Vermont.*)
11 FREDERIC GRIFFIN,[9] (*Hubbard*, 2,) b. October 11th, 1869.
12 FREDERIC,[10] (*Samuel*, 26,) b. September 30th, 1867.
12¼ Frederic B. Hansen,[8] (*Abbiette Corwin*, 1.)
12½ Frederic W. Noyes,[9] (*Louisa Corwin*, 3,) b. February 5th, 1859.
        (*Vermont.*)
13 Frederic Vine,[9] (*Sarah E. Corwin*, 30½,) b. May 15th, 1870.

14 Frederic W. Kyle,[9] (*Clarinda Corwin,* 1,) b. August 6th, 1870.

15 Frederic Noyes,[9] (*Maria L. Smith,* 17,) b. September 10th, 1868.

(*Vermont.*)

16 Freddie Thorne,[9] (*Diantha Smith,* 1,) b. ——, 1860.   (*Vermont.*)

17 Frederic E. Gray,[10] (*Sarah Corwin,* 56,) b. November 16th, 1835.

(1) * FREDERIC MORTIMER,[9] [*Amos,* (2),] b. at New-Rochelle, N. Y., August 25th, 1864.

(2) Frederick Mortimer Fountain,[9] [*Sarah H. Corwine,* (6),] b. July 3d, 1844. He enlisted in the 13th regiment, N. G. S. N. Y., in spring of 1862, as a three months' volunteer. Was stationed at Suffolk, Va. He contracted typhoid fever, and died (a few days after his return) September 7th, 1862.

1 Frederica Willinghouse,[8] (*Hannah J. Corwin,* 9.)

Gabriel, (Hebrew,) man of God.

1 GABRIEL,[7] (*Joseph,* 9,) b. ——, 1808 ?

M. Irene Murray.

Ch. Louis.   (*Middletown, N. Y.*)

2 GABRIEL L.,[7] (*Benjamin,* 5,) b. ——.

M. ——

Ch. Theodore, Albert, Caroline.   (*Marathon, Cortland Co., N. Y.*)

3 GABRIEL,[7] (*Phineas,* 3,) b. January 15th, 1815, d. August 19th, 1850.

M. (1) Hannah Campbell, February 20th, 1841 ; (2) Clarissa Carroll, March 12th, 1846. She (2) b. 1821, d. December 27th, 1853. She left a will.

Ch. Bethia M., Letitia J., Harriet A.   (*Cato, Cayuga Co., N. Y.*)

4 GABRIEL,[7] (*James,* 5,) b. October 10th, 1795, d. November 2d, 1820.

M. Olivia Gladen. She afterward m. Dr. Cary. She d. about 1862.

Ch. Anna Maria.

A cabinet-maker.   (*Luzerne Co., Pa.*)

5 REV. GABRIEL SMITH,[8] (*Daniel,* 11,) b. February 27th, 1802, near Middletown, N. Y.

M. (1) Adaline Houston. She d. April, 1861. (2) Sarah A. Corwin, No. 95.

Ch. (2) Emma M., G. Tracy, G. Eugene.

He was graduated from the Medical College of New-York, in 1825. At first a physician in Bloomingburg, N. Y. Licensed to preach the Gospel by the Presbytery of Hudson, September, 1841. Pastor of Presbyterian Church of Elba, N. Y., 1842–67; of Pembroke and Batavia, N. Y., 1867—.

6 GABRIEL D.,[8] (*John,* 16,) b. April 7th, 1824.

M. Harriet Coleman.

No Ch.   (*Port Jervis, N. Y.*)

`                              * Corwine.`

7 Gabriel Everett,[7] (*Deborah Corwin*, 1,) b. ——, 1810–20.

8 Gabriel Pellott,[8] (*or Reeve*,) (*Mary Corwin*, 24.)
   M. Mary McCarter.   (*Weston, Orange Co., N. Y.*)

9 Gabriel Mullock, (*Mary Corwin*, 22.)

1 Gaddis Blinn,[9] (*Jesse C. Blinn*, 13,) b. August, 1865, d. ——.

Gamaliel, (Hebrew,) a reward from God.

1 GAMALIEL,[7] (*Nathaniel*, 5,) b. ——, 1770–90.

1 GARDINER,[8] (*John*, 25.)

George, (Greek,) a landholder, or husbandman.

1 GEORGE,[4] or [5]? (*Theophilus*, 4,) b. ——, 1740–50 ? d. ——, 1806, at Mont-
gomery, Orange Co., N. Y.   (*See will at Goshen.*)
   M. Margary Williams ?
   Ch. John, George, James, Daniel, William, Nancy, Mary, Barney.
   (The name is spelled Curwin and Curran. He is thought to be the son
of Theophilus, 4, as he also spelled his name in both these ways.)

2 GEORGE,[6] (*Phineas*, 1,) b. November 10th, 1768, d. November 18th, 1831.
   M. Mary McCarty.
   Ch. Mehetable, Eleazer, Fanny, Phineas, Mary, George E., Arvilla,
Noahdiah, Christiana, Lewis W.
               (*Orange Co., N. Y., till 1822, then Central N. Y.*)
   There was a George Corwin in Jericho, Chemung Co., N. Y., 1836.
(*Records*, vol. viii. p. 442.)

3 GEORGE,[5] or [6], (*George*, 1,) b. ——, 1770–80.   (*Curwin or Curran.*)

4 GEORGE II.,[6] (*David*, 5,) b. January 7th, 1822, d. August 5th, 1829.

5 GEORGE,[6] (*James*, 1,) b. at Southold, L. I, June 27th, 1766, d. December
2d, 1834.
   M. Betty Brush, October 10th, 1805. She was b. at Greenwich, Ct.,
February 25th, 1772, and d. in New-York City, May 15th, 1852.
   Ch. Edward Brush.   (*New-York City*, 1796–1834.)

6 GEORGE,[6] (*John*, 6,) b. November 25th, 1776, d. young.

7 GEORGE W.,[7] (*John*, 9,) b. October 10th, 1801, d. May 27th, 1809.
Was run over by a wagon.

8 GEORGE STUART,[7] (*Moses*, 1,) b. at Middletown, N. Y., December 29th,
1821, d. March 5th, 1863, in Jersey City.
   M. Leah Margaret Corwin, (1,) March 24th, 1852, in Jersey City, N. J.
   Ch. Edward Tanjore, George Edwin.
         (Fruit merchant in New-York. Residence, *Jersey City, N. J.*)

9 GEORGE,[7] (*James*, 1,) b. ——, 1800, d. ——, in Mobile. Unmarried.
                             (*St. Louis, Missouri.*)

10 GEORGE,[7] (*Jabez*, 1,) b. August 29th, 1803, d. ——, 18—.
   M. Eliza Terry, November 29th, 1826; (2) Eliza Mandeville.
   Ch. Daniel T., George II., Emeline E.; by 2d wife, George II.

11 George,[7] (*Abel*, 1,) b. ——, 1797 ——, d. ——, 1819.

12 George S.,[7] (*Jesse*, 3,) b. December 15th, 1806, at Middletown, N. Y. *ʌ⋅ʰc*
   M. Eliza Wood, December 22d, 1829. She was b. February 27th, 1810. *ʌ*
   Ch. Francis, Harriet A., Charles, Annie, Henry B., Edward M., Jane A.
   (For eight years (1853–61) he was cashier of the Iron Bank at Morristown, N. J.; afterward moved to Catasauqua, Lehigh Co., Pa., 1861–70; now of Newark, N. J.

13 George Elkanah,[7] (*George*, 2,) b. ——, 1809.
   M. (1) Eliza Jane Smith; (2) Olive Smith.
   Ch. Franklin, Mary, Sevellen, Patia E., Emma George.
                          (*Reading*, *N. Y.*, 1836 ; *Batavia*, *Ill.*)
   (In May, 1835, a George Corwin and Eliza his wife were dismissed from the church of Aquebogue to the Presbyterian church of Westtown, N. Y.)

14 George Little,[7] (*Silas*, 2,) b. February 5th, 1811.
   M. (1) Harriet Roe, (2) Eunice Peck, (3) Mary Tompkins.
   Ch. Moses Young.   (*New-Milford*, *Pa.*)

15 George,[7] (*Daniel*, 5,) b. ——, d. in infancy.

16 George Wells,[7] (*Daniel*, 9,) b. ——, 1795–1805.

17 George D.,[7] (*David*, 10,) b. August 10th, 1855, d. October 15th, 1855.

18 George W.,[7] (*Thomas*, 2,) b. November 1st, 1814, d. in California, April 16th, 1850.
   M. Clarissa M. Sisson, June 3d, 1839. She b. April 10th, 1817.
   Ch. Cornelius G., Charles L., Iduella G., George W.
                          (*Greenport*, *L. I.*)

18½ George,[7] (*James*, 6,) b. July 26th, 1805, d. May 10th, 1836, in St. Louis, Mo.

19 George,[7] (*Richard*, 3,) b. May 1st, 1840, d. May 2d, 1844.

20 George,[7] (*Asa*, 3,) b. – —, 1804.
   M. Susan ——.
   Ch.   (*Flint Village*, *Mich.*)

21 George Washington,[8] (*Daniel*, 11,) b. ——, 1812.
   M. Lamina Horton.

22 George,[8] (*Isaac*, 4,) b. March 1st, 1820.   Unmarried.

23 George Washington,[8] (*Anson*, 1,) b. January 18th, 1826, / 8 Z 7
   M. Mary Ophelia Howell, February 11th, 1846.
   Ch. Mary E., Theodore F., Rose, Gertrude, Everett E., Evelina, Fanny.
                          (*Riverhead ; Greenport, L. I.*)

24 George Hiram,[8] (*David*, 12,) b. August 27th, 1828.
   M. Jane Terry, August 15th, 1850.
   Ch. Elbert O., Morrison M.   (*Greenport*, *L. I.*)

24½ George Washington,[8] (*George W.*, 18,) b. November 9th, 1849, d.
   September 13th, 1852.

25 George,[8] (*Eleazer*, 1,) b. ——, 1835–45.

26 GEORGE,[8] (*George E.*, 13,) b. 1830–40.

27 GEORGE HERVEY,[8] (*Hervey*, 1,) b. ——, 1858.

28 GEORGE HENRY,[8] (*George*, 10,) d. in infancy.

29 GEORGE HENRY,[8] (*George*, 10,) d. aged about 17.

30 GEORGE,[8] (*John*, 22,) b. October 12th, 1812.
   M.
   Ch. John, George.   (*Brooklyn, N. Y.*)

31 GEORGE,[8] (*Stephen*, 8,) b. ——, 1810–20.

32 GEORGE BRAINARD,[8] (*Edward C.*, 5,) b. in New-York City, June 12th,
   1836, d. March 22d, 1839, of measles. Buried at Preakness, N. J.

33 GEORGE EDWIN,[8] (*George S.*, 8,) b. in Jersey City, April 20th, 1858.

34 GEORGE F.,[8] (*Lemuel*, 1,) b. January 27th, 1870.

35 GEORGE,[8] (*Nathaniel*, 3.)

35½ GEORGE,[8] (*Henry*, 11,) b. April 18th, 1837, d. December 4th, 1855.

36 GEORGE CHAUNCEY,[8] (*Gilbert*, 4,) b. September 1st, 1842.

37 GEORGE W.,[8] (*Daniel*, 16,) b. ——, 1819, d. January 6th, 1871 ?
   M.
   Ch. Julia, Joseph, James, George.   (*Brooklyn.*)

38 GEORGE HOMAN,[8] (*Hudson*, 1,) b. July 13th, 1825.
   M. Elizabeth J. Miller, November 3d, 1853.
   Ch. Frederick, George H.   (*Greenport, L. I.*)

39 GEORGE,[8] (*Daniel*, 19,) b. May 9th, 1819, d. February 12th, 1849.

40 GEORGE J.,[8] (*Parker*, 1,) b. January 20th, 1838.
   M. Adelia C., (daughter of Loren Root and Sarah Porter,) February 29th,
   1860, at Victor, Ontario Co., N. Y. Adelia was born August 9th,
   1835. Loren Root, b. 1789, in Stockbridge, Mass., was a soldier in
   war of 1812. Died 1867. Sarah Porter, born 1794, near Hoosic
   Falls, N. Y. They were married April 6th, 1815, and resided at
   Victor, N. Y.
   Ch. Minnie E., (adopted,) Sarah F. S., Frank L. George studied at
   Oberlin, Ohio, and his wife Adelia in the Collegiate Institute, Rochester, N. Y. He served in Michigan Fourth Cavalry for three years, returning wounded and disabled.

41 GEORGE WASHINGTON,[8] (*De Witt C.*, 2,) b. September 8th, 1841.

41½ GEORGE A.,[8] (*James*, 16,) b. May 28th, 1854.

42 GEORGE,[8] (*Noah*, 2,) b. February 28th, 1826.
   M. Louisa M. Maine, September 24th, 1849. She born 1827.
   Ch. Lucy A., William, George.   (*Havana, N. Y.*)

43 GEORGE E.,[8] (*James H.*, 13,) b. March 31st, 1864. –

44 GEORGE,[8] (*James T.*, 21,) b. ——, 1840–50.

44½ GEORGE WHIPPLE,[8] (*Rev. Ira*, 4,) b. at Marietta, O., February 21st,

1848. Graduated at Granville, Ohio; now studying law at Ann Arbor, Mich.

45 GEORGE,[9] (*George W.*, 28.)

45½ GEORGE,[9] (*George*, 42,) b. March 16th, 1865.

46 GEORGE,[9] (*William*, 34.)

47 GEORGE F.,[9] (*William II.*, 38,) b. January 16th, 1840, in Vermont.

48 GEORGE W.,[9] (*John*, 44,) b. December ——, 1834.
    M. Frances Howell.   / / 4 / De au
    Ch. Mortimer M., Lucy A., Dwight, Grover.   (*Brooklyn, N. Y.*)

49 GEORGE,[9] (*George*, 30.)

49½ GEORGE HENRY,[9] (*James*, 36½,) b. about 1859.

50 GEORGE HARVEY,[9] (*Silas G.*, 12,) b. July 23d, 1850, d. September 21st, 1854.

51 GEORGE W.,[9] (*Silas G.*, 12,) b. March 14th, 1863.

51¼ GEORGE II.,[9] (*George II.*, 38,) April 16th, 1864.

51½ GEORGE R.,[9] (*William A.*, 41½,) b. March 19th, 1867.

(Another George Corwin married Isabel Rogers, January 6th, 1822, but neither are yet identified.)

52 George Everett,[7] (*Deborah Corwin*, 1,) b. ——, 1815–25.

53 George C. Little,[7] (*Desire Corwin*, 1,) b. May, 1814.
    M. Almira Hardy, October, 1839.
    Ch. George L., John C., Lucretia A., Lloyd B., Floy L. (*Wisconsin.*)

54 George P. Taylor,[7] (*Fanny Corwin*, 1,) b. January 1st, 1812, d. February 7th, 1832.

55 George Philips,[7] (*Linea Corwin*, 1,) b. ——, 1800 ?

56 George Duckworth Blinn,[8] (*Lucinda Thatcher*, 5,) b. July 9th, 1837.
    M. (1) Catharine Pence, March 17th, 1861. She d. May, 1864. (2) Nancy S. Elliott, January 28th, 1867.
    Ch. Manson, Josephine ; (2) Nettie, Frank.

57 George W. Millspaugh,[8] (*Gideon H. Millspaugh*, 1,) b. October 1st, 1835.

58 George Hunt,[8] (*Experience Corwin*, 1,) b. 1815–25.   (*Newark, N. J.*)

59 George Corwin Wetherby,[8] (*Eliza Corwin*, 3,) b. February 13th, 1838, d. March 11th, 1854.   (*Luzerne Co., Pa.*)

60 George Lewis Taylor,[8] (*Elizabeth Corwin*, 17,) b. at Rochester, Ohio, November 5th, 1848.

61 George Reading Reed,[8] (*Sophia Corwin*, 1,) b. July 2d, 1806.
                          (*Walnut Grove, Morris Co., N. J.*)

62 George C. Mitchell,[8] (*Maria Corwin*, 3.)

63 George Aldrich,[8] (*Deborah Corwin*, 6.)

64 George Chase,[8] (*Bethia Corwin*, 5.)

65 George Shannon,[8] (*Mehetable Corwin*, 5,) b. ——, 1820–30.

66 George Baldwin,[8] (*Jane Corwin*, 1,) b. at Goshen, d. young.

67 George Cox,[8] (*Mary E. Corwin*, 44.)

67½ George C. Shepherd,[8] (*Charlotte Corwin*, 6,) b. July 11th, 1846, d. November 9th, 1847.

68 George Clarence Cooper,[8] (*Martha C. Corwin*, 4½,) b. ——, 1852.

69 George W. Peters,[8] (*Mary Corwin*, 30,) b. ——, 1835–45.

70 George Loren Little,[8] (*George C. Little*, 53,) b. April, 1841.

71 George Taylor,[8] (*John C. Taylor*, 70,) b. April, 1841.

72 George R. Crego,[8] (*Clarissa Taylor*, 4,) b. August, 1853.

73 George Thatcher,[8] (*Jesse Thatcher*, 10,) b. October 6th, 1809.

74 George Smith,[8] (*Sarah Corwin*, 28,) b. in Tunbridge, Vt., in 1824.
    M. Ellen B. Post, 1852.
    Ch. Amanda F., Herbert P., George T., Judson L., Edmund E., Royal M., Emerett.   (*St. Lawrence Co., N. Y.*)

75 George G. Byram,[8] (*Mary A. Corwin*, 34,) b. October 27th, 1856.

76 George Spencer Hackett,[8] (*Mary Corwin*, 14,) b. January 24th, 1820, d. September 21st, 1841.   (*Vermont.*)

77 George Dunlevy,[8] (*Lucinda Corwin*, 4,) b. February 8th, 1838, d. January, 1840.

78 George W. Smith,[8] (*Philenia Corwin*, 1,) b. August 2d, 1837, d. May 29th, 1864. He died in the army.
    M. Luella Elimer, 1860.

78½ George Call,[8] (*Bethia Corwin*, 8,) b. May, 1827, d. ——,1829.

79 George W. Wetherby,[9] (*James H. Wetherby,* 88) b. September 26th, 1863.

80 George W. Gordon,[9] (*Sarah E. Corwin*, 43.)

81 George S. Garrigues,[9] (*Frances Corwin*, 4.)

82 George Murray,[9] (*Lucinda Corwin*, 1.)

83 George S. Allen,[9] (*Susannah Hobart*, 3,) b. October 19th, 1852.

84 George T. Smith,[9] (*George Smith*, 74,) b. ——, 1855–65.

85 George Huff,[9] (*Caroline Horton*, 13.)

86 George L. Kyte,[9] (*Clarinda Corwin*, 2,) b. January 6th, 1852, d. February, 1853.

87 George Davis,[9] (*Rachel Corwin*, 4,) b. 1851.

88 George W. Chamberlin,[9] (*Jennette Corwin*, 1,) b. February 13th, 1853.
                 (*Vermont.*)

89 George Henry Hackett,[9] (*Samuel B. Hackett*, 32,) b. September 4th, 1841.
    M. Ellen Parker, in Tunbridge, Vt., July 4th, 1867.
    Ch. Frank A.

89½ George Bigelow Colgrove,* (*Rev. Clinton Colgrove*, 1,) b. September 16th, 1860, d. January 14th, 1862.

90 George Nicholas Crater,[10] (*Mary Corwin*, 83.)

91 George Gray,[10] (*Sarah Corwin*, 56,) b. May 26th, 1868.

1 Georgiana Clark,[9] (*Cecilia Corwin*, 1,) b. July 4th, 1845. *M.* Frank Jeffrey, of New-London, Ct., March 29th, 1869.

(*Galveston, Texas.*)

(1) GEORGE CORWIN or CURWIN, (Captain,) b. in England, December 10th, 1610.   Came from Northampton, Eng., and settled at Salem, Mass., in 1638; d. January 3d, 1685. *M.* (1) Elizabeth, daughter of John Herbert, Mayor of Northampton, and had five children. *M.* (2) Elizabeth,* widow of John White.   She d. September 15th, 1668. *M.* (3) Elizabeth Brook, widow of Robert Brook, and daughter of Governor Edward Winslow,† of Plymouth Colony, on September 22d, 1669.   She d. after 1694.

Ch. Abigail, John, Jonathan, Hannah, Elizabeth; by third wife, Penelope, Susannah, George.

He settled at Salem in company with the celebrated Hugh Peter, and laid the foundation of the commercial enterprise of that place.   He first began the building of vessels in that port, and was afterward extensively engaged in commerce during the whole of his long life.   His books of account and his mercantile correspondence with Sir William Peake (Lord Mayor of London in 1666) show that he had embarked in the London trade previous to 1658.   The late Rev. Dr. Bentley thus notices him in his *Sketch of Salem*, published in the collections of the Massachusetts Historical Society, in 1800 : "This year, (1685,) Salem lost another eminent man, Captain George Curwin, who came here in 1638 with his family, and was rich.   He was often engaged in town affairs, and commanded a troop of horse.   He was also a representative in the General Court.   There is a three-quarter portrait‡ of him in the hands of Samuel Curwen, Esq., son of the Rev. George Curwin, and his great-grandson.   He had a fine, round forehead, large nostrils, high cheek bones, and gray eyes.   His dress, a wrought flowing neckcloth, a sash covered with lace, a coat with short cuffs, and reaching half-way between the wrist and elbow, the shirt in plaits below ; an octagon ring and cane, which still remain."   He left no debts, and one of the largest estates that had been administered in the colony, amounting to £5964 19s. 7d.   Besides the homestead, he left four dwelling-houses, four warehouses, and two wharves in Salem ; three farms in the vicinity, containing fifteen hundred acres ; a warehouse and wharf in Boston ; four ketches,

---

* She was the mother of two children by Mr. White ; Elizabeth, married Samuel Andrew, and Mary, married Samuel Gardner.

† She was daughter of Governor Edward and Susannah Winslow, of Plymouth Colony, and was twice married: (1) to Robert Brook, by whom she had one son, Robert, who died 1687, aged 31, and was buried at Charlestown.   Tombstone still standing.

‡ This portrait is now in possession of George R. Curwen, of Salem, Mass.

valued at £1050 ; merchandise valued at £2232 ; coin £93 7s., and 621 ounces of silver plate. Among the wearing apparel inventoried are a silver-laced cloth coat, a velvet ditto, a satin waistcoat embroidered with gold; a troping scarf and silver hat-band, golden-topped and embroidered gloves, and a silver-headed cane, which still remains.

In the settlement of the estate his widow claimed some plate given her by the lord mayor, by her father, and by the governor; together with eight pounds in gold, which her husband had received from Mr. Pope, being the "produce" of an Indian boy sent her by the governor and council from Plymouth.

In 1657, George Corwin and many others petitioned the government for a plantation of ten miles square between Springfield and Fort Awrania, (Orange, now Albany;) the petition was granted, if the plantation were begun within eighteen months. (*Mass. Bay Col. Recs.* iv. 374.)

In 1659, the court judged that a present claim of our just rights upon Hudson River, in the Fort of Awrania, by a letter from this court to the Dutch Governor, etc., etc.

(See Curwen's Journal and Letters, published by Little, Brown & Co., Boston, last edition, 1864. *N. E. Gen. & Ant. Reg.* iv. 299–301. *Mass. Bay Col. Records* iv. 187.)

(2) * GEORGE,[2] [*George Curwen*, (1),] b. 1674, d. before 1684.

(3) * GEORGE,[3] [*John*, (1),] b. February 26th, 1665–6, d. April 12th, 1696.
M. (1) Susannah, daughter of John Gedney; (2) Lydia, daughter of Hon. Bartholomew Gedney ; Lydia d. ——, 1700.
Ch. Bartholomew.

He is mentioned in Hutchinson's History of Massachusetts as a captain in the expedition against Canada under Sir William Phipps, in 1690. He was the unfortunate sheriff of Essex Co., Mass., May 7th, 1692, and for officiating in the witchcraft delusion was severely persecuted by the friends of the sufferers till his death, before which, however, he removed to Winthrop's Neck, Ct. His only son, on reaching maturity, removed to Amwell, Hunterdon Co., N. J.

That one so young as George (only 27) should have been appointed sheriff, may be explained by the circumstances that his uncle, Jonathan Corwin, his maternal uncle, Wait Winthrop, and his father-in-law, were among the judges of the Court of Oyer and Terminer.

(4) * REV. GEORGE,[3] [*Jonathan*, (1),] b. May 21st, 1683, d. November 3d, 1717.
M. Mehetable, daughter of Deliverance Parkman and Mehetable Waite, (daughter of Hon. John Waite, Speaker of the House of Representatives of Massachusetts,) on July 27th, 1711. She d. November 13th, 1717–18.
Ch. Jonathan, Samuel, George.

He graduated at Harvard College, 1701, and became pastor of 1st Congregational Church at Salem, as a colleague of Rev. Nicholas Noyes, 1714.

* Curwen or Corwin.

In this relation he continued till death. His portrait is in possession of George R. Curwen, of Salem. He died of a sudden cold. He left a reputation in the church of great philanthropy, good address, and excellent pulpit talents. A Boston paper of the day says, "He was highly esteemed in his life, and very deservedly lamented at his death, having been very eminent for his early improvements in learning and piety, his singular abilities and great labors, his remarkable zeal and faithfulness in the service of his Master. He was a great benefactor to our poor."

(5) * GEORGE,[4] (*Rev. George*, 4,) b. December 4th, 1717, d. at St. Eustatia, West-Indies, while engaged in a commercial enterprise there, June 7th, 1746.

M. Sarah, daughter of Benjamin and Abigail (Lindall) Pickman, March 18th, 1738. She d. in Salem, January 3d, 1810, aged 91.

Ch. George, Sarah, Mehetable.

He graduated at Harvard College, 1735; was engaged with success in mercantile pursuits till interrupted by the French war. He was commissary in the expedition against Louisburg, 1745-6, and a captain. His commission as commissary is signed by Governor Shirly, bearing date of March 18th, 1738.

(6) † GEORGE,[5] [*Bartholomew*, (1),] b. July 12th, 1718, d. 1780, (or 1783.)

M. Abigail Hixon. She born 1726, d. in Mason Co., Ky., 1810.

Ch. Esther, John, Margaret, Richard, Alice, Ruth, Ann, Amos, Elizabeth.

His name appears on the Hunterdon Co. (N. J.) records as purchasing land in 1760. His family removed to Kentucky, soon after his death. (1794 ?)

(7) * GEORGE,[5] [*George*, (5),] b. June 4th, 1739, drowned April 2d, 1762, while on a voyage to the West-Indies.

7½ † GEORGE,[6] [*Samuel*, (3),] b. about 1760-5, d. ——, 1785-90.

M. Ruth Corwine, (1) about 1788.

Ch. George.

He was married soon after the close of the Revolution, but d. shortly. When the son was about five years old, his mother removed to Kentucky with him. (1794.)

7¾ † GEORGE,[7] [*George*, (7½),] b. about 1789, d. in Indiana.

M. ——, in Kentucky.

Ch. Richard, Benjamin, George, Mary.

He moved after marriage to Indiana.

(There was a George Corwine who bought land, and lived in Fabius, N. Y., in 1801. Who?)

(8) * GEORGE REA,[7] [*Samuel*, (6),] b. July 4th, 1823.

He possesses the original portraits of George Curwen, (1,) the original settler; of Samuel Curwen, (2,) the Editor of the Journal and Letters; of Rev. George Curwen, (4,) and many others.   (*Salem, Mass.*)

* Curwen.   † Corwine.

(8½) + GEORGE,⁷ [*John*, (3),] b. ——, d. young.   (*West-Jersey.*)

(9) * GEORGE,⁷ [*Samuel*, (4),] b. June 11th, 1800, living.

    M. (1) Rachel Vossler, (2) Charity Stryker, (3) ——

    Ch. Rachel A., Abigail.   (*Flagtown, N. J.*)

(9½) * GEORGE,⁸ [*George*, (7¾),] b. ——, 1810–20, in Indiana.

(10) * GEORGE,⁸ [*Gideon R.*, (1),] b. October 10th, 1818.

    M. Catharine M., daughter of Jeremiah Van Dyke, of Hopewell, N. J.
    She b. May 4th, 1824.

    Ch. John V., Jennie A.   (*Pennington, N. J.*)

(11) † GEORGE,⁷ [*Richard*, (2),] b. in New-Jersey, August 6th, 1780, d. at
    Portsmouth, Scioto Co., O., May, 1852.

    M. Elizabeth Wilson, May 5th, 1800. She d. May, 1850. *No Ch.*

He came to Ohio for the purpose of constructing a mill, about 1798, being engaged in learning the business, and became acquainted with Miss
Wilson. He returned to Kentucky, and upon visiting her at one time, he
used a willow switch for a riding-whip. It was planted by them in the
father's yard, and grew to be a large tree, and was looked upon by them in
after years as typical of their own lives. A few years prior to their death,
the willow showed signs of decay, and the old people looked upon it with
feelings that can not be expressed. And it is rather a strange coincidence
that, when one side was lifeless, Mrs. Corwine had passed from this "tree
life" to immortality; and about the time Mr. Corwine yielded to the grasp
of the hand of death, the dead willow was prostrated by the storm.

They were frugal, industrious, and benevolent. In early life they were
both members of the Baptist Church. She remained so until her death.
But Mr. Corwine, at middle life, became a Universalist. By his generosity
and benevolence he won the esteem of all. He was active in business, and
amassed a very considerable fortune. He was one of the first associate
judges of the Common Pleas Court of Pike County, A.D. 1815, which office
he held for several terms. He took a lively interest in the political affairs of
the country, and in the welfare of the Indians, and in disposing of his
estate devised a considerable sum to aid in educating the "North-American
Indians." The bequest was received by a New-York society to be applied
as directed. The residue of his property was distributed to his relatives.

(11¼) * GEORGE,⁸ [*Amos*, (1½).]   (*Broadwell, Ill.*)

He has a ring containing the Curwen coat-of-arms, which has been
handed down through the Georges since the first settlement of the family
at Salem, Mass.

(11½) * GEORGE,⁸ [*Samuel*, (6½),] b. March 18th, 1818, in Pike Co., O.

    M. Lydia, daughter of Judge Charles McCollister, November 28th, 1844.

    Ch. Charles, Alice, Martha, Clara, George, Emma, McCollister.

    Judge McCollister was a prominent citizen of Pike Co., Ohio. He now

* Corwine.

6

resides in Iowa City, Iowa. George Corwine resides at Carthage, Jasper Co., Mo., whither he removed in 1871. He was clerk of the Court of Common Pleas for many years, a member of the State Senate, and was also engaged in banking for several years. He was a prominent member of the Democratic Party, and was very influential as a citizen. He is now engaged in farming.

(11¾) † George,⁹ [*Jesse*, (1),] b. February 23d, 1866.

(12) * George Endicott,⁹ [*James B.*, (1),] b. March 8th, 1861.

(12¼) † George,⁹ [*George*, (11½),] b. ——.   (*Missouri.*)

(13) George Atkinson Ward,⁷ (*Samuel C. Ward*, 5,) b. March 29th, 1793.

   M. Mehetable, daughter of James and Sarah (Ward) Cushing, October 5th, 1816. She was born February 28th, 1795, and died at New-Brighton, Staten Island, October 4th, 1862.

   Ch. George R., Sarah J., James C., Frank.

   See *Ward Genealogy, Hist. Collections of Essex Institute*, Vol. V. No. 5, 1863.

   He edited Samuel Curwen's Journal and Letters, in 1842, which has passed through several editions. See Samuel Curwen, (2.)

(*a*) * George,³ [*John*, (*a*),] b. ——, 1790, d. ——, 1848.

   M. ——

   Ch. John, Mary, Marshall F., George F.

(*b*) * George F.,⁴ [*George*, (*a*),] b. ——, 1810–20.   (*Near Philadelphia.*)

Gershom, (Hebrew,) an exile.

1 Gershom,⁵ (*Matthias*, 3 ?) b. ——, 1737, d. July 22d, 1756.

2 Gershom,⁵ ? (*Matthias*, 3,) b.

   M. Margaret —— —.

   Ch. Aaron ? Lucy ?

   Received a grant of 450 acres in Sterling township, Cayuga Co., N. Y., in 1806, being lot No. 3. He also bought land in same county, 1816. In 1825–35, his name occurs in the records, as being of the town of Sterling. In 1836, he is of the town of Hannibal, Oswego Co., N. Y. In Naomi Corwin's will, 1833, a Gershom is mentioned as yet living, probably her brother. In 1797, the name occurs in New-York City Directory. Perhaps he also lived in Paterson, N. J., for a time.

Gertrude, (Old High-German,) a spear-maiden.

1 Gertrude,⁹ (*Charles*, 21,) b. December 28th, 1858.

2 Gertrude,⁹ (*George W.*, 23,) b. November 22d, 1856.

3 Gertrude Knox,⁸ (*Julia Corwin*, 7,) b. January 30th, 1853.

4 Gertrude Cropper,⁹ (*Caroline Corwin*, 8,) b. November 22d, 1870.

5 Gertrude Robinson,¹⁰ (*Hannah Corwin*, 22.)

1 Ghordes,⁶ (*Edward*, 3,) b. September 3d, 1801.

M. Arminia Sartwell.

No Ch.   (*Smithport, Pa.*)

Gideon, (Hebrew,) a destroyer.

1 Gideon Harvey Millspaugh,[7] (*Cynthia Corwin*, 1,) b. October 9th, 1808.
M. Sarah Ann Kyle.
   Ch. George W., John Randolph, James, Mary J., Electa, William,
   Elizabeth.   (*Lockington, O.*)
   (Was killed in a saw-mill.)

2 Gideon,[9] (*Noah*, 3,) b. ——, 1840–50.

(1) * Gideon Reed,[7] [*John*, (3),] b. at Snydertown, N. J., February 8th,
   1786, d. May 28th, 1869.
   M. Jane, daughter of Captain Timothy Titus, January 11th, 1815. She
   born, September 20th, 1788, d. October 25th, 1868.
   Ch. Rebecca, George, Phebe, John, John, Cornelius. (*Hopewell, N. J.*)

Gilbert, (Old High-German,) yellow-bright, famous.

1 Gilbert,[6] (*Matthias*, 3,) b. at Southold, about 1742.
   M. Amy Knapp, of Haverstraw, N. Y. *see suplement—*
   Ch. Joseph, Elizabeth, and others. *Gershon, Bethia,*
He went to Nantucket when about nineteen years of age, and became
a whalesman, which business he followed for many years. He was a
soldier in the Revolution, having charge for a long time of King's Ferry, at
Peekskill. About 1768, he removed to Haverstraw, Rockland Co., N. Y.

2 Gilbert,[7] (*William*, 4,) b. October 25th, 1830.
   M. Ann M. Carpenter. She d. September 18th, 1856.
   Ch. Emma C.

3 Gilbert B.,[7] (*Jason*, 1,) b. May 11th, 1825.
   M. Catharine McWilliams.   (*Scotchtown, N. Y.*)

4 Gilbert,[7] (*Matthias*, 6,) b. May 7th, 1809.
   M. Sarah Corwin, No. 95, in 1832.
   Ch. Milla F., Sarahetta, George C., Maria M., Daniel, Charry A.,
   Sophia A., Joshua.

5 Gilbert Hopkins,[8] (*Samuel H.*, 13,) b. September 15th, 1842.
                                               (*Howells, N. Y.*)

6 Gilbert Corwin Millspaugh,[7] (*Cynthia Corwin*, 1.) b. August 18th, 1806.
   M. Lucy Williams, October 26th, 1826.
   Ch. Harvey, Ezra, Oliver, James L., Franklin, William, Peter, Catha-
   rine E., Leander N.   (*Nall's Mills, Ind.*)

7 Gilbert Harrison Ketchum,[8] (*Sophia Corwin*, 2.)

8 Gilbert H. Hood,[9] (*Caroline L. Corwin*, 4,) b. May 11th, 1866. (*Vermont.*)

Grace, (Latin,) favor, kindness.

1 Grace A.,[9] (*James*, 16,) b. September 25th, 1861.

* Corwine.

2 GRACE LOUISA,[9] (*Charles*, 21,) b. January 17th, 1868.

3 Grace E. Dunham,[10] (*Alida Durkee*, 1,) b. March 16th, 1863. (*Vermont.*)

4 Grace Jewett,[9] (*Mary F. Colgrore*, 141½,) b. about 1863.

(1) Grace Miller,[9] [*Sarah H. Corwine*, (6),] b. September 28th, 1855, d. December 15th, 1857.

1 GILLESPIE BIRNEY,[8] (*Samuel B.*, 16.)

1 GROTIUS SABINA,[8] (*Anson*, 1,) b. May 30th, 1838, at Riverhead.
   M. Emma B. Strong.

1 GROVER,[7] (*Abel*, 1,) b. ——, 1800 ?   (*Baiting Hollow, L. I.*)

2 GROVER,[10] (*George*, 48,) b. ——, 1866.

1 HALSEY,[8] (*Henry*, 10,) b. November 10th, 1806, d. April 10th, 1837.
   M. Emeline Fash.
   Ch. Halsey, Mary A.   (*New-York City.*)

2 HALSEY,[9] (*Halsey*, 1,)   (*New-York City.*)

3 HALSEY,[10] (*Oliver*, 8,) b. January 25th, 1864.

4 Halsey Tuthill,[9] (*Ann Corwin*, 22.)

1 Hamilton A. Millspaugh,[8] (*John M. Millspaugh*, 72,) b. July 27th, 1845, d. August 5th, 1846.

1 HAMPTON,[8] (*Daniel*, 19,) b. February 24th, 1822, d. September 3d, 1825.

                Hannah, same as Anna, (Hebrew,) grace.

1 HANNAH,[3] (*John*, 1,) b. ——, 1660–70.

Not married in 1690, when her father's will was written, nor in 1698, when her name occurs as Anna Corwin on census list.
                                              (*Doc. Hist. N. Y.* i. 450.)

2 HANNAH,[6] (*William*, 2,) b. March 28th, 1773.
   M. Jeremiah Woodhull, January 8th, 1795. He was b. January 22d, 1770, and was son of William Woodhull * and Elizabeth Hedges, of East-Hampton, L. I.
   Ch. Elizabeth S., Sarah, William H., Nancy, Mary S.   (*Chester, N. J.*)

3 HANNAH,[6] (*Jesse*, 2,) b. November 30th, 1774, d. ——, 18—.
   M. Elijah Philips, M.D., 1792.
   Ch. William, Mary, Sarah, Eliza, Jesse.

Mr. Phillips resided in Bourbon Co., Ky. He died about 1802, from the effects of falling through a hatchway while walking in his sleep. His wife then removed to Ohio, near Lebanon, and after five years to Champaign County. About 1816 she married Philip Kenton.

3½ HANNAH,[6] (*Abner*, 1,) b. ——, 1780–5 ?
   M. Ebenezer Kellum.   (*Huntington, L. I.*)

4 HANNAH,[6] (*Jonathan*, 1,) b. ——, 1750–60.
   M. Josiah Goodale, February 27th, 1784.   (*Aquebogue Records.*)

———————
* A record of his descendants may be found at Chester, N. J., in possession of Mrs. Temperance Topping.

5 HANNAH,⁶ (*Stephen*, 1,) b.
M. Jonas Ferrago. (*Ohio.*)

6 HANNAH,⁷ (*Richard*, 1,) b. ——, 1790–1800.
M. ——
Ch. Catharine Ellen, Augusta.

7 HANNAH,⁷ (*John*, 13,) bapt. at Aquebogue, December 2d, 1789.
M. —— Youngs.

8 HANNAH,⁷ (*Richard*, 2,) b. ——, 1810–20.

9 HANNAH JANE,⁷ (*Silas*, 5,) b. March 19th, 1819.
M. Frederick Willinghouse.
Ch. Mary J., Anna, Catharine, Margaret, Frederica.

10 HANNAH,⁷ (*Jonathan*, 2,) b. ——, 1790? d. in Connecticut.

10½ HANNAH,⁷ (*Phineas*, 1½,) b. May 26th, 1818, d. February 24th, 1819.

11 HANNAH,⁷ (*Benjamin*, 4,) b. ——, 1791?

12 HANNAH,⁷ (*James*, 6,) b. September 2d, 1818.
M. Thomas Petit, 1844.
Ch. Isabella, Charles, Charlotte, Emma B. (*Brooklyn.*)

13 HANNAH COOK,⁷ (*Stephen*, 4,) b.
M. ——
Ch. Marcellus. (*Wyandotte Co., O.*)

14 HANNAH,⁸ (*Stephen*, 8,) b. ——, 1810–20.

15 HANNAH,⁸ (*Hector*, 1,) b. ——, 1830–40.

16 HANNAH ELIZABETH,⁸ (*Horton*, 2,) b. January 20th, 1832.
M. James Easton, February 24th, 1849.
Ch. Thomas H., Cornelia, Ada. (*Howells, N. Y.*)

16½ HANNAH CHARLOTTE,⁸ (*Hazen*, 1.) b. August 7th, 1829, d. May 9th, 1842, at Hopkington, N. Y.

17 HANNAH JANE,⁸ (*Daniel*, 16,) b. ——, 1817? d. ——, 1829.

18 HANNAH ADELIA,⁸ (*Alsop H.*, 2.)

19 HANNAH E.,⁸ (*Cortlandt Elias*, 1.)

20 HANNAH MARIA,⁸ (*Manasseh, R. I.*,) b.
M. Jeremiah Beals.

21 HANNAH MARIA,⁸ (*Robert*, 1.)

22 HANNAH,⁹ (*Henry*, 20,) b. December 31st, 1840.
M. William H. Robinson, December 16th, 1857.
Ch. Alice, Elizabeth, Fanny, Gertrude. (*Flushing, L. I.*)

23 HANNAH M.,⁹ (*Charles L.*, 19,) b. May 9th, 1851.

23½ Hannah Case,⁴ (*Henry Case*, 1.)
M. Philemon Dickerson? 1709. He d. March 14th, 1718.
(*See Moore's Index of Southold*, p. 65.)

24 Hannah Osman,⁴ (*Sarah Corwin*, 1,) b. ——, 1690–1710.

86   THE CORWIN GENEALOGY.

24½ Hannah Maria Duckworth,[7] (*Sarah Corwin*, 9,) b. July 3d, 1822..

25 Hannah Thatcher,[7] (*Jemima Corwin*, 1,) b. 1796 ? Yet living, (1870,) in Dark Co., Ind.

26 Hannah Hackett,[8] (*Mary Corwin*, 14,) b. August 29th, 1808.
M. Daniel Cram, March 7th, 1832.
Ch. Mary H., Hannah H., Martha A., Lucy M., Susan E., Sarah E., Laura A.   (*Chelsea, Vt.*)

27 Hannah W. Woodruff,[8] (*Erastus Woodruff*, 1,) b. ——, 1825–35.

28 Hannah Smith,[8] (*Sarah Corwin*, 28,) b. in Tunbridge, Vt., 1832.
M. Carlos Colton, in 1859.   (*Pierrepont, St. Lawrence Co., N. Y.*)

29 Hannah M. Aldrich,[8] (*Deborah Corwin*, 6.)

30 Hannah E. Knox,[8] (*Julia Corwin*, 7,) b. June 12th, 1839.

31 Hannah E. Brown,[8] (*Silas C. Brown*, 16.)

32 Hannah Brown,[8] (*Elizabeth Woodhull*, 45,) b. October 21st, 1825.
M. Peter Jurnee.   (*Old Bridge, N. J.*)

33 Hannah H. Cram,[9] (*Hannah Hackett*, 26,) b. November 5th, 1835.

(1) * HANNAH,[2] [*George Curwen*, (1),] b. January 1st, 1645, d. November 21st, 1692.
M. Hon. Major William Brown, Jr., December 29th, 1664.
                                      (*Salem, Mass.*)

(2) † HANNAH,[3] [*John*, (1).] b. February 14th, 1671–2.

(3) ‡ HANNAH,[7] [*Samuel*, (4),] b. July 27th, 1795, d. June 20th, 1799.

(4) ‡ HANNAH CATHARINE,[5] [*Richard*, (4).]   (*Raritan Landing, N. J.*)

(5) ‡ HANNAH JANE,[8] [*Joab*, (1),] b, June, 1820.
M. G. L. Dart.   (*Peru, Ind.*)

Hannibal, (Punic,) grace of Baal.

1 HANNIBAL,[8] (*Nathan*, 5,) b. ——, 1825–30.

Harbert or Herbert, (Anglo-Saxon,) glory of the army.

(1) HARBERT,[5] [*Jonathan*, (1),] b. December 15th, 1690, d. February 10th, 1691.

1 Harmony Jennings,[6] (*Sarah Corwin*, 4,) b. ——, 1780–90.

1 Harlie A. Whitney,[9] (*Helen M. Corwin*, 4,) b. May 5th, 1859. (*Vermont.*)

Harriet, feminine of Henry, (Old High-German,) the little head of the house.

1 HARRIET N.,[7] (*Elisha*, 1,) b. ——, 1810–20. 1816, d. July 22
M. James J. Borgy.
Ch. Emile, Jenny H., William C.

2 HARRIET,[7] (*Phineas*, 3,) b. March 8th, 1820.
M. Robert P. L. Higby, October 5th, 1856.
Ch. Alfred, Sarah F., Milton.

* Curwen.   † Curwin.   ‡ Corwine.

3 HARRIET N.,[7] (*Jemuel*, 1,) b. January 5th, 1819.
M. M. G. Howell, November 15th, 1839.
Ch. Edwin S., William Y., Charles M.

4 HARRIET FORDHAM,[7] (*Joshua G.*, 3,) b. October 9th, 1828.
M. Marcus Meeker.
Ch. Henry F., Martha E., John M., Anna M., Harriet J., John E.

5 HARRIET NEWELL,[8] (*Daniel*, 11,) b. ——, 1816.

6 HARRIET J.,[7] (*Amaziah*, 3,) b. ——, 1820–30.

6½ HARRIET AVIS,[8] (*George S.*, 12,) b. October 19th, 1832. (*Newark, N. J.*)

7 HARRIET MARTHA,[8] (*John*, 16,) b. January 27th, 1832. (*Owego, N. Y.*)

8 HARRIET,[8] (*Archibald*, 1,) b. ——, 1840–50.

9 HARRIET,[8] (*Gabriel*, 3,) b. February 18th, 1850, d. August 24th, 1858.

10 HARRIET B.,[8] (*William H.*,) b. December 18th, 1854.

11 HARRIET AUGUSTA,[8] (*John*, 21,) b. October 28th, 1856.

12 HARRIET,[8] (*Matthias*, 8.)

13 HARRIET,[8] (*Ebenezer*, 3.) b. May 12th, 1846.
M. Linas Smith.

14 HARRIET,[8] (*Polydore B.*, 1,) b. September 10th, 1847.

15 HARRIET JANE,[8] (*Benjamin*, 14,) b. July, 1825, d. October 1st, 1853.
M. Burran Rogers.

16 HARRIET,[8] (*John*, 30,) b. ——, 1830, d. ——, 1833.

17 HARRIET ELIZA,[9] (*Henry W.*, 19,) b. December 29th, 1837, d. May 3d, 1867.

17½ HARRIET L.,[9] (*William H.*, 38,) b. June 13th, 1862.

18 HARRIET ESTELL,[9] (*Joseph*, 20,) b. November 21st, 1848.

19 Harriet A. Millspaugh,[7] (*Cynthia Corwin*, 1,) b. April 12th, 1816.
M. J. T. Milnor, October 14th, 1837.
Ch. Theodore, Cortlandt W., Samuel, Leander, William S.
(*Covington, O.*)

20 Harriet Everett,[7] (*Abigail Corwin*, 2.) b. May 17th, 1806.

21 Harriet Little,[7] (*Abigail Corwin*, 3,) b. November 13th, 1816, d. ——, 1857.
M. George Tryon. (*Illinois.*)

22 Harriet Olivia Young,[7] (*Charlotte Corwin*, 1,) b. May 1st, 1821, d. February 21st, 1840.

23 Harriet Pellet,[8] (*Mary Corwin*, 24,) b. ——, 1820–30.
M. Gilbert Moore. (*Mount Hope, N. Y.*)

24 Harriet Eliza Weed,[8] (*Charity Corwin*, 1,) b. ——, 1820–30.
M. William Marvin.

25 Harriet E. Byram,[8] (*Mary A. Corwin*, 34,) b. February 26th, 1846, d. June 20th, 1847.

26 Harriet Horton,[8] (*Eliza Corwin*, 2,) b. ——, 1830-40.
    M. ——— Jackson. ...
    Ch. ———

27 Harriet Cathlina Young,[5] (*Daniel Young*, 35,) b. April ——, 1848.
    M. ——— McCaw.   (*Morning Sun, Iowa.*)

28 Harriet Hulse,[8] (*Lucetta Corwin*, 1.)

29 Harriet Squires,[5] (*Seth ? Corwin*, 1.)

30 Harriet W. Howell,[8] (*Sophia Woodruff*, 5,) b. November 23d, 1843.

31 Harriet Craus,[8] (*Cynthia Little*, 7,) b. ——, d. February 17th, 1852.

32 Harriet J. Meeker,[5] (*Harriet F. Corwin*, 4,) b. August 3d, 1865, d.
    May 3d, 1867.

33 Harriet Smith,[9] (*Loren Smith*, 1,) b. ——, 1850-55.
                                (*St. Lawrence Co., N. Y.*)

34 Harriet Davis,[9] (*Rachel Corwin*, 4.)

35 Harriet Mapes,[9] (*Mary Corwin*, 49.)

36 Harriet D. Halleck,[10] (*Elizabeth M. Downs*, 53,) b. December 13th, 1865.
  (1) * HARRIET ELIZABETH,[8] [*Joab*, (1),] b. ——.
  (2) * HARRIET,[9] [*Jesse D. H.*, (1),] b. in Kentucky, October 4th, 1852.
  (3) * HARRIET LOUISA,[9] [*Amos*, (2).] b. in Panama. August 28th, 1858.
  (4) Harriet Stevenson,[9] [*Sarah A. Corwine*, (5,)] b. at Shelbyville, Ky.,
      December 11th, 1860.

HARRY, see HENRY.

1 HARRY,[8] (*Jesse*, 6,) b. ——.   (*Ohio.*)

2 HARRY JOHN,[9] (*John A. R.*, 60,) b. ——, 1852-60. (*Chelsea, Vermont.*)

3 Harry Houston Wilcox,[9] (*Mary Corwin*, 47.)

1 HARRISON,[7] (*Matthias*, 6,) b. June 19th, 1819.
    M. Betsy Ann Aldrich, September 7th, 1837.
    Ch. Morgiana, Thaddeus, Mary E., Arabella, Annie, Louisa.
                                 (*Riverhead, L. I.*)

2 Harrison Mapes,[8] (*Anna Corwin*, 7,) b. ——, 1820-30.
    M. Mary Corwin.

3 Harrison E. Mapes,[9] (*Mary E. Corwin*, 54,) b. January 31st, 1859.

1 HARVEY,[7] (*David*, 8,) b. ——.

2 HARVEY,[8] (*John*, 30,) b. May 24th, 1824.
    M. Amanda Barrett, July 19th, 1851.
    No Ch.   (*Penn Yan, N. Y.*)

3 HARVEY,[9] (*Alsop L.*, 3,) b. ——.

4 Harvey Skidmore,[8] (*Mary Corwin*, 28,) b. ——, 1810-20.

* Corwine.

5 Harvey, Jacques,[3] (*Julia Corwin*, 3,) b. ——, 1820–30.

6 Harvey Millspaugh,[8] (*Gilbert C. Millspaugh*, 6,) b. August 19th, 1827.

7 Harvey M. Clark,[8] (*Sarah Jane Millspaugh*, 66,) b. July 31st, 1834.

1 Hazen,[7] (*Hubbard*, 1,) b. August 1st, 1799, in Tunbridge, Vt., d. January 18th, 1866.
    M. Asenath Smith, of Tunbridge, Vt., February 2d, 1823. She b. 1802, d. December 18th, 1861.
    Ch. Hubbard, Hannah C., Rosina J., Fullam.   (*Hopkinton, N. Y.*)
    (He was killed by being run against by a runaway horse.)

1 Hector,[7] (*Alexander*, 1,) b. ——, 1810.
    M. Eliza Thompson.
    Ch. Sarah A., Hannah, David H., Susan.   (*Washingtonville, N. Y.*)

2 Hector,[8] (*William*, 31,) b. ——, 1840–50.

1 Hekshedah Lippincott,[5] (*Elizabeth Corwin*, 18.)

· Helen, (Greek,) light.

1 Helen,[8] (*William S.*, 9.) b. ——, 1845–50.

2 Helen,[8] (*Henry W.*, 8,) b. ——, 1825–30.
    M. Isaac Stashur ?

3 Helen Mary,[8] (*Benjamin Wickham*, 11,) b. September 10th, 1831.
    M. Lewis Penney. December 14th, 1853.
    Ch. Lewis, Mary A., Thomas C.

4 Helen M.,[8] (*John*, 32,) b. July 18th, 1838.
    M. Carlos Whitney, December 29th, 1857.
    Ch. Harlie A., Carrie A., Elmer F.   (*Vermont.*)

5 Helen,[8] (*Charles*, 3,) b. August 16th, 1855, d. June 3d, 1864.

5½ Helen Mary,[8] (*Rev. Jason*, 4,) b. at Shelburne, N. Y., October 19th, 1835.
    M. (1) ——— Fisher.
    Ch. Calista.
    M. (2) General H. C. Pierce, February 7th, 1871.

(The following is taken from the New-York Weekly.)

DISTINGUISHED WEDDING.

"Under the above caption, the *Washington Herald* (of Tazewell Co., Ill.) gives the particulars of the marriage of Mrs. Helen Corwin Fisher (well known to the writers of the New-York Weekly as a powerful and fascinating romance writer) and General Pierce, of Pine Bluff, Ark. This is what the *Herald* says :

"'On Tuesday evening last, February 7th, the Christian Church was crowded with the beauty and *élite* of our town, to witness the marriage of General H. A. Pierce, of Pine Bluff, Ark., and Mrs. Helen Corwin Fisher, of this city.

" ' The hour set for the ceremony was eight o'clock, and expectation was on tiptoe, for General Pierce was a stranger to our people, and we wanted to know who the fortunate Southerner was, who had come to take away one of our brightest ornaments.

" ' Mrs. Fisher's name has become familiar throughout our whole land as an authoress of rare ability, and her charming productions have done much to render the NEW-YORK WEEKLY so highly prized. She has been familiar here among us since her girlhood, and we feel more than an ordinary interest in her, for in a measure her honor and praise are reflected upon her many friends here.

" ' The parties appeared shortly after the time indicated, and proceeded to the altar without waiting-men and maids. The ceremony was short, simple, and unostentatious, and at its close the parties retired from the church to the residence of the bride's mother, Mrs. D. C. Corwin, where they received those who wished to pay their respects personally.

" ' General Pierce, we learn, is a lawyer of ability, an eloquent speaker and writer, and the editor of a very prosperous paper, called the *Jefferson Republican*, published at Pine Bluff, Ark.

" ' The happy pair will leave in a day or two for their sunny home, and the best wishes of all go with them. We congratulate Arkansas upon this acquisition.'

" And this is what *we* say : We have known Mrs. Helen Corwin Pierce (*née* Fisher) for some eight or ten years, during which time she has written for us constantly, and we have always found her not only an accomplished writer, but a most estimable lady. General Pierce, her husband, we have never seen but once, and then for a few moments only ; but if there is any thing in physiognomy, he is just the man to make a true woman happy. Sincerely do we congratulate the newly-married pair, and wish them all the happiness that can attend the union of

> ' Two minds with but a single thought—
> Two hearts that beat as one.' "

(*Now of Fort Smith, Ark.*)

6 HELEN FLORENCE,[9] (*Henry W.*, 19,) b. June 18th, 1849.

7 Helen Hunsike,[b] (*Julia Corwin*, 1,) b. ——, 1825–35.
   M. E. Davaw.
   Ch. Delaphine, Charles J., Julia, Mary.

8 Helen G. Millspaugh,[8] (*Benjamin F. Millspaugh*, 31,) b. February 15th, 1847.

9 Helen M. Sperry,[8] (*Mary Corwin*, 40,) b. May 30th, 1842.
   M. Charles A. Willetts.

10 Helen A. Colgrove,[8] (*Bela H. Colgrove*, 1,) b. July 30th, 1842.
    M. Charles A. Clark, of Buffalo, November 27th, 1862.
    Ch. Seth C., Charles A., Catharine, Thomas D.

(*Akron, Erie Co., N. Y.*)

11 Helen Eliza Colgrove,[9] (*Rev. Clinton Colgrove*, 1,) b. about 1863.

12. ..

Hendrik, the Dutch for Henry.

1 HENDRIK,[7] (*Daniel*, 9,) b. ——, d. ——, 1824?

2 Hendrik Taylor,[7] (*Fanny Corwin*, 1,) b. July, 1830.
M. Erin Thompson, April, 1853.
Ch. Charles, Florence. (*Michigan.*)

Henrietta, a feminine diminutive of Henry, (Old High-German,) a little chief.

1 HENRIETTA ELIZABETH,[x] (*John*, 21,) b. October 15th, 1852.

2 HENRIETTA,[9] (*Anson L.*, 2.) b. ——, 1855–65.

3 Henrietta Hunsike,[x] (*Julia Corwin*, 1,) b. ——, 1825–35.

4 Henrietta Sophia Howell,[x] (*Sophia Woodruff*, 5,) b. February 4th, 1830.
M. Rev. Eli Corwin, July 16th, 1851.

5 Henrietta Meyers,[x] (*Arminda Brown*, 2.)

6 Henrietta Crans,[8] (*Cynthia Little*, 9.)

7 Henrietta Sophia Williams,[9] (*Sarah T. Howell*, 87,) b. January 1st, 1856.

Henry. (Old High-German,) the chief of a house.

1 Henry Case,[3] (*Martha Corwin*, 1.) b. ——, 1659?
M. Tabitha ——. She d. December 16th, 1735.
Ch. Henry, Samuel, Benjamin, Tabitha, Mary. (*In Doc. Hist. N. Y.*, i.
453, *this family is thus found in census list*, 1698.) In 1683, Henry
Case was rated at £35. In 1686, had 3 males and 3 females in his
family. In 1694, his mother (Martha Hutchinson) deeds him land,
etc. (See *Moore's Index of Southold*, 64, No. 111.)

2 HENRY,[4] (*Daniel*, 1?) b. about 1700, d. May 22d, 1735.

3 HENRY,[5] (*Samuel*, 2,) b. October 12th, 1735, d. October 12th, 1756?

4 HENRY,[5] (——,) b. ——, 1760–70, d. ——.
M.
Ch. Henry.

5 HENRY,[6] (*Henry*, 3,) b. ——.
M. Levina Fanning? December 21st, 1822.
Ch. John, David, Mary, Jane. (*Brooklyn, N. Y.*)

6 HENRY,[6] (*Daniel*, 3,) b. June 10th, 1752, d. January 25th, 1833.
M. Bethia Reeve, November 11th, 1773, at Aquebogue. She b. Sep-
tember 10th, 1744, d. September 3d, 1826.
Ch. Henry, Daniel, Benjamin, Temperance, Betsy, Sarah, Bethia,
David, Patty. (*Riverhead, L. I.*)
He signed the engagement to support Congress; in 1776, his name
occurs on census list 411, having in his family one male, one female over
16, and two children.

7 HENRY B.,[6] (*David*, 3,) baptized at Mattituck, November 13th, 1763,
d. ——.

M. Elizabeth Allen?

Ch. Henry, Jerusha?

8 HENRY WISNER,[7] (*James,* 5,) b. (near Middletown, N. Y.) October 5th, 1794, d. July, 1842.
   M. Eliza Ann Kelley.
   Ch. Arminda, Mary A., James E., Henry, Helen, Emerson.
   A cabinet-maker.   (*Luzerne Co., Pa. New-York City.*)

9 HENRY,[7] (*Daniel,* 7,) b. ———, 1800?
   (A Harry Corwin died at Riverhead, 1824. Who?)

10 HENRY,[7] (*Henry,* 6,) b. October 21st, 1774, d. June 27th, 1847.
   M. Mehetable Howell, February 8th, 1797. (*Aquebogue Records.*) She
      b. February 3d, 1775, d. August 18th, 1850.
   Ch. Oliver, Alvah, Halsey, John H., Ann, Maria.
                                  (*Riverhead and Brookhaven, L. I.*)

11 HENRY,[7] (*Samuel,* 11,) b. August 7th, 1802.
   M. Oriet Richmond, January 9th, 1827.
   Ch. Daniel, Denniston, William H., James, Albert, George, Charles,
      Theodore, Orin, Warren.
                          (*Southold, L. I., till* 1860, *then Lade City, Minn.*)

12 HENRY,[7] (*Henry,* 7,) bapt. November 4th, 1798, at Aquebogue.
   M. Phebe, Cleves.
   Ch. Jerusha, Huldah.

13 HENRY K.,[8] (*John,* 16,) b. February 21st, 1821.
   M. ——— Coleman.   (*Mongaup, N. Y.*)

14 HENRY,[8] (*Benjamin Harvey,* 10,) b. ———, 1835-45.

15 HENRY HORTON,[8] (*Isaac S.,* 5,) b. January 1st, 1836.

16 HENRY,[8] (*Henry W.,* 8,) b. about 1825.
   M. Annie Van Duyn, August, 1868.   (*Greenville, N. J.*)

17 HENRY ROBERTS,[8] (*Benjamin W.,* 11,) b. in Orange County, N. Y., April
      17th, 1837.   (*Cincinnati, O.*)

18 HENRY,[8] (*William,* 17,) b. ———, 1834, d. ———, 1839.

19 HENRY WICKHAM,[8] (*Benjamin,* 14,) b. June 2d, 1802.
   M. Miriam Raynor, about 1836. She d. October ———, 1861, aged about
      46 years.
   Ch. Harriet E., Henry H., William M., Helen F., Mary M.
                                           (*Riverhead, L. I.*)

20 HENRY,[8] (*Daniel,* 19,) b. February 19th, 1811.
   M. Roxana Concklin, December 27th, 1836.
   Ch. Hannah, Ida, Augusta, Henry H.   (*Riverhead, L. I.*)

20½ HENRY CLAY,[8] (*Ira,* 4½.) b. May 29th, 1844, killed May, 1862. En-
      listed in federal army in 1861, Company H, 11th Illinois.

21 HENRY B.,[8] (*George S.*, 12,) b. June 4th, 1839.
   M. Mary Davis. *June 14, 1866 ...*
   Ch. Charles. *Henry, ... 1867.*   (*Newark, N. J.*)

22 HENRY,[8] (*David*, 18,) b. ——, 1815-25.

23 HENRY E.,[8] (*Salem*, 1.)

23½ HENRY WISNER,[8] (*David*, 18,) b. January 19th, 1827.
   M. Sarah E. Brown, April 10th, 1851. She d. May 11th, 1854.
   Ch. Ella, (b. April 1st, 1852,) Lydia S., (b. March 28th, 1854, d. March 28th, 1854.)
   Ch. of second wife: Henry P., (b. in Tennessee, October 8th, 1859, d. there August 14th, 1862,) William Beson, (b. at Alleghany City, Pa., August 31st, 1863,) Carrie Robinson, (b. October 24th, 1869.)
                  (*Pittsburg, Pa.*, 1844.)

24 HENRY HAMILTON,[9] (*Henry*, 20,) b. January 5th, 1845.

25 HENRY HARRISON, M.D.,[9] (*Nicholas*, 1,) b. ——, 1842? (*Stamford, Ct*)

26 HENRY HARRISON,[9] (*Henry W.*, 19,) b. ——, 1843.
   M. Sarah Ellen Terrill, December 29th, 1866.
   Ch. Miriam G., Dwight T.

27 HENRY,[9] (*Amos*, 1.)

27¼ HENRY FRANK,[10] (*Oliver*, 8,) b. March 29th, 1869.

27½ Henry Case,[4] (*Henry Case*, 1,) b. ——, 1685? d. April 16th, 1720.

28 Henry Little,[7] (*Abigail Corwin*, 3,) b. July 1st, 1801.
   M. Elizabeth Millspaugh, June 8th, 1824.
   Ch. Abigail J., Bittridge V., Dorothy, Edwin.   (*Middletown, N. Y.*)

29 Henry Horton,[8] (*Eliza Corwin*, 2,) b. ——, 1830-40.
   M.
   Ch. Ada, Sarah.

30 Henry Campbell,[8] (*Josiah Campbell*, 1,) b. ——, 1835-40.

31 Henry Wood,[8] (*Mary Corwin*, 16.)

32 Henry Shepherd,[8] (*Charlotte Corwin*, 6,) b. November 6th, 1852.

33 Henry F. Meeker,[8] (*Harriet Corwin*, 4,) b. April 12th, 1858.

34 Henry Elwood Wells,[8] (*Puah Corwin*, 1.)

35 Henry Baldwin,[8] (*Jane Corwin*, 1,) b. at Goshen, d. at sea. Unmarried.

36 Henry Beach Windsor,[8] (*Sarah Vance Beach*, 77.)

37 Henry A. C. Noyes,[9] (*Louisa Corwin*, 3,) b. April 27th, 1840.   (*Vt.*)

38 Henry Austin Crater,[10] (*Mary Corwin*, 83.)

(1) * HENRY,[9] [*Samuel R.*, (7),] b. September 9th, 1847. (*Salem, Mass.*)

       Herbert, (Anglo-Saxon.) glory of the army.

1 HERBERT R.,[8] (*Albert*, 2) b. April 29th, 1859.

2 HERBERT M.,[9] (*Jabin*, 2,) b. August 28th, 1861, at Chelsea, Vt.

3 Herbert P. Smith,[9] (*George Smith*, 74,) b. ——, 1855–65.

(*St. Lawrence Co., N. Y.*)

4 Herbert Baker,[9] (*Rosalinda Corwin*, 1,) b. March 29th, 1845.
M. Mary Potter.
Ch. Allen E., Viola K.   (*Linden, Genesee Co., Mich.*)

5 Herbert Tinney,[9] (*Lucilia Corwin*, 1.)

1 HERVEY,[7] (*Phineas*, 3,) b. May 5th, 1816.
M. Sarah M. Prince, November 14th, 1839.
Ch. Julia A., Angelina, Mary J., Alice M., Clarissa R., George H.,
Caroline C.

Hester, same as Esther.

1 HESTER,[3] (*John*, 2,) b. about 1685. On census list of Southold, Suffolk
County, N. Y., 1698, under that of John Corwin, No. 2, and Sarah
Corwin.   (*Doc. Hist. N. Y.* i. 450.)

2 Hester Osman,[4] (*Sarah Corwin*, 1,) b. ——, 1690–1700.
M. Z. Hallock, 1720.

3 HESTER J.,[9] (*William*, 41½,) b. January 22d, 1860, d. March 2d, 1860.

1 HEWLETT CONCKLIN,[8] (*Benjamin*, 14,) b. January, 1807, d. March 25th,
1852.
M. Phebe Case, January 7th, 1832. She b. May, 1807, d. Sept. 11th,
1868.
Ch. Phebe A., Samuel A., Mary E., Sarah E.   (*Riverhead, L. I.*)

2 Hialwar Knox,[5] (*Julia Corwin*, 7,) b. May 13th, 1855.

Hiram, (Hebrew,) most noble.

1 HIRAM,[7] (*Rev. Joseph*, 5,) b. about 1796, d. about 1811.

2 HIRAM,[7] (*Moses*, 1,) b. in Middletown, N. Y., September 15th, 1812, d.
November, 1868.
M. Elizabeth Anderson, April 7th, 1842.
No Ch.
An Alderman.   (*New-York City.*)

3 HIRAM BURTON,[9] (*Benjamin*, 23,) b. about 1850.

4 HIRAM,[8] (*John*, 22,) b. June 4th, 1823.
M. (1) —— Tuthill; (2) —— Vail?
Ch.   (*Cutchogue, L. I.*)

5 Hiram Howell,[6] (*Jemima Corwin*, 7.)

6 Hiram Carr,[8] (*Aznbah Corwin*, 3.)
M. Agnes Brown.   (*McGrawville, N. Y.*)

7 Hiram Vail,[8] (*Elizabeth Corwin*, 28.)

(A Hiram Corwin is buried at Upper Aquebogue, who died October
13th, 1822, aged 22.)

1 Hobart North,[8] (*Sarah Hobart*, 65,) b. ——, 1834, d. September, 1856. *M.* Mary Smith.

2 Hobart Eugene Allen,[9] (*Susannah Hobart*, 3,) b. August 1st, 1854.

Homer, (Greek,) a pledge.

1 Homer Campbell,[8] (*Jonah Campbell*, 1,) b. ——, 1835–40.

Horace, (Greek,) visible.

1 HORACE,[8] (*Isaac*, 4,) b. April 11th, 1833.
M. Elizabeth King, June 1st, 1854.   (*Cattarangus Co., N. Y.*)

2 Horace Halsey Peters,[8] (*Mary Corwin*, 80,) b. 1830–40. Was in the federal army.

3 Horace Little,[7] (*Abigail Corwin*, 3,) b. January 7th, 1809.
M. Maria Case.   (*New-Milford, Pa.*)

4 Horace Sellen Peters,[9] (*Eleazer Peters*, 3.)

5 Horace Drake,[3] (*Mary Corwin*, 20.)   (*Ohio.*)

Horatio, same as Horace.

1 Horatio A. Crego,[5] (*Clarissa Taylor*, 4,) b. May, 1856.

1 HORTON,[6 or 7] (*Phineas*, 1½,) b. ——, 1780–90, in Lisbon or Franklin, Ct. *M.* Elizabeth Armstrong.
Ch. Ira, etc.
In 1816, bought land in Camillus, N. Y.; living there yet, 1831.

2 HORTON,[7] (*Martin Luther*, 1,) b. July 31st, 1800.
M. Jane Wheat, December 31st, 1825. She b. March 13th, 1806, d. June 3d, 1862.
Ch. Martin, Rebecca J., Salmon W., Hannah E., Martin, William F., Abbey C.
A justice of the peace.   (*Howells, N. Y.*)

3 HORTON,[9] (*Martin*, 5,) b. October 5th, 1853.

4 Horton Porter,[9] (*Adelia Horton*, 1.)

1 HOWARD,[7] (*John II.*, 8,) b. July 28th, 1831.
M. Emma C. Conger, January 12th, 1853. She was b. March 21st, 1832.
Ch. Andrew K. C., Robert W., Elmer E.

2 Howard Chappel,[8] (*Charlotte Hobart*, 9,) b. August, 1844.   / 8 Ʋ. ~

3 Howard Dunlevy,[8] (*Lucinda Corwin*, 4,) b. May 5th, 1836.

Howell, (British,) sound, whole.

1 HOWELL,[7] (*Nebat*, 1?) b. 1810–20.

2 Howell Reeve,[5] (*Mary Corwin*, 24,) b. 1820–30.
M. Sarah Decker.

1 HUBBARD,[6] (*John*, 5,) b. ——, 1759, bapt. at Mattituck, August 16th, 1761, d. ——, 1833.

M. Lydia Hazen, of Norwich Landing, Connecticut. She b. 1760, d. April 13th, 1833.

Ch. Mary, William, John, Jabin, James, Sarah, Philenia, Hazen, Russell, Lydia, Spencer.   (*Chelsea, Vt.*)

He removed from Long Island to New-England about 1790, or before. +

2 HUBBARD,[8] (*John*, 22.) b. March 3d, 1822.

M. Emeline Aldrich, 1842. She b. October 30th, 1825.

Ch. Alice V., James H., Eveline B., Cora E., Katy L., Carrie M., Emeline M., Leah B., Fanny H., Fred. G.   (*Riverhead, L. I.*)

3 HUBBARD,[8] (*Hazen*, 1,) b. January 2d, 1824, at Tunbridge, Vt., d. October 2d, 1827.

1 HUDSON,[7] (*Abel*, 1,) b. March 23d, 1791, d. January 29th, 1864.

M. (1) Mary Homans, (2) Dolly Homans, (3) Lucy A. Boerum, b. May 12th, 1820.

Ch. George H., Samuel B., Charles.

(*New-York City*, 1824–31 ; *Greenport, L. I.*)

1 HULDAH,[4] (*Theophilus*, 2, ? She had a brother David,) b. ———, 1700 ?

M. (1) Gershom Brown ; (2) Jacob Hawkins, of Brookhaven.

2 HULDAH,[7] (*Asa*, 8,) b. ———, 1798.

M. Daniel Knapp.

3 HULDAH,[8] (*Henry*, 12.)

4 HULDEY E.,[9] (*Barnabas*, 6,) b. March 8th, 1837.

M. Egbert C. Corwin, No. 1.

5 HULDEY T.,[9] (*Charles L.*, 19,) b. July 3d, 1860.

6 Huldey Penney,[7] (*Annie Corwin*, 3,) b. June 29th, 1798.

M. James R. Mapes, November 13th, 1822.

7 Huldey Tuthill,[9] (*Ann Corwin*, 22.)

(A Huldah Ann Corwin married Caleb S. Ruspar, August 7th, 1830.)

Ichabod, (Hebrew,) where is the glory?

1 ICHABOD,[6] (*Jesse*, 2,) b. February 29th, 1768, d. ———, 18 .

M. Sarah Griffin, January 20th, 1789. She was the daughter of Zadok Griffin, who was killed by the Indians in Fayette Co., Pa.

Ch. Moses B., William G., Mary, Eliza, Jesse B., Lucinda, James H., Elvira, Ichabod, Julia A., Sarah G., Robert G., Matthias.

He moved from Flanders, Morris Co., N. J., about 1790, (stopping for a while in Pennsylvania,) to Bourbon Co., Ky. While pursuing the Indians into the new State of Ohio, when they made incursions in Kentucky, he discovered the large and beautiful tract of land between the two Miami rivers, and since land-titles were bad in Kentucky, he made his home there, as soon as peace was secured with the Indians, March, 1796. It is the spot where Lebanon, Ohio, now stands.

2 ICHABOD,[7] (*Ichabod*, 1,) b. March 30th, 1808, d. September, 1843.

M. Catharine Dunlevy, March, 1836.

3 Ichabod,[7] (*Joseph*, 10,) b. ——.

4 Ichabod,[8] (*Moses B.*, 4,) b. ——.
Was elected, in 1807, Judge of Common Pleas.

5 Ichabod Case,[4] (*Theophilus Case*, 6.)
M. (1)? Mary Terrill, 1715, who d. 1716.   (2)? Abigail Mapes, 1717,
who d. 1724.   (3)? Hannah Goldsmith, 1725.
(*See Moore's Index, Southold*, p. 63.)

Ida, (Old-German,) God-like.

1 Ida V. R.,[8] (*Benjamin V. R.*, 9,) b. ——, 1845–55.

2 Ida W.,[8] (*Charles*, 3,) b. November 15th, 1850.

3 Ida,[8] (*Henry*, 20,) b. April 7th, 1843.

4 Ida,[8] (*John*, 30.)

5 Ida,[9] {*Charles*, 7,) b. October 6th, 1862. *m. June 7 / 2 / A. Williams*

6 Ida Little,[8] (*John A. Little*, 51,) b. ——, 1855.

7 Ida Concklin,[9] (*Lucretia R. J. Corwin*, 3,) b. ——.

1 Idomay Cox,[8] (*Mary E. Corwin*, 44.)

1 Iduella G.,[8] (*George W.*, 18,) b. July 5th, 1843, d. in Greenport, October 2d, 1847.

1 Ilia Ann Call,[8] (*Bethia Corwin*, 8,) b. November 7th, 1823.
M. Daniel Bennett, 1843.   (*Northport, L. I.*)
Four Ch.

1 Imogene Cox,[8] (*Mary E. Corwin*, 44.)

2 Imogene Davis,[9] (*Polly Corwin*, 2,) b. ——, 1845.
M. James Lewis.
Ch. Franklin L., William, Jennie.

Ira, (Hebrew,) watchful.

1 Ira Case,[7] (*Jabez*, 1,) b. August 6th, 1827, d. June 2d, 1870.
M. Aminda Louisa Mayo.
Ch. F., Ira L., Thomas M.   (*Jamesport, L. I.*)

2 Ira,[7] (*John*, 13,) b. ——, 1780–90, d. ——, 1853.   (*Southold, L. I.*)

2½ Ira,[7] (*Phineas*, 1½,) b. August 5th, 1789, d. September 16th, 1796, at
Franklin, Ct.

3 Ira M.,[8] (*William*, 11,) b. February 11th, 1834.
M. Jennie Swadin.
Ch. William. *Stephen 7½*

4 Rev. Ira,[7] (*Phineas*, 1½,) b. December 12th, 1809.
M. Mary Annie Baker, September 3d, 1840. She was born August
21st, 1817, in Herkimer, N. Y.
Ch. Timothy B., William H., George W., Mary E., Charles B. N.,
George.

He spent seven years at Hamilton Theological School, now Madison University, 1831–1838. He settled in Medina, Ohio, where he was ordained, 1839. He removed to Erie, Pa., 1840. His chief pastorates were Marietta, Ohio, and South-Bend, Ind. He is now of Norwalk, Ohio.

4¼ Ira,[7] (*Horton*, 1,) b. ——, 1809, d. November 1st, 1867.

    M. (1) Jennette Dutcher, about 1842; (2) ——, 1854.

    Ch. Henry C., Myron, Elvira, Thomas C., Ira.   (*La Salle, Ill.*)

4½ Ira,[8] (*Ira*, 4¼,) b. ——, 1857.

4¾ Ira G.,[8] (*Parker*, 1,) b. September 27th, 1835, d. September 8th, 1846.

5 Ira Linwood,[8] (*Ira C.*, 1,) b. ——, 1862, d. October 2d, 1866.

6 Ira W.,[9] (*Charles*, 21,) b. March 3d, 1857.

7 Ira Swina,[8] (*Bethia Hobart*, 11,) b. March, 1828.

8 Ira Baldwin,[8] (*Jane Corwin*, 1,) b. at Goshen.

9 Ira Harding,[9] (*Polly M. Corwin*, 4.)

1 Irad,[8] (*Abel*, 3.)

Irena, (Greek,) peaceful.

1 Irena Tuthill,[7] (*Azubah Corwin*, 1,) b. ——, 1805.

    M. Dr. E. L. Blakeley.   (*Wyoming Co., N. Y.*)

2 Irene Crans,[8] (*Cynthia Little*, 9.)

Isaac, (Hebrew,) laughter.

1 Isaac,[6] (*John*, 5,) b. on Long Island, April 7th, 1759, d. November 1st 1830.

    M. Experience Reeve. She b. April 13th, 1759, d. August 15th, 1839.

    Ch. Deborah, Martha, Isaac H., Joseph, John, Manasseh R., James Y., Sarah, Jerusha, Nancy, Zechariah, Elizabeth, Experience, Stephen O.

He left Long Island during or at the beginning of the Revolution, and settled near Chester and Flanders, Morris Co., N. J., on the Black River. Perhaps he also lived at Elizabeth, N. J., for a time. Will at Morristown, N. J. Spells his own name Curwin, in will.

2 Isaac,[6]? (*Jonathan*, 1,) bapt. at Mattituck, 1759, d. September 10th, 1777, according to *Aquebogue Records*.

3 Isaac,[6]? (*John*, 5½,) bapt. at Mattituck, May 14th, 1758. His mother's name was Mary.

There was an Isaac who died October, 1777; probably this one.

4 Isaac,[7] (*John*, 7,) b. December 29th, 1789, d. December 16th, 1857.

    M. (1) Clarissa Mullock, who was b. February 24th, 1796; (2) ——.

    Ch. Elizabeth, George, Adaline, Ann M., Samuel, Sarah R., Julia F., Horace, William L., Albert.   (*Mt. Hope, N. Y.*)

(An Isaac H. Corwin and Belinda, his wife, sell land to Allerton in 1854. They were of Arcadia, Wayne Co., N. Y. *Records, lib.* 107, *p.* 132.)

(An Isaac L. Corwin, (or Cowan, as also written,) of Phelps, Ontario Co., N. Y., sells land in 1847.)

5 Isaac Shultz,[7] (*Silas*, 2,) b. May 29th, 1809.
M. Valeria Ann Crane, November 26th, 1829.
Ch. Silas B., Eli E., Henry H., Rosetta A., Archibald L., Valeria C., Eunice O., Mary O. (*Providence, Pa.*)

6 Isaac Hubbard,[7] (*Isaac*, 1,) b. February 4th, 1782, d. ——, 1814.
M. Lydia, daughter of Silas and Susannah Horton. She b. September 26th, 1788, d. November 28th, 1808.
Gives his lands to father, Isaac. Will at Morristown, N. J.

7 Isaac,[7] (*Stephen*, 2,) b. ——, 1800 ?
M. Nancy Dods. (*Plymouth, Mich.*)

8 Isaac Little,[7] (*John II.*, 8,) b. February 3d, 1829.
M. Margaret Jemima Marquis, September 5th, 1854. She was b. January 28th, 1836.
Ch. David J., Jane E., John J.

9 Isaac Hubbard,[8] (*James Y.*, 11,) b. ——, 1810–20. (*Niagara Co., N. Y.*)

10 Isaac,[8] (*Peter*, 4,) b. ——, 1835–40.

11 Isaac Bryant,[8] (*John*, 23,) b. October 27th, 1804.
M. Maria H. Hillard, January 16th, 1827. She b. February 28th, 1805.
Ch. Charles, Elizabeth. (*See Jesse*, 1.) (*Near Chester, N. J.*)

12 Isaac,[8] (*Manasseh R.*, 1.)

13 Isaac,[8] (*Noah*, 2,) b. August 3d, 1836, d. March 12th, 1863.

14 Isaac T,[9] (*Charles L.*, 19,) b. March 9th, 1857.

15 Isaac,[9] (*William*, 39,) b. ——, 1855–60.

16 Isaac Reeve,[7] (*Mary Corwin*, 12,) b. ——, 1810–20. (*Middletown, N. Y.*)

17 Isaac L. Millspaugh,[7] (*Dorothy Corwin*, 2,) b. February 1st, 1827.
M. Deborah E. Munday, a physician. (*Staten Island.*)

18 Isaac Little,[8] (*Benoni B. Little*, 1,) b. December, 1828.

19 Isaac Rieger,[8] (*Nancy Corwin*, 2,) b. 1810–20.

20 Isaac H. Leck,[8] (*Deborah Corwin*, 4,) b. ——, 1815–25. (*Newark, N. J.*)

(An Isaac died in Suffolk Co., N. Y., in 1829, whose mother's name was Mary.)

(1) Isaac Bellos,[7] [*Alice Corwine*, (1).]

Isabel, (Hebrew,) same as Elizabeth.

1 Isabel,[8] (*Charles*, 3,) b. March 31st, 1857.

2 Isabella,[8] (*Polydore B.*, 1,) b. May 22d, 1843.
M. Edson Pike. (*McGrawville, N. Y.*)

3 Isabella,[9] (*Charles T.*, 17,) b. ——, 1849, d. January 11th, 1865.

4 Isabella,[9] (*James Horton*, 34,) b. at Bergen, N. J., November 2d, 1858.

5 ISABELLA,[9] (*William*, 39,) b. ——, 1855-65.

6 Isabel H. Lankton,[9] (*Phebe A. Padden*, 10,) b. May 15th, 1854, d. February 12th, 1856.

7 Isabella Petit,[8] (*Hannah Corwin*, 12.)

Isaiah, (Hebrew,) salvation is of the Lord.

1 ISAIAH,[7] (*Jabez*, 1,) b. ——, 1815, d. May 21st, 1850, at Providence, R. I.
M. Huldah Howell.
Ch. Isaiah S.   (*Jamesport, L. I.*)

2 ISAIAH SEYMOUR,[8] (*Isaiah*, 1,) b. ——, 1830-50.
M. Josephine Reeve.

3 Isaiah Mapes,[8] (*Anna Corwin*, 7,) b. ——, 1820-30.

4 Isaiah Corwin Wells,[8] (*Puah Corwin*, 1,) b. ——, d. young.

Israel, (Hebrew,) soldier of God.

1 ISRAEL,[5] (*Jesse*, 1,) b. about 1730-40, on Long Island.

2 ISRAEL YOUNGS,[7] (*Peter*, 2,) b. August 29th, 1791, d. November 24th, 1846.
Unmarried. A devoted Christian, of the Presbyterian Church.
(*Will at Goshen, N. Y.*)   (*Minisink, N. Y.*)

3 Israel Case,[5] (*Samuel Case*, 29.)
M. ? —— King, 1761.   (*Moore's Index of Southold*, p. 64.)

4 Israel Youngs Carr,[8] (*Azubah Corwin*, 3.)
M. Margaret Rowe.   (*Genoa, N. Y.*)

5 Israel Concklin,[9] (*Sarah M. Corwin*, 34,) b. February 14th, 1849.

Jabez, (Hebrew,) pain-causer.

1 JABEZ,[6] (*Silas*, 1,) b. October, 1770, d. September 29th, 1836.
M. (1) Patience Tuthill, October 12th, 1794, who was b. 1773, d. April 14th, 1822 ; (2) Christiana Skidmore, May 7th, 1823. She b. 1786.
Ch. Tuthill, Jabez, Silas, Daniel, George, Puah, Daniel, Elizabeth, Isaiah, Patience M., Ezra F., Ira C., Martha C.
He united with church of Aquebogue, May 7th, 1809.
(*Aquebogue and Riverhead, L. I.*)

2 JABEZ,[7] (*Silas*, 6,) b. ——.
M. Ann Hulse. No *ch.*   (*West Town, Orange Co., N. Y.*)

3 JABEZ,[7] (*Jabez*, 1,) b. May 19th, 1797, d. May 17th, 1868.
M. (1) Karan Moore, b. ——, 1797, d. September 29th, 1831 ; (2) Melinda Cook, sister of Rev. N. B. Cook.
Ch. Addison M. ; (2) Lucy K., Emily E., Rosetta M., Francis P., Jabez B.   (*Franklinville, L. I.*)

4 JABEZ,[8] (*Tuthill*, 1,) b. ——, d. when a young man.

5 JABEZ BALDWIN,[8] (*Jabez*, 3,) b. ——.   (*Northville, L. I.*)

⌈ Jabin, (Hebrew,) whom God observes.

1 JABIN,[7] (*Hubbard*, 1,) b. at Tunbridge, Vt., August 7th, 1791; living
   M. Abigail Burroughs, April 7th, 1818. She was b. at Sorrell,
   Canada, April 22d, 1792.
   Ch. Ephraim II., Reuben, Abigail, Philenia, Jabin II., Charles D.

2 JABIN HUBBARD,[8] (*Jabin*, 1,) b. August 11th, 1827.
   M. Dolly Bean, January 1st, 1856.
   Ch. Herbert M., Eliza J., Solon C., Lura A.   (*Chelsea, Vt.*)

Jacob, (Hebrew,) a supplanter.

1 REV. JACOB,[5] (*Matthias*, 3,) b. ——, 1747, d. September 20th, 1833 or
   183 .
   M. Mehetable ————.
   Ch. Sarah, Mary Ann.

This is, no doubt, the Jacob who signed the agreement to support
Congress in 1775.   He was pastor of the strict Congregational Church at
Riverhead, Long Island, November, 1787–1800.   This church had been
organized in 1785, with a few members, by Rev. Daniel Youngs, with the
style of "The Second Strict Congregational Church of Riverhead." Rev.
David Wells, nephew of Jacob Corwin, became pastor of this church in
1809.   He died in 1821, aged 46.   It had no settled pastor after this.   It
belonged to *The Separated Churches* of New-England.   Those on Long
Island formed a convention with the kindred churches of Connecticut,
called "The Strict Congregational Convention of Connecticut," and in 1781
published a *Confession of Faith* and *Form of Government*, which were re-
published on Long Island in 1823.   This also contains a brief history of
their separation.   In 1791, "The Strict Congregational Convention of Long
Island " was established.   The original members were Revs. Daniel
Youngs, Jacob Corwin, and Noah Hallock, all of whom had been ordained
by the Connecticut Convention.   As early as 1793, however, ministerial
exchanges began to take place with the other body.   On April 2d, 1815,
Jacob Corwin organized the Third Congregational church at Brookhaven,
Long Island, and supplied it for several years.   He also prepared a genea-
logy of the Corwin family, but his papers were destroyed.

2 JACOB,[6] (*Jeremiah*, 1,) b. November, 1777, d. January 10th, 1849.
   M. Deborah Petty.   She b. July, 1777, d. April 18th, 1844.
   Ch. Maria, Deborah, Mary, Azubah, Jemima.   (*Aquebogue, L. I.*)

3 Jacob Bryant Reed,[6] (*Sophia Corwin*, 1,) b. July 27th, 1812.

4 Jacob Blinn Thatcher,[8] (*Jesse Thatcher*, 10,) b. May 3d, 1835.

5 Jacob Blinn,[9] (*Adam Blinn*, 3,) b. ——, 1862 ?

6 Jacob Paris,[9] (*Julia A. Blinn*, 18,) b. ——, 1855 ?

(1) JACOB,[7] [*Richard*, (2),] b. ——, 1780–90, d. young.

James, (Hebrew,) a supplanter.

1 JAMES,[5] (*John*, 3,) b. August 22d, 1741, d. November 9th, 1791.
   M. Mehetable Horton, ——, 1763. (She was b. September 29th, 1743,
and d. October 27th, 1795. She was a daughter of William Horton
and —— Wells. The ancestors of this William Horton were successively William, Jonathan, and Barnabas, who was the first emigrant of the name. Barnabas was born at Mousely, England, about
1600, and d. July 13th, 1680, at Southold. Jonathan was b. 1648,
and d. February 23d, 1708. William was b. 1685, d. September,
1728. See *Moore's Indexes of Town of Southold.*)
   Ch. William, George, Martin Luther, James, Mehetable, Benjamin,
Elizabeth, John, Moses.
   The record gives the following parcels of land as belonging, in 1776, to
" James Corwin, son of John, of Mattituck :"
   (1) " 12½ acres of land, bought in 1763, lying on the south side of the
King's road, bounded easterly by the land of Storrs Hubbard, and north-
erly by the road, being about 27 poles on the highway; westerly and
southerly by land of Timothy Corwin, which land was purchased of Timothy Corwin, (No. 1,) for £122."
   (2) "¼ acre of land at Mattituck, purchased of John Corwin in 1763,
probably his brother."
   (3) " A parcel, purchased of Harmony Pike, at Mattituck, containing
6½ acres, 1766. These deeds were all recorded April 29th, 1776."
   He lost his father when only fourteen years of age, and when in the
prime of life was obliged to leave his native place (Mattituck) because of
the Revolutionary war, having signed the association to support Congress.
He settled, finally, near Middletown, Orange Co., N. Y. Owing to his
early orphanage and subsequent removal to a then distant and new country,
his descendants lost the name of James's father, (John, 3,) for many years.
The writer, a great-grandson, was unable to discover this name, by most
diligent inquiry, among any of the descendants. In 1863, however, he
visited Mattituck and adjoining places, and in the town and county
records, in the grave-yards, and by tradition, the connecting link with the
earlier generations was soon indisputably established. The father's tombstone (John's) is yet standing at Mattituck, (see John, 3.) James is said
to have been buried at Warwick, Orange Co., N. Y. Another tradition assigns his burial place at Millo, between Middletown and Goshen.
   1½ JAMES,[6 or 6] (——,) b. ——, d. September 4th, 1789.
   His mother's name was Mary.   (*Aquebogue Records.*)
   2 JAMES,[6] (*Samuel*, 3,) b.——, 1741, d. June 24th, 1796.
   M. Lydia ——.
   Ch. Prudence, Lydia, Beulah, Fanny?
   In 1775, he signed the engagement to support Congress. In 1776, on
census list 241, having in his family one male, one female over sixteen,
and two children. He was known as "The Deacon."   (*Peconic, L. I.*)

3 JAMES,[5 or 6] (*George*, 1,) b. ——.
He wrote his name Curran.
4 JAMES,[6] (*John*, 5,) bapt. at Mattituck, August 15th, 1762, d. March 9th 1832.
M. Catherine Halleck, who was b. ——, 1774, d. May 28th, 1824.
Ch. Charlotte, Joseph, Laura, Sarah.   (*Southold, L. I.*)
5 JAMES,[6] (*James*, 1,) b. at Mattituck, L. I., July 14th, 1771, d. in New-York City, July 8th, 1848.
M. (1) Mary Simrall, December 17th, 1793. She was b. September 10th, 1769, d. in New-York City, July 22d, 1825. [She was daughter of William Simrall (b. May 5th, 1743, d. July 2d, 1814) and Bethia Owen, (b. January 9th, 1749.) The Simralls probably came from New-Hampshire. See note on Eliza, 3.] *M.* (2) Mary Carland, December 10th, 1833, in New-York City. She d. in 1869, aged about 85.
Ch. Henry W., Gabriel, (a daughter, b. April 9th, d. April 24th, 1797, Laura, William O., (a child, b. and d. April 9th, 1802,) Eliza, Maria, James H., Edward C., Lewis B., (and another, b. and d. —— 1809.)
He removed with his father to Orange Co., N. Y., when five or six years of age, about the opening of the Revolution. When nineteen years of age, his father died. They lived at what was called White Oak Bridge, or Monhagen, on a farm of forty or fifty acres, in Walkill township, near Middletown. About 1820, he removed to the village of Middletown, and in 1824 to New-York City. The farm was bought (it is said) by David Reeve. He remained in the city till his death. Was a deacon in the Spring Street Presbyterian church. Was buried at Preakness, N. J. (*See Edward C.*, 5.)
6 JAMES,[6] (*William*, 2,) b. April 21st, 1779, d. October 10th, 1844, at Piketon, Ohio.
M. (1) Margaret Cameron, of Scotland. She d. October 3d, 1822. (2) Elizabeth Smith, widow of James Mallory, of New-York City. She subsequently married a Demorest.
Ch. George, Ann E., William, Margaret, James, Charlotte, Hannah.
(Kept a shoe-store at 94 Broadway, New-York City, from 1805–20.)
7 JAMES,[6] (*Thomas*, 1,) b. ——, 1790.
M. —— Appleby.
Ch. John, Mary, James C., Joseph.   (*Near Southold.*)
8 JAMES,[6] (*Phineas*, 1,) b. May 25th, 1773, d. ——, 18—.
M. Mary Bull.
Ch. Charlotte, Mehetable, John, Annie, Louisa.
(*Camillus, Onondaga Co., N. Y.*)
9 JAMES,[6]? (*Nathan*, 1½,) baptized at Aquebogue September 18th, 1785.
M.
Ch. Jesse, James, Nathan, Stephen, Joanna.

(This whole family moved from Chester, Morris Co., N. J., to Central New-York, about 1820-30. Nathan, 1½, had a son James, but it is not certain that this is the same James.)

10 JAMES HERVEY,[7] (*Jemuel*, 1,) b. August 24th, 1829, d. November 18th, 1829.

11 JAMES YOUNGS,[7] (*Isaac*, 1,) b. November 11th, 1789.
   M. Sarah Stout.
   Ch. Stephen F., Jerusha, Isaac H., Benjamin, Christopher, James T., and three daughters.   (*Central New-York.*)

12 JAMES HERVEY,[7] (*Jemuel*, 1,) b. July 23d, 1838.

13 JAMES HORTON,[7] (*John*, 10,) b. October 15th, 1812, in Walkill, Orange Co., N. Y.
   M. Emma Hawkins, September 25th, 1851. She b. December 4th, 1826.
   Ch. Elizabeth, Coe Hawkins, Clarinda, George E.
                     (*Howells, Orange Co., N. Y.*)

14 JAMES,[7] (*Moses*, 1,) b. at Middletown, N. Y., July 22d, 1830, d. ——, 1860 ?   (*New-York City.*)

15 JAMES,[7] (*James*, 6,) b. August 12th, 1812, d. September 13th, 1870, in Memphis, Tenn.
   M. (1) Sarah L. Sinnerd, of Hamilton, Ohio, September 8th, 1842. She d. July 14th, 1843. (2) —— ——, September 13th, 1853.
   Ch. James, Pauline.   (*Indiana.*)

16 JAMES CLARK,[7] (*Thomas*, 2,) b. March 16th, 1820.
   M. Grace A. King, July 15th, 1851. She b. September 6th, 1826.
   Ch. Ella J., George A., Alice M., Flora E., Grace A., Laura B., Betsy A.
                     (*Greenport, L. I.*)

17 JAMES C.,[7] (*James*, 7,) b. ——, 1810-30.

18 JAMES,[7] (*Benjamin*, 5,) b. April 9th, 1813.
   M. (1) Emeline Botsford, December 13th, 1840. She b. 1824, d. July 8th, 1854. (2) Elizabeth Hulse. She b. 1822, d. September 23d, 1859. (3) Susan N. Hulse.
   Ch. Lucina, Emily, Emeline ? James F., Ellen C., Benjamin O.
                     (*Mt. Hope, Orange Co., N. Y.*)

19 JAMES,[7] (*Joseph*, 10,) b. ——.

20 JAMES,[7] (*Stephen*, 4,) b. ——, 1824, d. September 21st, 1863.
   M. Margaretta Barchus, June 25th, 1844, in Richland Co., Ohio.
   Ch. Edith R., Elmore H., Alice E., Wolverton M., James P., Alonzo J., Leonard S.
   He removed to Fulton Co., Ill., in 1854, enlisted in 103d regiment Illinois infantry, and died in camp, before Vicksburg.

21 JAMES TOWNSEND,[7] (*Alexander*, 1,) b. ——, 1817?
   M.
   Ch. George, Daniel.
   (Was in the Federal army.)   (*Chester, Orange Co., N. Y.*)

22 JAMES HARVEY,[7] (*Ichabod*, 1,) b. January, 1803, d. December 25th, 1867.
M.
Ch. William II.

23 JAMES,[7] (*Nathaniel*, 5.)

24 JAMES,[7] (*Hubbard*, 1,) b. August 19th, 1793.

25 JAMES,[7] (*James*, 9,) b. ——.   (*Central N. Y.*)

26 JAMES HORTON,[7] (*James*, 5,) b. May 5th, 1806, d. April 29th, 1830, in
Virginia.   (*A grocer in New-York City*, 1826–30.)

27 JAMES,[8] (*John*, 30,) b. ——.

28 JAMES PRESTON,[8] (*James*, 20,) b. October 5th, 1852.

29 JAMES,[9] (*Jesse*, 5,) b. ——.

29½ JAMES ADDISON,[8] (*Albert*, 2,) b. May 24th, 1842.

30 JAMES F.,[9] (*James*, 18,) b. March 20th, 1858.

31 JAMES BARRETT,[8] (*Joseph W.*, 12,) b. October 22d, 1842.

32 JAMES YOUNGS,[8] (*James Y.*, 11,) b. ——, 1810–20.
                                        (*Niagara Co., N. Y.*)

33 JAMES,[8] (*Henry W.*, 8,) b. about 1825, d. ——.

34 JAMES HORTON,[8] (*Edward C.*, 5,) b. in New-York City, March 17th, 1830,
d. at Bergen, N. J., February 2d, 1861.
M. Annie M. Garretson, January 31st, 1854. She was daughter of John
Garretson and Catharine Ann Riker, and was born May 9th, 1833.
Ch. Emma, Mary, Isabella, James Horton.
Removed to Jersey City with his father, May 1st, 1847; to Bergen,
1856. Timber merchant. After his death, his family soon removed to
New-Brunswick, N. J.

35 JAMES,[8] (*Daniel*, 16,) b. ——, d. young.

36 JAMES,[8] (*David M.*, 13.)

36½ JAMES,[8] (*Henry*, 11,) b. February 11th, 1834.
M. Helen Reeves, June 22d, 1856.
Ch. Effie L., George H., Lucy E.   (*Peconic, L. I.*)

37 JAMES ADELMER,[8] (*Ebenezer*, 3,) b. July 1st, 1834.
M. Helen M. Taylor.

38 JAMES,[8] (*Timothy*, 6.)

38½ JAMES,[8] (*James*, 15,) b. about 1855.   (*West.*)

39 JAMES MADISON,[9] (*Barnabas*, 6,) b. October 2d, 1834.
M. Nancy E. Jennings, December 29th, 1857.
Ch. Theresa B., Frank J., Lizzie P., Carrie M.

40 JAMES HOWELL,[9] (*Hubbard*, 2,) b. June 29th, 1845.

40¼ JAMES,[9] (*George W.*, 28.)

40½ JAMES,[9] (*Ezra*, 3,) b. ——, 1855–60.

41 JAMES HORTON,[9] (*James H.*, 34,) b. at Bergen, N. J., January 17th, 1861.

42 James Jennings,[6] (*Sarah Corwin*, 4,) b. ——, 1780?

43 James P. Taylor,[7] (*Fanny Corwin*, 1,) b. June, 1815, d. September 16th, 1816.

44 James Little,[7] (*Abigail Corwin*, 3,) b. November 3d, 1818.
   M. Sarah Jessup.   (*Middletown, N. Y.*)

45 James Corwin Halsey,[7] (*Prudence Corwin*, 1,) b. ——, 1790–1800.

45½ James B. Colgrove,[7] (*Bela II. Colgrove*, 1,) b. at Sardinia, N.Y., October 12th, 1831.
   M. (1) Theodora Clark, November 17th, 1858, of Ellicotville. (She died about 40 days after marriage.) (2) Mary Belle M. Giffin, in 1864.
   Ch. Russel, Bela II., James B.
   He graduated as medical doctor at the University of Buffalo, N. Y. He emigrated to St. Louis in 1861; served as an assistant surgeon in a government hospital in that place; was appointed examining surgeon of recruits for the Union army, and continued in some similar capacity throughout the Rebellion. He removed to Lincoln, Benton Co., Mo., in 1865.

46 James M. Little,[8] (*William C. Little*, 65,) b. ——, 1839.

47 James Lyons Millspaugh,[8] (*Gilbert C. Millspaugh*, 6,) b. January 19th, 1834.

48 James Millspaugh,[8] (*Gideon II. Millspaugh*, 1,) b. January 7th, 1839, d. January 20th, 1841.

48½ James Dunlevy,[8] (*Lucinda Corwin*, 4,) b. May 8th, 1829, d. February, 1831.

49 James Horton Wetherby,[8] (*Eliza Corwin*, 3,) b. May 4th, 1836.
   M. Sarah Maria Campbell, of Herrick, Susquehanna Co., Pa., October 16th, 1862.
   Ch. George W., Franklin N., Vangelia M.
                                   (*Green Grove, Luzerne Co., Pa.*)

50 James B. Odell,[8] (*Bethia Corwin*, 4.)

51 James Hackett,[8] (*Mary Corwin*, 14,) b. January 5th, 1812, d. September 8th, 1840.
   M. Hannah Richardson.
   Ch. James, Cornelius.

52 James W. Brown,[8] (*Dorastus Brown*, 1.)

53 James Arthur Taylor,[8] (*Elizabeth Corwin*, 17,) b. at Rochester, Ohio, August 22d, 1852.

54 James Lewis Ketcham,[9] (*Rebecca J. Corwin*, 3,) b. August 21st, 1861, d. April 18th, 1862.

55 James Lewis,[9] (*Emeline Corwin*, 2.)

56 James Stryker,[9] (*Mary Corwin*, 48,) b. February 18th, 1851. Entered St. Stephen's College, Annandale, 1865.

57 James Paris,[9] (*Julia A. Blinn*, 18,) b. ——, 1862?

Irene Amelia Corwin (George 1 12)
b. April 18. 1845, m. Oct 13   874
Milton Faust   (Bernardsville, N. J.)

issue –

George Wilbur Corwin, b. March 5. 1875
              died Feb 12. 1877.
Jennie Eliza Corwin, b. Dec 7. 1877.
Hattie Martha Corwin, b. June 28, 1879,
              died Dec 23. 1885.
Else Flora Corwin, b. Nov 16. 1881,
              died Dec 10. 1887.
Edna Mary Corwin, b. Feb 23. 1884,
Roger Corwin, b. Aug 1. 1886.

James Hackett,[9] (*James Hackett*, 51.)   Lives in the West.

58½ James B. Colgrove,[9] (*James B. Colgrove*, 45½,) b. ——, 1869 ?

59 James Winfield,[10] (*Elizabeth M. Downs*, 65,) b. March 4th, 1868.

(A James C. Corwin, and Emily, his wife, were of Sterling, Cayuga Co.,
N. Y., 1825–37.   She was a widow in 1837.)

(1) *JAMES BARR,[8] [*Samuel Curwen*, (6),] b. in Salem, Mass., December
20th, 1818.

M. Rebecca Hovey, daughter of Samuel and Caroline (Collins) Endicott,
February 3d, 1848.

Ch. Samuel E., Caroline E., James E., George E.   (*Salem, Mass.*)

(2) *JAMES ENDICOTT,[9] [*James B.*, (1),] b. January 14th, 1856, d. March
19th, 1857.

(3) †JAMES,[9] [*John*, (11),] b. ——, 1849.   (*Waverley, Ohio.*)

Jane, feminine of John, (Hebrew,) a gracious gift of God.

Jane, see Jenny, Jennette.

½ JANE,[6] (*David*, 5,) b. August 16th, 1827, d. June 15th, 1837.

1 JANE,[7] (*Jesse*, 3,) b. ——, 1810 ?

M. Daniel J. Baldwin.

Ch. Albert, Mary, George, Henry, Jane, Ira.   (*Goshen, N. Y.*)

2 JANE E.,[7] (*David*, 7,) b. April 29th, 1830.

3 JANE,[7] (*Henry*, 5,) b. ——, 1829 ?

M. —— Luce.

4 JANE AMELIA,[8] (*George S.*, 12,) b. April 18th, 1845. *[handwritten: Oct 13 1874]*

5 JANE,[8] (*Anson*, 1,) b. ——, 1820–30.   *[handwritten]*

6 JANE,[8] (*Lewis W.*, 1,) b. ——, 1825–35.

M. Benjamin Merry.   (*Seneca Falls, N. Y.*)

7 JANE ELIZABETH,[8] (*Isaac L.*, 8,) b. November 2d, 1858.

8 JANE,[8] (*Nathan*, 5,) b. ——, 1825–30.

M. Nathaniel A. Griffin.

9 JANE,[8] (*Abel*, 3.)

10 JANE,[8] (*Jesse*, 6.)   (*Ohio.*)

11 Jane Little,[7] (*Abigail Corwin*, 8,) b. January 12th, 1813.

M. James W. McWilliams, April, 1832.

12 Jane E. Porter,[7] (*Elizabeth Corwin*, 8.)

13 Jane Leonora Hill,[8] (*Margaret Corwin*, 3.)

M. —— Burton.

14 Jane Baldwin,[8] (*Jane Corwin*, 1,) b. at Goshen.

M. —— McCoy.

15 Jane Wood,[8] (*Mary Corwin*, 16.)

16 Jane Concklin,[9] (*Sarah M. Corwin*, 34,) b. August 3d, 1841.

* Curwen.   † Corwine.

Jason, (Greek,) a healer.

⚮ JASON,⁶ (*Jonathan*, 1,) bapt. May 2d, 1773, d. April 12th, 1775.

(Aquebogue Records.)

1 JASON,⁶ (*Eli*, 1,) b. September 1st, 1784, d. October 2d, 1860.

    M. (1) Ruth Brown, May 26th, 1805. She was b. June 19th, 1787, d. April 12th, 1806. (2) Jennette Millspaugh, September 22d, 1808. She was b. July 22d, 1787.

    Ch. Adam, Dolly J., Susan C., Mary, William H., John E., Ruth, Gilbert, Jason W., Oliver.   (*Scotchtown, Orange Co., N. Y.*)

2 JASON W.,⁷ (*Jason*, 1,) b. February 19th, 1828.

    M. Sarah Howell.

3 JASON,⁷ (*Jonathan*, 2,) b. ——, 1788.   (*Central N. Y.*)

4 REV. JASON,⁷? (*Phineas*, 1½,) b. at Franklin, Ct., February 16th, 1792, d. May 15th, 1860, at Washington, Ill.

    M. Deborah Calista, daughter of Thomas Orton, of Hamilton, N. Y., on January 6th, 1824.

    Ch. Charlton J., Eugene P , Orton, Helen M., Charles L., Thomas D.

He was for several years a pupil of Dr. Nott, pastor of the Presbyterian church in his native place. He removed with his family to Cazenovia, N. Y., and united with the Baptists, at Woodstock, when at the age of twenty-five. He studied at Hamilton Literary and Theological Institution, graduating in 1822. He first preached a few months at Coventry, in Chenango Co., N. Y., and then settled successively at the following places: Deposit, Delaware Co., N. Y., 1823–6 ; Penfield, Monroe Co., N. Y., 1826–31; Webster, Monroe Co., N. Y., 1831–3 ; Binghamton, Broome Co., N. Y., 1833–5 ; Great Bend, Susquehanna Co., Pa., 1835–7 ; Earlville, Madison Co., N. Y., 1837–9 ; Bridgewater, Oneida Co., N. Y., 1840–1 ; Augusta, Oneida Co., N. Y., 1841–2 ; Clinton, Oneida Co., N. Y., 1839–43 ; Webster, Monroe Co., N. Y., 1843–5 ; supplied destitute churches, 1845–8 ; Belleville, St. Clair Co., Ill., 1848–9 ; Bloomington, McLean Co., Ill., 1849 ; Washington, Tazewell Co., Ill., 1850–4. The last three appointments in Illinois were from the American Baptist Home Missionary Society. He subsequently became an agent of the American Bible Union, traveling in Indiana, Illinois, Iowa, Missouri, and Kentucky. His widow afterward lived in Brooklyn, N. Y., and in Washington, D. C., and now at Fort Smith, Arkansas.

1 Jasper Smith Reed,⁸ (*Sophia Corwin*, 1,) b. May 19th, 1808.

Jedediah, (Hebrew,) beloved of the Lord.

⚮ JEDEDIAH,⁴ (*Samuel*, 1,) b. about 1700.

    Ch. John ?

1 JEDEDIAH,⁴ (*Theophilus*, 2 ?) This name is given, as there is evidently some confusion between the names of Jedediah, 1, and Daniel, 2. The next Jedediah (namely, No. 2) is called, in the letters of administration,

given his administrators, in 1799, the son of Jedediah. Yet he and all his brothers and sisters are elsewhere called children of Daniel. Hence I have supposed that Daniel, 2, had also the second name of Jedediah, though I have no proof. (*See Daniel*, 2.)

2 JEDEDIAH,* (*Daniel*, 2; *see Jedediah*, 1,) b. ——, 1728, d. May 7th, 1799.
   M. Abiah Sweezy, November 23d, 1752. She was b. 1729, d. August 8th, 1795. Both of Mattituck, when married.
   Ch. Daniel, Joshua, Richard, Sand, Rhoda, Abiah, Anna, Nathan, Jedediah.

In 1775, he signed engagement to support Congress. In 1776, on census list 137, having in his family 2 males, 2 females over 16, and 5 children.   (*Riverhead and Orient, L. I.*)

3 CAPTAIN JEDEDIAH,* (*Jedediah*, 2,) b. ——, 1755, bapt. at Mattituck, February 4th, 1756, d. February 24th, 1824.
   M. Polly Luce. She b. 1770, d. January 27th, 1854. She was a church member for 72 years.
   Ch. Mary, Elury, Jedediah.   (*Riverhead, L. I.*)

4 JEDEDIAH,[7] (*Jedediah*, 3,) b. —-, 1792, d. February 18th, 1796.

Jehiel, (Hebrew,) treasured of God.

1 JEHIEL, (*Nathaniel*, 9,) b. ——, d. ——, 1827.   (*Brookhaven, L. I.*)

Jehu, (Hebrew,) Jehovah is He.

(1) *JEHU,[7] [*John*, (3),] b. about 1780, d. July 6th, 1823, in Baltimore, Md.
   M. Margaret ——.
   Ch. Mary Ann?
A letter of his to his father in 1806, (yet preserved,) written in Baltimore, seems to refer to a Margaret as his wife, and a "little Mary Ann."

1 Jefferson Halsey,[7] (*Elizabeth Corwin*, 2,) b. ——, 1814?
   M. Elizabeth Burt.
   Ch. Caroline, Burt, Frank.

Jeffrey, same as Godfrey, (Old High-German,) at peace with God.

1 Jeffrey Howell,[8] (*Jemima Corwin*, 7,) b. ——.

Jemima, (Hebrew,) a dove.

½ JEMIMA,* (*Simeon*, 1,) bapt. at Mattituck, L. I., October 29th, 1752.

1 JEMIMA,* (*Jesse*, 2,) b. April 26th, 1765, d. January 18th, 1845.
   M. Amos Thatcher, in Kentucky, 1788. He was b. 1752, d. June 25th, 1834, in Darke Co., Ohio.
   Ch. Jesse, Elizabeth, Mary, Joseph, Elijah, William, Sarah, Hannah, Alexander, Lydia, Lucinda, David.

* Corwine.

Removed to Bourbon or Campbell Co., Ky., 1785; to Lebanon, Ohio, 1798; to Preble Co., O., 1830. After his death, the widow removed to Frankfort, Indiana, where she died. Mr. T. was eccentric, enjoying fun very much. He was also a Revolutionary soldier, and in many engagements, but was never injured. The children live near Frankfort, Indiana.

2 JEMIMA,[7] (*Joshua*, 1,) b. August 25th, 1774, d. August 19th, 1776.

3 JEMIMA,[7] (*Abner*, 1,) b. ——, 1792, d. ——, 1859.
    M. —— Tuthill.
    No Ch.

4 JEMIMA,[7] (*Joseph*, 4,) b. July 22d, 1806.

5 JEMIMA P.,[7] (*Jacob*, 2,) b. October 15th, 1811.

6 JEMIMA,[7] (*Samuel*, 9,) b. ——, 1810–20.
    M. Gilbert Williams, April 11th, 1832.
    No. Ch. Farmer.   (*Middle Hope, Orange Co., N. Y.*)

7 JEMIMA,[7] (*Peter*, 2,) b. March 16th, 1794.
    M. Jeffrey Howell.
    Ch. Hiram, Mary A., William, Jeffrey, Lewis, Sarah J., Marietta H.
                             (*Ridgebury, Orange Co., N. Y.*)

7½ JEMIMA,[7] (*Seth*, 1,) b. about 1809.
    M. Silas Whitney.
    Ch. Hannah, Mary, (who *m.* James M. Payne,) Silas, (who has 7 children,) Charles, (who has 3 children,) Harriet, (who *m.* John Benjamin,) Elizabeth, (who *m.* William Goodale,) George, John, Sarah, (who *m.* Abijah Smith,) and Monroe.

8 JEMIMA,[7] (*Joshua*, 2,) b. ——, 1792.
    M. William Coleman.
    Ch. William L.

8¼ JEMIMA JANE,[8] (*David*, 18,) b. March 8th, 1830.
    M. Adolph J. You, January 23d, 1851. He d. August 17th, 1853, aged 25 years.
    Ch. Adolph Jackson, Sarah Elizabeth, (twins.)   (*Pittsburg, Pa.*)

8½ Jemima Case,[6] (*Benjamin Case*, 27,) b. ——, 1723, d. January 20th, 1745.

9 Jemima Hulse,[8] (*Lucetta Corwin*, 1.)

10 Jemima Thatcher,[5] (*Jesse Thatcher*, 8,) b. January 10th, 1832.

11 Jemima Blinn,[6] (*Lucinda Thatcher*, 5,) b. January 23d, 1834, d. August 12th, 1836.

12 Jemima Harding,[8] (*Polly M. Corwin*, 4.)
    (There was a Jemima Corwin, in 1853, of Genoa, N. Y.)

Jemuel, (Hebrew,) a day of God.

1 JEMUEL,[6] (*Eli*, 1,) b. June 15th, 1788, d. November 19th, 1849.
    M. (1) Susan Young, March 20th, 1815. She d. September 27th, 1825;

(2) Esther Young, December 21st, 1828. She b. January 26th, 1805.
Ch. Harriet, Luther, Cynthia, James H., John M., Susan, Amelia, Daniel T., James H., Dorothy J., Charlotte. (*Pine Bush, N. Y.*)

Jennette, Jenny, same as Jane.

1 JENNETTE,[8] (*John*, 32,) b. May 2d, 1819, in Vermont.
M. William Chamberlin, March 30th, 1841.
Ch. Sarah J., William M., Mary E., John F., George W., Alice K.

2 JENNETTE M.,[8] (*William H.*, 12,) b. November 13th, 1845.
M. Robert Henderson, of Williamsburg, L. I.
Ch. William T.

1 JENNIE SCOTT,[9] (*Barnabas*, 5,) b. January 23d, 1859.

2 JENNIE,[10] (*Bartholomew*, 1,) b. May 1, 1870, d. July 17th, 1870.

3 JENNY,[10] (*John*, 67,) b. November 17th, 1860.

3½ Jenny Henrietta Borgy,[8] (*Harriet N. Corwin*, 1,) b.——, 1825–35.

4 Jenny Ketcham,[9] (*Nancy Ann Corwin*, 3,) b. October 23d, 1848.
M. Charles W. Hartly, October 27th, 1870. (*Greenport, L. I.*)

5 Jennie Lewis,[10] (*Imogene Davis*, 1,) b. ——, 1865–70.

(1) * JENNIE,[9] [*George*, (10),] b. June 20th, 1847, d. November 4th, 1868.
Graduated at Lasher's Seminary.

Jeremiah, (Hebrew,) exalted of the Lord.

1 JEREMIAH,[5] (*Matthias*, 3,) b. ——, 1735, d. October 22d, 1821.
M. (1) Mary ——, who was b. ——, 1732, d. October 22d, 1798 ; (2) Abigail ——.
Ch. Elizabeth, Bethia, Keturah, Jacob, Matthias, Jeremiah, Benjamin.
(*Riverhead, L. I.*)

In 1775, he signed the engagement to support Congress. In 1776, census list 412 shows that he had 1 male, 3 females over 16, and 4 children in his family.

2 JEREMIAH,[6] (*Jeremiah*, 1,) b. about 1765.
M. Jerusha Edwards.
Ch. Jerusha, Jeremiah, Daniel, Abigail, Mary.
He went away from his family about 1800, and was never heard of again.

3 JEREMIAH,[7] (*Matthias*, 6,) b. about 1803, d. 1828.
M. Mary Ann Helm, about 1825. She afterward married a Tuthill.
Ch. John V., Calcina.
He once lived, perhaps, in Cuba, N. Y., also for a short time in Massachusetts, and then at New-Windsor, N. Y. His widow (now Mrs. Tuthill) lives (1871) in Jersey City.

4 JEREMIAH,[7] (*Jeremiah*, 2,) b. May 8th, 1796.

* Corwine.

M. Susan Sanxy.

Ch. Elizabeth, Mary, Charlotte, Charles.

5 Jeremiah Terry,[7] (*Keturah Corwin,* 1,) b. ——, 1780–90.

6 Jeremiah Terry,[8] (*Julia Ann Corwin,* 5.)

Jerome, (Greek,) holy name.

1 Jerome Swina,[8] (*Bethia Hobart,* 11,) b. March ——, 1837.

Jerusha, (Hebrew,) possessed, (by a husband,) married.

1 JERUSHA,[7] (*Isaac,* 1,) b. December 6th, 1793, d. ——, 1814.

(*Chester, N. J.*)

2 JERUSHA,[7] (*Jeremiah,* 2,) b. December 10th, 1786, d. ——.

M. Nathaniel Girard, (or Garret,) of Middletown, N. Y., February 21st,
1807.   (*Aquebogue Records.*)

Ch. Franklin, Catharine, Anna, William, Wessell, Mary, Sarah, John.

(*Newburg, N. Y.*)

3 JERUSHA,[7] (*Henry,* 7,) bapt. at Aquebogue, March 10th, 1800.

4 JERUSHA,[8] (*James Y.,* 11,) b. ——, 1810–20.

5 JERUSHA,[8] (*Henry,* 12,) b. ——.

6 Jerusha Galloway,[8] (*Amanda Corwin,* 2,) b. ——.

7 Jerusha Hunt,[8] (*Experience Corwin,* 1,) b. ——, 1822, d. May 28th,
1849.

M. Samuel H. Rorick.   (Her tombstone at Chester, N. J.)

(1) *JERUSHA,[8] [*Samuel,* (6½),] b. October 8th, 1824.

M. William Marshall, October 28th, 1856.

Ch. Clarence, Edward.   (*Omega, Ohio.*)

Jesse, (Hebrew,) wealth.

1 JESSE,[4] (*Matthias,* 2 ?) b. ——, 1700–1710.

M.

Ch. Israel, Jesse, Amy.   (*Long Island and Connecticut.*)

David Corwin, (No. 8,) of Lebanon, Ohio, has a tradition that this Jesse
emigrated from Wales about 1718, coming to Long Island, and that Jesse's
grandfather came from Hungary, and was a son of Matthias II., who
reigned in Hungary about ten years. This Matthias II. was Archduke of
Austria, 1608–18, but history declares he died childless. He granted religious toleration throughout Austria. It does not appear, as far as the
writer is aware, that he possessed the *Corwin* name at all. He was apparently no relation to Matthias Corvin, I., 1458–90. The tradition seems
therefore inaccurate. (*But see Introductory Chapter.*) Isaac B. Corwin
(No. 11) also informed the writer that he had received a tradition from his
ancestors that the family came from Wales. But this Isaac (whose
grandfather was an own cousin of the writer's grandfather) is a direct de-

* Corwine.

scendant, beyond dispute, of Matthias, No. 1, who came to America in 1638, and the writer has not met with a line as to a Welsh extraction of this Matthias. Isaac Corwin's grandfather, (Isaac,) and the Jesse under consideration, both settled in Flanders, Morris Co., N. J., about 1760–70. (At about the same time, settled in the same county, William, 2, Benjamin, 2, and James, 9.) This Welsh tradition, therefore, being found only in Morris County branches, had probably a common origin. Now, there is a town called *Corwen* in the north-eastern part of Wales, and noticing this on the map, probably gave rise to the tradition. The writer has learned from Rev. Mr. Richardson, the clergyman of that place, with whom he has corresponded, that none of the Corwin name live there, and that the etymology of the word is very different from the surnames Corwin, in America. (*See Letter in the Introduction.*) Traditions are very uncertain things, unless they can be substantiated by documentary evidence. There was little emigration to this country in the early part of the last century. The probable ancestor of the above Jesse, the writer has put down as Matthias, 2, whose descendants were lost track of. If Baptists at that early day, as many of Jesse's descendants have been since, it will explain the apparent absence of names on the baptismal records. Jesse possibly had the middle name of William. (*See William, 1.*)

2 JESSE,⁵ (*Jesse,* 1,) b. on Long Island, 1736, d. in Kentucky, 1791.

M. Keziah Case, on Long Island, probably, but about the time of removal to New-Jersey. The Cases also removed to New-Jersey about the same time as the Corwins, (1750–70.)

Ch. Matthias, Mary, Jemima, Ichabod, Joseph, Hannah, David, Sarah.

He removed from Long Island to Morris Co., New-Jersey, about 1760, (some say 1767;) in 1776 he removed to Fayette Co., in south-western part of Pennsylvania, with a family of six children. He emigrated thence to Bourbon Co., in north-eastern Kentucky, in 1785. (*See Richard Corwine, No. 2.*)

3 JESSE,⁶ (*David,* 4,) b. September 12th, 1768, d. August 9th, 1831.

M. Mrs. Wakeman, originally Sabrina Sherwood. She was of Fairfield, Connecticut, b. ——, 1769, d. ——, 1832.

Ch. George S., Mary, Jane, Samuel W.   (*Middletown, N. Y.*)

He became blind when about eighteen years of age, from inflammation. He was, notwithstanding, the original inventor and patentee of a mill for cleaning wheat.

4 JESSE,⁷ (*Rev. Joseph,* 5,) b.——, 1790–1800.

M. Polly Benjamin.

Ch. Nancy J.   (*Greenport, L. I.*)

5 JESSE,⁷? (*James,* 9,) b.——, 1790–1800?

M. Sarah Terry.

Ch. James, and several others, names unknown.

He moved from Chester, Morris Co., New-Jersey, to Central New-York about 1825–30. There was a Jesse Corwin, of Sullivan, Tioga Co., New-

York, who sold land to J. H. Corwin, of Chemung Co., about 1830–40.
Possibly the same as No. 5.

6 JESSE,[7] (*Matthias,* 5,) b. June 30th, 1794, *1797*
   M. Jane McMahon, ——,1829.
   Ch. Jane, Erin, Harry, Warren, Thomas.
   A lawyer, and a member of the Ohio Legislature.

7 JESSE B.,[7] (*Ichabod,* 1,) b. October 14th, 1798.
   M. (1) Nancy Gilchrist, August 19th, 1823 ; (2) Rebecca Knox, April
     11th, 1826.
   Ch. Sarah, John K.   (*Ohio.*)

8 JESSE,[9] (*William,* 39,) b. ——, 1853–60.

9 JESSE,[9] (*Anson L.,* 2,) b. ——, 1855–65.

10 Jesse Thatcher,[7] (*Jemima Corwin,* 1,) b. December 23d, 1785, d.
   March 7th, 1863.
   M. Anna Maria Painter, September 9th, 1807. She b. September 5th,
     1791, d. October 7th, 1844.
   Ch. George, Joseph, John, Sophia, Lucinda, Anna, Dilly, Amos,
     Jemima, Jacob B.

11 Jesse C. Penny,[7] (*Annie Corwin,* 3,) b. November 30th, 1801.
   M. Elizabeth Montgomery, October 3d, 1822.

12 Jesse Philips,[7] (*Hannah Corwin,* 2,) b. ——.

13 Jesse Corwin Blinn,[8] (*Lucinda Thatcher,* 5,) b. January 4th, 1840.
   M. Ellen Gaddis, November, 1864.
   Ch. Gaddis, Ora.
   He enlisted in August, 1861, for three years, in Co. K, 10th Indiana,
under General Manson. Army of the Cumberland.

14 Jesse C. Duckworth,[7] (*Sarah Corwin,* 9,) b. February 24th, 1819.

15 Jesse Wood,[9] (*Mary Corwin,* 16.)

(1) * REV. JESSE DANIEL HITT,[8] [*Rev. Richard,* (5),] b. June 7th, 1825, at
   Maysville, Ky.
   M. Susan Respess, of Kenton Co., Ky.
   Ch. Harriet, Jesse R., Corliss H., William A., Conde M., Thomas R.
   He graduated at Transylvania University, Lexington, Ky., in 1846. At
first a Methodist minister. After a few years united with the Universa-
lists. Settled for a time at North-Vernon, Indiana. In 1870, removed to
Mason, Ohio.

(2) * JESSE RICHARD,[9] [*Rev. Jesse D. H.,* (1),] b. in Kentucky, June 2d,
   1854.

(3) Jesse Frances Stevenson,[9] [*Sarah A. Corwine,* (5),] b. August 31st,
   1862, at Shelbyville, Ky.

1 Jessie Fremont Boisseau,[9] (*Bethia,* 9½,) b. March 14th, 1862.

1 J. FRANKLIN,[9] (*John,* 56,) b. August 25th, 1862.

                 * Corwine.

Joab, (Hebrew,) whose father is Jehovah.

(1) * JOAB HOUGHTON,[7] [*Amos*, (1),] b. in Hunterdon Co., New-Jersey,
——, 1788, d. in Louisville, Ky., May 20th, 1837.

M. Elizabeth Lucas, 1810. She was daughter of General Samuel Lucas,
who emigrated from Pennsylvania to Kentucky in 1789.

Ch. Richard M., Amos B., Samuel L., David M., Joab II., John F.,
Aaron II., William A., Margaret L., Hannah J., Sarah II., Anna E.,
Hattie E.

He was a printer and editor till within a few years of his death. He
established the first newspaper, *The Maysville Eagle*, in the town of
Maysville, Ky. He also established the first newspaper in Washington,
Mason Co., Ky. It was called *The Dove*. *The Eagle*, in its day,
became a prominent organ of the Whig Party in that portion of Kentucky.
It is yet published. In 1833, he removed to Cincinnati, Ohio, and became
editor and proprietor of the Cincinnati *Courier*, but soon after removed
again to Kentucky, and settled at Louisville. In 1830, his family removed
to Benton, Miss. His father had removed to the Tucahoe and Jersey
Ridges, Mason Co., Ky., from New-Jersey, soon after Joab's birth.

(2) * JOAB HOUGHTON,[8] [*Joab II.*, (1),] b. July, 1826, d. ——, 1849.

(3) * JOAB HOUGHTON,[9] [*Samuel*, (12),] b. and d. March 4th, 1866.

Joanna, (Hebrew,) a gracious gift of the Lord.

1 JOANNA,[6] (*Amaziah*, 1,) b. August 31st, 1792.

M. Bethuel Howard, February 14th, 1816.

(*New-York City and Mattituck.*)

1½ JOANNA,[7] (*John*, 9,) b. August 30th, 1783, d. July 24th, 1853.
M. Benjamin Reeve.

2 JOANNA,[7] (*Moses*, 1,) b. March 7th, 1825, d. April 19th, 1851.

3 JOANNA,[7] (*Selah*, 1,) b. ——, 1783?
M. Jeremiah Young, June 13th, 1801. (*Aquebogue Records.*)

3½ JOANNA ELIZABETH,[7] (*Amaziah*, 3,) b. ——, 1820–30.

4 JOANNA [7] ? (*James*, 9.)

5 JOANNA REYNOLDS,[8] (*Daniel*, 11,) b. ——, 1822.

6 Joanna A. Flannery,[8] (*Mary Corwin*, 33,) b. at New-London, Ohio,
June 28th, 1854. (*Cleveland, Ohio.*)

Joel, (Hebrew,) Jehovah is God.

1 Joel L. Davis,[9] (*Polly Corwin*, 2,) b. ——, 1839.

M. Sarah Crane, of Wellsville, N. Y. She was graduated from the
Musical Institute of Friendship, N. Y.

Accountant of Morris Run and Fall Brook Coal Co., Pa. (*Blossburg, Pa.*)

* Corwine.

1 John,[2] (*Matthias*, 1,) b. ——, 1630 ? d. September 25th, 1702.
  M. Mary, daughter of Charles Glover, February 4th, 1658.   She d.
  before 1690, probably.
  Ch. John, Matthias, Samuel, Sarah, Rebecca, Hannah, Abigail, Mary.
  In 1661, bought land and meadow at Oyster Pond and Aquebogue.
Admitted as a freeman of Connecticut, for Southold, 1662.   In 1675, rated
for 2 heads, 21 acres, 16 cattle, 9 horses, 5 swine, 6 sheep, £228 10s.
(*Doc. Hist. N. Y.* ii. 259.)   In 1678, united in deed for common lands.
1679, deed to John Sweezy, for lot in brush meadow.   1683, John Carwine,
of Southold, is estimated to be worth £131.   (*Doc. Hist.* ii. 310.)   1686,
had 4 males, 1 female in his family.   1689, deed to John Paine, 1 acre
meadow at Goose Creek.   1691, deed to R. Howell, meadow in Aquebogue.
1695, deed to eldest son, John, Jr., 50 acres.   1696, deed to Simon
Grover, 10 acres, on Mill Pond Creek.   His name occurs in census list
1698, (*Doc. Hist. N. Y.* i. 450,) with names of all his children, excepting
Mary and Rebecca, who were already married.
  The following is his will :
  In ye name of God, Amen.   I, John Corwin, Senr., of Southold, in ye
County of Suffolk & Province of New Yorke, being aged, but in health of
body, & of a sound & pfect memory, doe make this my last will & testamt,
as followeth.   First, I bequeath my spirit to God that gave it & my body
to be decently buried in hope of ye resurrection again with Jesus Christ,
at ye last day, &c.   And for my worldly estate, after my just debts and
funerall charges paid, I doe bequeath as followeth :   Impmus, I give &
bequeath to my eldest son, John Corwin, one hundred acres of ye land
lying and being at Mattchucke, be it more or less, ye land of Jonathan
Reeve on ye West ; also one first lot of Meadow at ye deep Creeke & all
my meadow at ye other side of Peaconnucke river, and also a third lot
of upland agt ye Indian neld, and one first lot of upland on ye north side
of ye high way near ye fresh meadow, and a third lot of upland on ye
south side of hogge neck, and one lot of Creek thatch at ye Indian Necke ;
all & every part of upland & meadow as above recited I doe give, grant, &
bequeath unto my said son John Corwin, his heirs & assignes for ever, to
Have & to Hold in his & their own proper right for ever, &c., as also one
lot of Commonage to him as abovesd, &c.   2dly I give & bequeath unto
my second son Matthias Corwin my dwelling house, barne, & home lot,
also one & twenty acres of woodland on ye north side of S-hold, near ye
Cleft, and eighteen acres of upland at ye backside lots, and a third lot of
meadow eastward from John Budds, and a third lot of meadow at ye head
of ye Millpond, & a second lot of meadow in Pine Necke.   Also one lot of
Meadow Creek thatch in Indian Necke, & one lot of commonage, all which,
housing, barne, & lands, & meadows I give & bequeath unto my second
son Matthias Corwin, & to his heires & assigns for ever, to possess &
enjoy in his & their owne proper right for ever, &c.

3rdly. I give & bequeath to my youngest son, Samll Corwin, one hundred acres of upland, more or less, lying & being at Mattchucke, east from ye land of Jonathan Reeve, as also one first lot of meadow at ye deep Creek, and a second lot of Creek thatch at Indian Necke, together with one lot of commonage to him my sd son, Samll Corwin, to have & to hold to him, his heires, & assignes for ever, &c. As also I give one first lot layd out at ye Wading river, with halfe ye commonage belonging, yet undivided, &c.

4thly. I doe give & bequeath to my daughter Sarah, now ye wife of Jacob Osman, one cow, to be delivered by my executor herafter mentioned.

5thly. I doe give & bequeath to my daughter Rebecca, now ye wife of Abram Osman, one cow, to be delivered after my decease.

6thly. I give & bequeath to my daughter, Hannah Corwin, one young cow or cow kind. 7thly. I give and bequeath to my daughter, Abigail Corwin, one young cow or cow kind. 8thly. I give & bequeath to my grand child Sarah, ye daughter of Jacob Mapes, one cow, all wch is to be delivered after my decease by my executor hereafter mentioned, & my horse & my plow & plow irons I give to my son Samll Corwin, and all my sheep to be equally divived between my two youngest sons and all my daughters, equally, &c., and what carpenters tools I leave to be equally divided between my three sons &c. I give my oxen, & carte, & wheels, & one plow with irons, & ye bed I ly on to my son, Matthias Corwin, &c. And all ye rest of my moveable goods I give to my two youngest sons. Matthias Corwin & Samll Corwin, equally to be divided. Also I doe make & ordein my second son, Matthias Corwin, my sole executor to see this my last will & testamt duly pformed, and this I declare to be my last will & testmt, as witness my hand & seale this 26 day of November, 1700.

*JOHN CORWIN. (Seal.)

Signed & sealed as witnessed by us—
JOSHUA WELLS, JAMES REEVE, STEPHEN BAYLEY.

By ye tenour of these pesents know ye yt on ye 14th day of Octob. Anno Dni, 1702, at ye Mannr of St. Georges, in ye County of Suffolk, before Coll. William Smith, Judge of ye Prerogative Court, in ye sd county, was proved & approved ye last will & testamt of John Corwin, late of Southold, in ye sd county, deceased on ye 25th day of Sept., 1702, who by his sd will did nominate & appoint Matthias Corwin, his son, his sole executor, to whom was granted ye adminsn of all & singular ye goods, chattels, & credits of ye sd deceased.

(Hinman, in his *First Settlers of Connecticut*, makes this John to be probably son of George Curwen, (1,) of Salem, Mass., but wrongly.)

2 CAPTAIN JOHN,[3] (*John*, 1,) b. ——, 1663, d. December 13th, 1729.

    M. Sarah ————, before 1698, when her name occurs on census list, as wife of John. (*Doc. Hist. N. Y.* i. 450.)

* Once his name is spelled Curwinne. *Southold Records*, B, p. 41.

Ch. Benjamin, John, David, Sarah, Elizabeth, Hester.

In 1692, received of his father a lot of woodland lying west of the town, (Southold,) and on the north side of the road, by Nathaniel Terry's land. (*Book C*, 124.) Same year, received a second lot of land at Aquebock. 1695, another lot of upland, west of the town, and bounded west by Benjamin Moore's land, and east by Mr. Hubbard's land. 1696, deed from John and Jemima Paine for 50 acres; in 1702, the same number of acres was deeded back to Paine. 1702, deed from Jonathan Horton for land between Duck Pond and Inlet; same year, sold 25 acres of fresh meadows to Caleb Horton. 1704, deed from W. Downs for land on Peconic River. 1705, bought land of Jonathan Horton. (*Book C*, 248.) 1708, Theophilus Curwin sells land to John Curwin for five shillings, namely, $\frac{1}{2}$ part of a meadow at Aquebogue, on west side of Vail's Brook. (Samuel *Curwin*, a witness.) 1708, buys the homestead and 18 acres at Southold, of his brother Matthias, for £76, who also sold him a certain lot of land in Indian Neck for 50 shillings. 1711, deed from John Youngs, 2 lots in Queen's Road. 1712, deed to Benjamin Youngs, 50 acres.

In 1712, the following exhibit of John Corwin's land is found in Southold town records.

His right of commonage is a third lot.

(1.) House lot, 4 acres, bounded west by Jonathan Horton's land and east by Sarah Young's.

(2.) 18 acres of land, west by the lane that goes to Jonathan Horton's land, bounded east by Thomas Moore's.

(3.) 22 acres lying west of Jonathan Horton's, a north sea lot. Bought of John Budd and Sam. Youngs.

(4.) 21 acres on north side; east is John Peken's, west, John Hubbard's.

(5.) A third lot of meadow at Tom's Creek, John Budd's land on the south and Benjamin Young's (?) meadow north.

(6.) A parcel of meadow at Tom's Creek Head, by the beach.

(7) A second lot of meadow in Pine Neck, Mrs. Elton's land west, and Mr. Hubbard's north.

(8.) 1 acre of meadow on south side of Goose Neck.

(9.) His land on south side of Hog Neck, a third lot of 9 acres.

(10.) A second lot of land at Mattituck, butting on the highway.

(11.) His meadow on south side of river is 4 shares.

(12.) A second lot of meadow at Deep Creek, Aquebogue, Jeremiah Vail's land on the north.

3 JOHN,[4] (*John*, 2,) b. July 10th, 1705, d. December 22d, 1755.

    M. (1) Perhaps Hester Clark, according to a book in possession of Augustus Griffin, of Orient, L. I.; (2) Elizabeth Goldsmith, 1732, yet living, probably, in 1776.

    Ch. John, James, William, Elizabeth, Sarah. Several children of a John Corwin, who died young, perhaps belonged to this John, namely, a child died 1735, another died 1738, his second daughter died 1746, and Elnathan died January, 1738.

A widow, Elizabeth Corwin, joined the church of Aquebogue, August 26th, 1764. Possibly John's widow. The family lived about a mile and a half east of Mattituck, in the town of Southold. The following is his will:

In ye name of God, Amen. I, John Corwin, of Southold, in ye County of Suffolk and province of New-York, being weak in body but of perfect & sound mind and memory, blessed be God, and calling to mind ye uncertainty of this life, do make this my last will & testament in manner following. First, I recommend my soul to Almighty God that gave it, hoping through ye *pations* of my Savior to have full pardon & remission of all my sins; and my body to ye earth to be buried according to ye descretion of my execurs hereafter named; & as touching of the disposition of all my worldly goods & estate, I dispose of in the following manner & form. First: I give & bequeath unto my beloved wife, Elizabeth Corwin, one third part of all my lands, to improve, so long as she shall remain my widow, & no longer; and one third part of all my living stock, excepting one yoke of oxen, and one third part of all my household goods, to dispose of as she shall think fit. Item: I give, devise, & bequeath unto my son, John Corwin, my lot of land whereon I now dwell, to his heirs and assigns; likewise one half lot of land which I purchased of my brother, David Corwin, unto his heirs & assigns; condition: my sd son John shall pay unto my son William Corwin, the sum of one hundred & fifty pounds, York currency, when my son William should arrive to twenty one years of age, & fifty pounds to my two daughters, to be equally divided between them, my sd two daughters; provided, nevertheless, that in case my sd son William should happen to die, I will that one hundred pounds should be paid to my son James Corwin, & ye other fifty pounds to my two daughters, Elizabeth & Sarah Corwin, or to either of them that shall be then living, (if in case that either of them should happen to die;) but if so be that my son John Corwin should refuse to pay the above payments as abovementioned, I order & direct that my sd half lot of land purchased of my sd brother David Corwin, be sold, & the money disposed of as abovesaid. Item: I give & bequeath unto my son James Corwin, two hundred pounds out of my personal estate, to be paid him by my executor, hereafter named, immediately after my decease. Item: I give & bequeath unto my son John Corwin one yoke of oxen, one mare, and one third part of all my farming utensils. Item: I give unto my two sons, James & William Corwin, ye other two thirds of all my farming utensils. Item: I give & bequeath unto my two daughters, Elizabeth & Sarah Corwin, all the rest, residue, & remainder of my moveable personal estate, to be divided equally between them, share & share alike. Lastly, I ordain & appoint my beloved wife, Elizabeth Corwin, & my son, John Corwin, Executricks & Executor of this my last will & testament, to see the same performed according to my intent & meaning. In witness whereof I, the said John Corwin, have to this my last will & testament set my

hand & seal this eighteenth day of December, in the year of our Lord
Christ seventeen hundred and fifty-four.

JOHN CORWIN. (Seal.)

Witnesses :

 SAMUEL CORWIN, Jr.,

 OBADIAH HUDSON,

 PARKER WICKHAM.

All yeomen of Southold ; proved will 7 January, 1755, in Suffolk Co., on
26 February, 1755, will approved and letters granted to Elizabeth and
John Corwin. Lib. 19, p. 206. Records now in New-York City.

The inscription on his tombstone, yet standing, (1872,) a little south of
the centre of the Mattituck grave-yard, Long Island, is as follows :

"Here lyes Buried ye Body of Mr. John Corwin, Who died Decem. ye
22d, 1755, aged 50 years, 5 months, and 12 days."

(A widow, Elizabeth Corwin, married Benjamin Brown, of Oyster Ponds,
L. I., 1763. (Mattituck Church Records.) Possibly the widow of John,
3, or John, 4.)

4 JOHN,[4] (*Theophilus*, 2,) b. ——, d. June, 1740.

 M. (1) Mehetable Clark, 1725 ; she d. August 16th, 1739 ; (2) Elizabeth (Terrill ?) November, 1739.

 Ch. Amaziah, Mehetable, Jonathan. Perhaps some of the small children mentioned under John, 3, belong here. He was a weaver. His
will (Lib. 13, p. 21, *in archives N. Y. C.*) speaks of father Theophilus and brother Samuel. In 1704, received a deed from his father
for land in Aquebogue. In 1736, gives a deed to S. Hutchinson. His
name occurs on census list, 1698. (*Doc. Hist. N. Y.* i. 454.)

             (*Southold, L. I.*)

4½ JOHN,[5] (*Jedediah,* ½,) b. ——, 1737, d. October 9th, 1738.

5 JOHN,[6] (*John*, 3,) b. ——, 1735, d. December 22d, 1817.

 M. (1) Sarah Hubbard, March 20th, 1755. She was born 1731, d.
December 28th, 1763. Tombstone at Mattituck ; (2) Deborah Brown,
September 9th, 1767. She was born 1733, d. February 22d, 1823.

 Ch. John, James, Joseph, Hubbard, Isaac. He inherited the homestead of his father at Mattituck. Was an eminent "deacon" in the
church there.

5¼ JOHN,[6] (*Jedediah,* ¼ ?) b. ——, 1730–35 ?

 M. Mary ——.

 Ch. Isaac.   (*Mattituck, L. I.*)

6 JOHN,[6] (*Daniel*, 2,) b. ——, 1732, d. November 8th, 1815.

 M. (1) Elizabeth Wells, November 18th, 1755, at Cutchogue. She was
born 1735, and was daughter of Captain Daniel Wells. She died
April 13th, 1799 ; (2) Anna Wells, who was b. 1749, and d. August
21st, 1821.

 Ch. John, Nathaniel, Abel, Deliverance, Lina, Phile, (or File,) Daniel,
George, Sarah, Elsie, Polly.   (*Will in Suffolk Co., N. Y.*)

.7 John,⁶ (*Joshua*, 1,) b. December 25th, 1769, d. May, 1856.
  M. Julia Vail.
  Ch. Isaac, Elizabeth, John, Samuel, Julia, Abram, Esther, Benjamin H.,
    David, Archibald, Peter, William, Julia ?   (*Orange Co., N. Y.*)
  A John Corwin bought land in Orange Co., N. Y., in 1807. Perhaps
this one.

7½ *John,⁶ or ⁶, (*George*, 1,) b. ——, 1770–80.

8 John Howell,⁶ (*Eli*, 1,) b. July 17th, 1793, yet living.
  M. Cynthia, daughter of Daniel Wells, May 17th, 1817. She was b.
    March 6th, 1796, d. December 27th, 1867.
  Ch. Elizabeth W., Enoch B., Daniel W., Rev. Eli, John, Isaac L.,
    Howard, Edward P., Albert.
  He bought land in Orange Co., N. Y., in 1820. In 1823, bought land
in Camillus, Onondaga Co., N. Y., and in Middlesex, Ontario Co., N. Y.
                                (*Newburg, N. Y.*)

9 John,⁶ (*John*, 5,) b. February 13th, 1756, bapt. at Mattituck, d. July
  19th, 1816.
  M. Joanna Mapes, August 14th, 1775, who was b. August 20th, 1758,
    d. January 1st, 1850.
  Ch. Sarah, Elizabeth, John, Johanna, Lydia, Abigail, Bethia, Dency, '
    Parnel, Parnel, George W.
  He was known as the Major, of Mattituck, in the Revolution. The patriotism of his wife, in compelling a British officer to refrain from turning
his horses into her husband's wheat-field, upon the threat of instantly
shooting the first horse that entered, is narrated in *Griffin's Journal*, p. 185.
They both united with the church of Aquebogue, August 1st, 1790. In
1775, a John Corwin, (probably this one,) is found in Captain Lupton's
Company, and signed engagement to support Congress. May, 1776, appointed First Lieutenant in Captain Paul Reeves's Company of minute men.
  (In 1776, one John died, and a John is on census list, 363, having in
family 2 males, 2 females over 16, 3 children, and 1 slave.)

10 John,⁶ (*James* 1,) b. September 11th, 1782, d. October 11th, 1850.
  M. Elizabeth Wickham. She b. August 28th, 1788—living.
  Ch. Matthew, Benjamin W., Samuel H., Mahala, James H.
                                  (*Howells, N. Y.*)

11 John Calvin,⁶ (*William*, 2,) b. October 21st, 1768, (bapt. at Mattituck,
  April 2d, 1769,) d. June 6th, 1849.
  M. (1) Deborah Terry. She born December 27th, 1767, d. January
    30th, 1791 ; (2) Elizabeth Vance, of Mendham, Morris Co., N. J.
  Ch. Elias, Nathan, John B., William, Eliza, Sarah.
  In 1803, bought 111 acres of his father, in Roxbury, Morris Co., for
$550. He subsequently removed to Ithaca, N. Y., where his will is found.
In his will he directs that certain sums of money shall be paid to Theodore

               * Curran.

Corwin, and to George Corwin, to school them. The will seems to indicate that Elias, his son, had four children, George, Theodore, Deborah Sterns, ? and Hannah Condict? It refers to a grandson, John C., son of William S., and a daughter of the same, Elizabeth. Also to a grandson, Joseph. A Joseph Corwin was executor. The items of this will are thus recorded, as the writer could not harmonize them with other material.

12 John C.,[6] (*David*, 5,) b. March 5th, 1820. (An invalid, 1871.)

13 John,[6] (*John*, 6,) b. August 5th, 1757, d. April 30th, 1839.
   M. (1) Julia Hedges. She b. 1761, d. January 21st, 1823 ; (2) Penelope ——. She b. 1767, d. February 22d, 1854.
   Ch. John, Frederick, Charles, Ira, Hannah, Josiah, Robert, Parker, Sarah C., Benjamin F.
   He, and his wife, and son John, were all baptized at Aquebogue, December 26th, 1781. (*Baiting Hollow, L. I.*)

14 John,[6] (*Timothy*, 2,) b. ——, 1792 ? d. about 1818.
   M. Abigail Bloomer. She d. about 1864.
   Ch. Mary, William, Elmira, Thomas, Ann E. (*Balmville, N. Y.*)

15 John,[7] (*Abner*, 1,) b. ——, 1782.
   Unmarried.

16 John,[7] (*John*, 7,) b. June 5th, 1791, d. July 29th, 1862.
   M. Mary Knight, November 14th, 1816. She was b. February 23d, 1796.
   Ch. Alfred, Henry, Gabriel D., Samuel, Abram, Emeline, John J.

17 John,[7] (*James*, 8,) b. ——, 1800–10. (*Camillus, N. Y.*)

18 John W.,[7] (*Eli II.*, 2,) b. January 31st, 1811, d. December 29th, 1812.

19 John Eli,[7] (*Jason*, 1,) b. November 26th, 1820, d. March 19th, 1863.
   M. Martha J. Wallace, October 14th, 1843.
   Ch. Theodore, Elizabeth, Martin, Matilda, John E.
   (*Middletown, N. Y.*)

20 John M.,[7] (*Jemuel*, 1,) b. August 24th, 1830.
   M. —— Schaffer, 1858. (*Hopewell, Orange Co., N. Y.*)

21 John,[7] (*John II.*, 8,) b. August 4th, 1826.
   M. Harriet E. Finch, December 24th, 1850.
   Ch. Henrietta E., Emma F., Harriet A., Mary F.

22 John,[7] (*John*, 9,) b. September 16th, 1781, d. April 30th, 1859.
   M. Bethia Griffin, November 12th, 1809, at Aquebogue, L. I., who was b. 1791, d. October 8th, 1855.
   Ch. John, George, Annie, Frances, William, Hubbard, Hiram, Charles, Oliver, Bethia J.
   In 1831, at Aquebogue, were baptized of these children, Fanny, Hubbard, Hiram, Charles, and Oliver. (*Aquebogue Records.*)

23 John,[7] (*Isaac*, 1,) b. January 22d, 1787, d. December 22d, 1859.
   M. Elizabeth M. Bryant, 1802. She b. January 15th, 1785.
   Ch. Isaac. (*Near Chester, N. J.*)

24 John Brown,[7] (*John C.*, 11,) b. ——, 1790–1800.

25 John,[7] (*Benjamin*, 8,) b. ——, 1790–1800.
M. ——.
Ch. N. Benjamin, John L., Shepherd, Gardiner, Frederick V., Lutheria.

26 John,[7] (*Henry*, 5,) b. ——, 1823?
Unmarried. (*Riverhead, L. I.*)

27 John,[7] (*John*, 13,) b. ——, 1781, d. May 3d, 1840.
M. Harma ——. She b. 1791, d. July 7th, 1842.

28 John Bloomer,[7] (*Richard*, 2,) b. ——, 1810–20.
M. Jane Hardenbergh.
Ch. Rachel J., John R.

29 John,[7] (*James*, 7,) b. ——, 1810–30.

30 John,[7] (*Stephen*, 2,) b. in Essex Co., N. J., 1786.
M. (1) Elizabeth French, 1807. She b. 1790, in Essex Co., N. J., d.
—— , 1846 ; (2) Hannah Rapalje.
Ch. Ezra, Sarah, Rachel, Phebe, Noah, Miranda, Polly, Harvey, William, Amos, Lyman, Harriet, John, Lucelia ; by second wife, Mary,
Martha, James, Frank, David, Emily, Ida.
He moved from Essex Co., N. J., in 1814, to Starkey, Yates Co., N. Y.
In 1826, he moved to Jerusalem, in same county, and settled on lot 27 of
the Beddoe tract, (French survey,) his farm containing 118 acres. He
afterwards moved to lot 41, (Guernsey survey,) about three quarters of a
mile north of the last farm. He had twenty-one children. Was a carpenter.

31 John,[7] (*Stephen*, 4,) b. ——.

32 John,[7] (*Hubbard*, 1,) b. September 4th, 1790, in Tunbridge, Vt., d.
May 10th, 1864, of bilious fever.
M. (1) Hannah Brown, March 5th, 1812. She d. November 15th, 1824 ;
(2) Clarissa Thompson, of Tunbridge, Vt., March 16th, 1826.
Ch. William H., Louisa, Lydia, Jannette, Mary ; Caroline, Emeline,
John F., Maria, Mahala, Helen.
(*From 1816–38, at Tunbridge, Vt. ; Chelsea, Vt., till 1842 ; then again
Tunbridge, Vt.*)

33 John,[7] (*David*, 8,) b. ——.

34 John,[7] (*Benjamin*, 4,) b. ——, 1797? d. a young man.

35 John,[7] (*Richard*, 3,) b. November 28th, 1850.

36 John,[7] (*William*, 7,) b. August 21st, 1844, d. October 15th, 1849.

36½ John,[7] (*Edward*, 3,) b. January 5th, 1817.
M. J. Robbins.
Seven Ch.

37 John Totten,[8] (*David,*[8,]) b. ——, 1815–25.

38 John Jackson,[8] (*John*, 16,) b. August 30th, 1835, d. July 29th, 1862.

39 John,[8] (*Benjamin H.*, 10,) b. ——, 1840–50.

40 Jonn C.,⁸ (*Lewis W.*, 1,) b. ——, 1835–45.

    M. Anna Ocobock.   (*Seneca Falls, N. Y.*)

41 John Eli,⁸ (*John E.*, 19,) b. ——, 1850–5.

42 John Howard,⁸ (*Rev. Eli*, 4,) b. July 5th, 1852, at San José, Cal.

43 John James,⁸ (*Isaac L.*, 8,) b. March 28th, 1861.

44 Jonn,⁸ (*John*, 22,) b. June 19th, 1810.

    M. Hannah Sweezey.

    Ch. George.   (*Riverhead, L. I.*)

45 John Eli,⁸ (*Benjamin W.*, 11,) b. May 26th, 1839.

    M. Alvira Jane, daughter of Allan and Nancy Makepeace, of Chesterfield, N. H., September 19th, 1867. She was b. March 8th, 1848, at Chesterfield, Ind.

    Ch. Ella G., Allen W.

Has lived, since 1860, successively at Cincinnati, O., Indianapolis, Ind., Philadelphia, Pa., and now, (since 1869,) at *Anderson, Ind.*

46 Jonn,⁸ (*Elias*, 1,) b. ——, 1820–5.   (*Jackson, Mich.*)

47 John D.,⁸ (*Benjamin*, 7,) b. August 10th, 1826.

    M. Louisa Alpock.

    Ch. Mary E., Charles, Catherine A., Margaret, Cynthia L.

                                     (*Succasunna, N. J.*)

48 John A.,⁸ (*Moses B.*, 4,) b. ——, 1815–25.

Was elected one of the Judges of Supreme Court of Ohio, and served five years with distinction; also Attorney-General, Ohio.

49 Jonn,⁵ (*Daniel*, 16,) b. ——, d. ——, 1856.

• 50 John Lewis,⁸ (*John*, 25.)

51 John Stringham,⁸ (*Silas*, 7,) d. in New-Orleans.

52 John R.,⁸ (*Robert*, 1,) b. April 13th, 1832.

    M. (1) Martha T. Jones ; (2) Carrie B. Robins.

    Ch. Martha J., Julia J.

53 John Morrison,⁸ (*Nathaniel*, 7,) b. February 21st, 1821.

    M. Emily Clark, November 9th, 1847.

    Ch. Edward L., John W., Frank L., Carrie, William L.

54 John Howell,⁸ (*Henry* 10,) b. June 8th, 1812.

    M. Rachel ——.

    No ch.   (*Brooklyn, N. Y.*)

55 John Richard,⁸ (*John B.*, 28.)

56 John F.,⁸ (*Parker*, 1,) b. June 2d, 1831.

    M. Margaret Stone, April 1st, 1857.

    Ch. Lillian, Emma J., Franklin, Edna, Frederic.

                             (*Lewiston, Niagara Co., N. Y.*)

57 John Vincent,⁸ (*Jeremiah*, 3½,) b. March 16th, 1828, d. Dec., 1871.

    M. Mary Electa Marsh, November 9th, 1850.

    Ch. Mary E., John V.

(Expressman; home in *New-York City, and Jersey City after* 1863.)

58 John,[8] (*John*, 30,) b. ——, 1832.
    M. Olive Tinney.
    Ch. Flora, Libbie, Ebenezer.
    Living N. W. of Penn Yan.   (*Livingston or Ontario Co., N. Y.*)

59 John K.,[8] (*Jesse*, 7.)
    A lawyer.   (*Cleveland, Ohio.*)

60 John Amos Russell,[8] (*Russell*, 1,) b. in Lowell, Mass., May 6th, 1831.
    M. Fanny J. Hatch, of Chelsea, Vt.
    Ch. Charles R., Harry J., Ellen M., Ernest A., Venton A., Flora M,

61 John W.,[8] (*William*, 23,) b. April 10th, 1858.

62 John Francis,[8] (*John*, 32,) b. June 25th, 1831.
    M. Esther Hunt, December 31st, 1857.
    Ch. Frank J., Frederic S.   (*Derry, N. H.*)

62½ John Abbott,[9] (*Major F.*, 1,) b. August 18th, 1862.

62¾ John E.,[9] (*Barnabas*, 6,) b. May 19th, 1841.
    M. Sarah F. Horton, January 4th, 1863.

63 John W.,[9] (*John*, 5,) b. September 11th, 1856.

63½ John,[9] (*Charles N.*, 15½,) b. January 12th, 1859.

64 John,[9] (*Noah*, 3.)

65 John,[9] (*William*, 39,) b. ——, 1855-65.

66 John,[9] (*Ezra*, 3,) b. ——, 1855-60.

67 John Henry,[9] (*Alva*, 1,) b. August 31st, 1832.
    M. Phebe Corwin, daughter of H. Conckling Corwin.
    Ch. Charles Henry, William Herbert, Jenny.

68 John,[9] (*George*, 30,) b. ——.
    (A John H. Corwin, and Elizabeth, his wife, bought land of Jesse Corwin about 1840-5, in Chemung Co., N. Y. They also sold land to Temperance Corwin. John H. Corwin was then of Liberty, Tioga Co., Pa. (Records at Elmira, N. Y., vol. xxviii. p. 242.) There was a John Corwin, 1857-64, in Cayuga Co., N. Y. Also a John Corwin, and wife Melissa, were of Lansing, Tompkins Co., N. Y., 1852-68, and bought land in Genoa, Cayuga Co., N. Y.)

68¼ John Vincent,[9] (*John V.*, 57,) b. July 28th, 1854, d. October 5th 1855.

68½ John Osman,[4] (*Rebecca Corwin*, 1,) b. before 1698.

68¾ John Case,[4] (*Theophilus Case*, 6,) b. ——, 1718, d. ——, 1775.
    M. Jemima Hulse, 1733.   (*See Moore's Index of Southold*, p. 64.)

69 John A. Little,[7] (*Desire Corwin*, 1,) b. May, 1823.
    M. Jane Stearns, September, 1849.
    Ch. Adell, Charles E., Ida.   (*Onondaga Co., N. Y.*)

70 John C. Taylor,[7] (*Fanny Corwin*, 1,) b. May, 1818.

M. Melinda Lampman, 1836.

Ch. Calista, George, Orlando, David.   (*Mich.*)

71 John Little,[7] (*Abigail Corwin*, 3,) b. March 1st, 1815.

M. Mary Ann Young.   .   (*Middletown, N. Y.*)

72 John W. Millspaugh,[7] (*Cynthia Corwin*, 1,) b. April 9th, 1816.

M. Harriet A. Armstrong, June 25th, 1843.

Ch. Albert E., Hamilton A., L. Ermina, Alice J.

(*Booneville, Ind.; afterward Mount Pleasant, Iowa.*)

73 John H. Millspaugh,[7] (*Dorothy Corwin*, 2,) b. May 29th, 1817.

M. Mary Ann Wells, February 23d, 1841.

Ch. Samuel W., Mary L., Julia.   (*Middletown, N. Y.*)

74 John Terry,[7] (*Keturah Corwin*, 1,) b. ——, 1780–90.

75 John H. Brown,[8] (*Silas C. Brown*, 16.)

76 John Bowers,[8] (*Elmira Corwin*, 1,) b. February 1st, 1831.

(*Cornwall, Orange Co., N. Y.*)

77 John C. Little,[8] (*George C. Little*, 53,) b. March, 1843.

78 John Wright,[8] (*Elizabeth Corwin*, 16,) b. ——, 1820–30.

79 John Flannery,[8] (*Mary Corwin*, 33,) b. New-York City, June 28th, 1845, d. May 15th, 1846.

80 John Taylor,[8] (*David J. Taylor*, 30,) b. February, 1840.

M. Harriet Reed, December, 1863.   (*Savannah, N. Y.*)

81 John Randolph Millspaugh,[8] (*Gideon H. Millspaugh*, 1,) b. May 28th, 1837.

82 John Brown,[8] (*Eliza Corwin*, 4,) b. ——, 1830–5.

83 John Albert Howell,[8] (*Susan Corwin*, 1,) b. ——, 1823.

84 John Reed,[8] (*Sophia Corwin*, 1,) b. November 28th, 1799, d. April 7th, 1814.

85 John E. Beebe,[8] (*Azubah Corwin*, 2.)

86 John Terry,[8] (*Julia Ann Corwin*, 5.)

87 John Ketcham,[8] (*Sophia Corwin*, 1.)

88 John Girard,[8] (*Jerusha Corwin*, 2.)

89 John F. Odell,[8] (*Bethia Corwin*, 4.)   •

90 John Swina,[8] (*Bethia Hobart*, 11,) b. February, 1840.

91 John Leek,[8] (*Deborah Corwin*, 4,) b. ——, 1815–25.

91½ John Clark,[8] (*Nathaniel C. Wells*, 10.)   (*Cold Spring, N. Y.*)

92 John C. Smith,[8] (*Sarah Corwin*, 28,) b. in Tunbridge, Vt., 1818.

M. Catharine C. Winne, 1844.

Ch. Emily, Salina, Josiah.   (*St. Lawrence Co., N. Y.*)

93 John M. Meeker,[8] (*Harriet F. Corwin*, 4,) b. February 20th, 1862, d. June 25th, 1862.

94 John Osborn,[8] (*Amelia Corwin*, 2,) b. May 13th, 1830, d. August 26th, 1832.

95 John C. Dunlevy,[8] (*Lucinda Corwin*, 4,) b. October 6th, 1822.

Was Probate Judge in Warren Co., Ohio. Appointed Assessor of internal revenue by President Lincoln, but resigned under President Johnson. He graduated at Denison University, Granville, Ohio. Now practicing law in Chicago, Illinois.

96 John Galloway,[8] (*Amanda Corwin*, 2,) b. ——.

97 John Thatcher,[8] (*Jesse Thatcher*, 10,) b. March 16th, 1815.

98 John L. Flannery,[8] (*Mary Corwin*, 33,) b. at Milltown, N. Y., April 20th, 1847.   (*Sewing-Machine Agent, Chicago, Ill.*)

99 John Mongown,[8] (*Elma Corwin*, 1,) b. ——, 1839, d. December 4th, 1858.

100 John Spencer Hackett,[8] (*Mary Corwin*, 14,) b. June 3d, 1817.

M. Sarah A. Noyes, March 7th, 1843.   (*Vermont.*)

101 John Edward Meeker,[8] (*Harriet F. Corwin*, 4,) b. August 8th, 1867.

102 John Drake,[8] (*Mary Corwin*, 20.)   (*Lebanon, Ohio.*)

103 John North,[9] (*Fitz-Alan North*, 2,) b. October 20th, 1860.

104 John H. Green,[9] (*Bethia Corwin*, 7,) b. and d. 1855.

105 John F. Chamberlin,[9] (*Jennette Corwin*, 1,) b. September 9th, 1850.
                    (*Vermont.*)

106 John A. Boyd,[9] (*Emeline Corwin*, 7,) b. October 28th, 1860.
                    (*Vermont.*)

107 John B. Sears,[9] (*Eliza N. Colgrove*, $26\frac{1}{2}$,) b. about 1855.

(1) * John,[2] [*George*, (1),] b. in Salem, Mass., July 28th, 1638, d. July 12th, 1683.

M. Margaret Winthrop, May, 1665, daughter of Governor John Winthrop,† Jr., of Connecticut. She d. September 28th, 1697.

Ch. George, Elizabeth, Lucy, Hannah, Samuel.

He was a merchant, and a deputy from Salem to the General Court; made a freeman, May 3d, 1665. He enjoyed by gift from Hugh Peters his houses and lands in Salem, Mass.

(2) ‡ John,[5] [*Bartholomew*, (1),] b. February 26th, 1722, d. September 26th, 1744.

(3) ‡ John,[6] [*George*, (6),] b. September 24th, 1744, d. July 3d, 1815.

* Curwin.

† John Winthrop's mother was daughter of Col. Read, of the English army. After Col. Read's death, his widow married Hugh Peters. She had no children by her second husband, but three by her first. One of these, Elizabeth, became the wife of John Winthrop Jr., and their daughter, Margaret, became the wife of John Curwin, as above. Mr. Read was attracted to England by the conflicts between Charles I. and his parliament, taking part in the great civil wars, in which he was colonel of a regiment. Fitz-John Winthrop, another of John Curwin's maternal uncles, was a captain in Read's Regiment.

‡ Corwine.

M. Rebecca Stillwell, 1768. She was b. February 5th, 1744, d. November 30th, 1807.

Ch. Phebe, George, Abigail, Alice, Jehu, Rebecca, Gideon R.

(Amwell, N. J.)

Several early letters to this John are still preserved by his granddaughter Phebe—one from Ruth Corwine, (John's sister,) of Mason Co., Ky., 1800 ; two from a John in Kentucky, 1804, 1806 ; and one from his son Jehu, dated Baltimore, Md., 1806. A deed is also preserved, bearing date, 1746, for land in Hunterdon Co., adjoining lands of Joseph Hixon and Richard Reed, to George Corwin, No. 6.

(4) * John,[6] (——,) b. May 23d, 1755.

M.

Ch. Mary A., William.   (*Baltimore, Md.*)

This John is said to have been a great-grandson of Bartholomew, (1,) but the intervening links have not been obtained.

(5) * John,[7] [*Richard*, (2),] b. ——, 1780–90, d. young.

(6) * John,[3] [*Jonathan*, (1),] b. July 9th, 1684, d. September 10th, 1684.

(7) *John,[7] [*Amos*, (1),] b. ——, 1790–1800.

(8) * John,[8] [*Gideon, R.*, (11),] b. ——, 1823, d. ——, 1825.

(9) * John Edgar,[8] [*Joab*, (1),] b. ——.

(10) * John,[8] [*Gideon R.*, (1),] b. ——, 1826, d. ——, 1830.

(10½) * John H.,[8] [*Amos*, (1½),] b. ——.   (*Illinois.*)

(11) * John W.,[8] [*Samuel*, (6½),] b. March 13th, 1822.

M. Margaret R., adopted daughter of James and Keziah Davis, February 7th, 1849.

Ch. James, Mary W., Keziah D., Rachel M.

Mr. Davis was one of the most prominent business men in the State, a large dealer in stock. He was a pioneer in the Scioto Valley. He possessed great executive ability. He died, 1854, leaving a large fortune, mostly to his wife, who lives near Waverley, Ohio.

(12) * John,[9] [*Richard M.*, (6).]

(13) * John,[9] [*George*, (10),] b. May 3d, 1845, d. March 1st, 1864, at Princeton, N. J.

He graduated at the High-School of Pennington, N. J., and died in the second year of his course in the College of New-Jersey.

(14) John Bellos,[7] [*Alice Corwine*, (1).]

(15) John T. Sharp,[9] [*Sarah Corwine*, (9),] b. ——.   (*Kingston, Ohio.*)

(A) John, (——,) b. ——, 1780–90.

M.

Ch. Malcolm.

John lived in London, England.

* Corwine.

(*a*) * JOHN,[2] [*Jonathan*, (a),] b. in England, about 1750, d. ——, 1825.
Approving of the principles of the American Revolution, he emigrated
from Keswick, England, to America, in 1784.
M.
Ch. Joseph, John, Jonathan, George, and a daughter who married a
Mr. Nicholson, and remained in England.

(*b*) * JOHN,[3] [*John*, (a),] b. ——, 1786.
M.
Ch.
He bought land and laid out the town of Curwinville, Clearfield Co.,
Pa., but the enterprise was afterward abandoned.

(*c*) * JOHN, M.D.,[4] [*George*, (a).] b. in Lower Marion township, Montgomery
Co., Pa., September 20th, 1821.
M.
Ch.
Physician and Superintendent of Lunatic Asylum, Harrisburg, Pa.,
since about 1850. He was graduated at Yale College, 1841; received his
degree of M.D. from the University of Pennsylvania, 1844, and of LL.D.
from Jefferson College, Pa., 1862.

Jonah, (Hebrew), a dove.

1 Jonah Campbell,[7] (*Sarah Corwin*, 6,) b. November 5th, 1816.
M. Melinda Bogardus, February, 1843.
Ch. Homer, Henry, Josephine.   (*Syracuse, N. Y.*)
2 Jonah C. Chappel,[3] (*Charlotte Hobart*, 9,) b. December ——, 1849.

Jonathan, (Hebrew,) gift of Jehovah.

1 JONATHAN,[5] (*Theophilus*, 4? *or Timothy*, 1?) b. December ——, 1721,
d. April 11th, 1798.
M. Rachel Howell, 1748.   She b. ——, 1729, d. May 14th, 1785.
Ch. Jonathan, Asa, Rachel, Selah, Jason, Hannah, John? Isaac,
Richard, Elizabeth.
His will is at Riverhead, (*Lib. A*, p. 528.) In 1776, a Jonathan is on
census list, 306, having in family, 1 male over 50, 2 males and 2 females
over 16. A Jonathan Corwin married Experience Howell, December 23d,
1784. Possibly a second wife of this one. (*Aquebogue Recs.*)
1½ JONATHAN,[5] (*John*, 4?) b. ——, d. ——, 1793?
M. ? Elizabeth Corwin, May 26th, 1774.
An infant daughter died September 11th, 1775.
(These names are Jonathan Corwin, Jr., married Elizabeth Corwin.
But the Jr. does not always imply in these records that the father's name
was the same.) (*Aquebogue Recs.*)
2 JONATHAN,[6]? (*Jonathan*, 1,) bapt. at Mattituck, January 28th, 1754, d.
January 4th, 1785.
                              * Curwen.
9

M. Hannah Reese, who afterward married John Howell, on June 19th, 1794. (*Aquebogue Recs.*)

Ch. Jason, Hannah, Rebecca.

In 1775, a Jonathan Corwin was in Captain Wallack's Company, and signed engagements to support Congress. In 1776, on census list, 100, having in family, 1 male, 2 females over 16, and no young children. In 1776–77, there was a Jonathan at Norwich, Ct., who had served in two campaigns. Probably Jonathan, 2. A widow, Hannah Corwin, was baptized at Aquebogue, October 30th, 1785, with her three children, Hannah, Jason, and Rebecca. The writer has also a memorandum that a Jonathan Corwin married an Abigail Howell in 1748, but he has been unable to locate him.

3 JONATHAN,[7] (*Asa*, 2,) bapt. at Aquebogue, May 14th, 1786.

The Ontario Co., N. Y., records show a Jonathan Corwin and Mary his wife, of the town of Lyon, selling land in 1816 to Griffin Hazard. William R. Corwin a witness. Possibly this was Jonathan, 3.

4 Jonathan Brown,[7] (*Mary Corwin*, 8,) b. ——, 1800 ?
    Unm.

5 Jonathan Meyers,[8] (*Arminda Brown*, 2,) b. ——.

(1) JONATHAN,[2] [*George*, (1),] b. at Salem, Mass., November 14th, 1640 d. July 9th, 1718.

 M. Elizabeth Gibbs, of Boston, March 20th, 1675–6, widow of Robert Gibbs, (who was son of Sir Henry Gibbs,) and daughter of Jacob and Margaret (Webb) Sheaf, of Boston. She d. August 19th, 1718.

 Ch. Elizabeth, Margaret, Sarah, Jonathan, George, John, Margaret, Anna, Jonathan, Harbert.

He was conspicuous for energy and influence, holding high judicial and political positions, and being all his life clothed with public trusts. In the Massachusetts civil list we find that Jonathan Corwin was a member of the Provincial Council of Massachusetts (having also been named such in the new charter) from 1692–1714. (He had been deputy to the General Court in 1682 and 1689.) The following extract from the Salem records shows much confidence in him, and, at the same time, the action of that town toward reëstablishing a government under the venerable Bradstreet, in opposition to the tyranny of Sir Edmund Andros :

" May 7th, 1689. Captain John Prince and Mr. Jonathan Corwin were chosen to assist in the Council at Boston, to be held on the 9th inst., and we desire that the honorable the governor, the magistrates, and deputies chosen in 1686, would (having always due respect to our dependence on the Crown of England, and the obligations we are under by the late declaration before the surrender of the last government) reassume our charter government, by taking their places and forming a General Court as soon as possible ; unto which we shall readily and cheerfully subject ourselves, and be always assisting to the utmost of our power, with our lives and estates, as formerly."

In 1701, he was negatived, but accepted by the governor. He was one of the judges of the Special Court of Oyer and Terminer, having taken the place of Richard Saltonstall, who resigned at the trial of the supposed witchcraft cases. This appointment was made June 13th, 1692, "to hear and determine all manner of crimes and offenses perpetrated within the counties of Suffolk, Essex, and Middlesex, Mass." He was judge of the Superior Court of Common Pleas of Essex from 1692–1708. He was Judge of Probate from June 3d, 1698–1702. He was appointed one of the Commissioners of Excise and Impost, June 24th, 1692. On November 22d, 1703, he was appointed, with others, for the trial of an Indian, at Salem, Mass. On February 20th, 1707-8, he was appointed justice in the Superior Court of Judicature, (caused by the resignation of Judge John Leverett, who was elected to the Presidency of Harvard College,) in connection with Winthrop, Sewall, Walley, and Hathorne. These appointments were made by the Governor and Council. He is referred to as Jona., in *N. E. Hist. Reg.* iv. 128.

He resided, after 1674–5, in the house which is still standing on the south-west corner of Essex and North streets, Salem, which he purchased of the administrators of the estate of Captain Richard Davenport, of Boston, and which the research of William P. Upham, Esq., shows to have been the identical house originally owned and occupied by Roger Williams, in the rooms of which the whole power of the Colony was employed to suppress this champion of religious liberty, before the persecutions to which he was subjected drove him to Rhode Island.

(2) * JONATHAN,[4] [*Rev. George*, (4),] b. May 6th, 1713, d. November 6th, 1718.

(3) JONATHAN,[3] [*Jonathan*, (1),] b. October 2d, 1681, d. August 12th 1682.

(4) JONATHAN,[3] [*Jonathan*, (1),] b. September 15th, 1689, d. December 25th, 1689.

(a) JONATHAN CURWEN was born in England about 1720. Was of Little Broughton, Cumberland Co., England. Lived and died in England. Had a son John, who came to America in 1784. See John, (a.)

(b) * JONATHAN,[3] [*John*, (b),] b. ——, 1788.

Joseph, (Hebrew,) he shall add.

1 JOSEPH,[5] (*David*, 2.) b. ——, 1730–40, d. ——, 1803.
    M. Anna Wells.
    Ch. William, Rev. Joseph, Mary, Sarah, Deborah.

In 1776, a Joseph on census list 334, having in family 1 male, 1 female over 16, and 4 children.

2 JOSEPH,[5] ? (——.)
    M. Zeruiah Case, April, 1739.

---

* Curwen. Jonathan (1) generally or always wrote Corwin.

3 Joseph,[5]? (*John*, 5,) bapt. at Mattituck, September 9th, 1759, d. July, 14th, 1793.

4 Joseph,[6] (*Joshua*, 1,) b. March 21st, 1765, d. May 12th, 1852.
    M. Jemima Ketchum, who was b. March 15th, 1772, d. April 20th 1852.
      Ch. Anna, Elizabeth, Marion, Naomi, Sarah, Prudence, Esther, Jemima, Charles S.

5 Rev. Joseph,[6] (*Joseph*, 1.) b. ——, 1768, d. January 29th, 1811.
    M. Mary Sweezy, b. ——, 1796, d. September 7th, 1846.
      Ch. Shubal, Joseph, David, Jesse, Hiram, Richard S., Mary, Bethia, Eleanor, Ann.
  He was never settled over a church.   (*Riverhead, L. I.*)

6 Joseph,[6] (*Eli*, 1,) b. August 24th, 1792, d. August 27th, 1792.

7 Joseph,[6] (*William*, 2,) b. July 6th, 1781, d. September 23d, 1800, in Chester, N. J.

8 Joseph,[6] (*Benjamin*, 2 ?) b. ——, 1750 ? d. ——, 1823.
    M.
      Ch. Sophia, Peter, Nathaniel, Margaret.
  He had land on Black River, near Chester, N. J., as early as 1776.
        (*Flanders, N. J.*)

9 Joseph,[6] (*David*, 4,) b. ——, 1781, d. May 4th, 1832.
    M. Hannah Finch.
      Ch. Gabriel, Eliza, Mary, Sarah J., Julia.

10 Joseph,[6] (*Jesse*, 2,) b. October 9th, 1771, d. ——, 1835, in Ohio.
    M. Susannah Wickham, in Kentucky.
      Ch. James, Matthias, Richard, Mary, Ichabod, Joseph, Keziah, William, Milton, Benjamin.
Moved from Pennsylvania to Kentucky in 1785, and to Lebanon, Ohio, in 1798. Was fleshy, eccentric, jovial, of fair skin, blue eyes, light hair.

10¼ Joseph R.,[6] (*Gilbert*, 1,) b. August 5th, 1781, at Haverstraw, N. Y.
    M. Phebe Johnston, July 16th, 1802. She b. July 14th, 1784, d. September 19th, 1834.
supplement   *Ch.* Gilbert, Jane, Elijah W., Elizabeth, Amy, Charity, William P., Harriet, John W., Abram, Joseph.   (*Quincy, Ill.*, 1828.)

10½ Joseph R.,[7] (*Seth*, 1,) b. November 19th, 1814.
    M. Antise N. White.
      Ch. Sarah.   (*San Francisco, Cal.*)

11 Joseph,[7] (*Benjamin*, 4,) b. ——, 1770–90.
    M. Jane Case.   (*Morris Co., N. J., and Ill.*)

12 Joseph William,[7] (*William*, 4,) b. June 8th, 1816 or 1817.
    M. Ann Maria Wells, January 4th, 1841.
      Ch. Epenetus L., Rose, James B., Frances A., William M.
        (30 *S. Tenth Street, Brooklyn, N. Y.*)

Joseph Thomson 27 (Mary Corwin 25)
  born July 21, 1838
  Married Nov. 4, 1867
    Jane L. Amidon, daughter of
    Francis H. Amidon of New York City -
  issue –
6 Josephine Thomson, born Sept 1. 1870
1 Amidon Thomson born Oct 30. 1878

13 JOSEPH,[7] (*Rev. Joseph*, 5,) b. ——, 1788, d. November 28th, 1811.

14 JOSEPH,[7] (*James*, 4,) b. ——, 1799, d. April 24th, 1839.
A Joseph, of Riverhead, L. I., d. 1840, aged 40. ·This one?

15 JOSEPH,[7] (*Isaac*, 1,) b. November 10th, 1795.
M. Mary Hopkins? *Two Ch.*   (*Michigan.*)

16 JOSEPH A.,[7] (*William*, 6,) b. in Sparta, N. J., May 17th, 1810.
M. (1) Tarquinia Kenney; (2) Emma Whybrew Baldwin, September
18th, 1856. She was b. July 29th, 1831.
Ch. Francis N. W., William A., Charles F., Mary G., Theodore W.,
Robert L.   (*A physician, Newark, N. J.*)
He received his degree of M. D. from Yale College, 1835.

17 JOSEPH,[7] (*Benjamin*, 8,) b. ——, 1790–1800. *Unm.*

17½ JOSEPH,[7] (*Joseph*, 10.)

18 JOSEPH,[7] (*Joseph*, 10,) b. ——.

18½ JOSEPH,[7] (*Joseph*, 10½,) b. 1810–20.   (*St. Louis, Mo.*)

19 JOSEPH HORTON,[8] (*Shubal*, 1,) b. October 9th, 1815.
M. Electa Edwards.
Ch. Joseph E., Oliver C.

20 JOSEPH,[8] (*Elias*, 1,) b. ——, 1824, d. April 10th, 1859.
M. Louisa F. Langstaff.
Ch. Harriet E., Louisa A., Josephine, Mary L., Walter F.
                                        (*Ithaca, N. Y.*)

21 JOSEPH,[8] (*William H. Harrison*, 18,) b. March 23d, 1839, d. February
4th, 1863.

22 JOSEPH,[6] (*Peter*, 3,) b. ——, 1810–20.
Unm.   (*Morris Co., N. J.*)

23 JOSEPH SYDNEY,[8] (*Alsop H.*, 2.)

23½ JOSEPH WALTER,[6] (*David*, 12,) b. April 6th, 1824, drowned November
28th, 1844.

24 JOSEPH,[9] (*George W.*, 28.)
There was a Joseph C. Corwin and Polly his wife, 1822–37, in Sterling,
Cayuga Co., N. Y. In 1845, they were of Hannibal, Oswego Co., N. Y.
A Joseph C. Corwin bought land in 1868, in Ira, Cayuga Co., N. Y. A
Joseph Corwin, Jr., bought land in Orange Co., N. Y., in 1810.

25 JOSEPH EGBERT,[9] (*Joseph*, 19,) b. December, 1845.

26 Joseph Osman,[4] (*Rebecca Corwin*, 1,) b. before 1698.

27 Joseph Thompson,[8] (*Mary Corwin*, 25,) b. July 29th, 1838. *m. Jane L. An*

28 Joseph Woodruff,[8] (*Silas Woodruff*, 15,) b. ——, d. young.

29 Joseph North,[8] (*Mehetable Campbell*, 10,) b. July 28th, 1831.

30 Joseph Wells,[8] (*Lydia Corwin*, 2,) b. ——, 1805–20.

31 Joseph Reed,[6] (*Sophia Corwin*, 1,) b. March 1st, 1804.

32 Joseph Dill,[8] (*Phebe Corwin*, 6.)   (*Lebanon, Ohio.*)

33 Joseph Thatcher,[7] (*Jemima Corwin*, 1,) b. ——, 1788 ? in Warren Co.,
       Ohio, d. in Vevay, (near Cincinnati, Ohio.) Was a soldier in war of
       1812.

34 Joseph Thatcher,[8] (*Jesse Thatcher*, 10,) b. April 6th, 1812, d. November
       9th, 1893.

35 Joseph Thomson (Cookingham (Nettie Thomson 2)6 March 31.189

   (1) † Joseph,[5] [*Bartholomew*, (1),] b. November 24th, 1724.
       M. ——.
       Ch. Naomi, Keziah.
       This family removed to Canada about the close of the Revolution.

   (a) * Joseph,[2] [*John*, (a),] b. about 1783, in England.
       Was brought to America by his father, (John,) in 1784. Joseph left
   one daughter.

Josephine, feminine of Joseph.

1 Josephine,[7] (*David*, 10,) b. October 25th, 1846, d. April 8th, 1848.

2 Josephine,[9] (*Joseph*, 20,) b. September 12th, 1852, at Ithaca, N. Y.
                                                   (*Elmira, N. Y.*)

3 Josephine Campbell,[8] (*Jonah Campbell*, 1,) b. ——, 1835–40.

4 Josephine Losee,[9] (*Ann Kate Corwin*, 21.)

5 Josephine Blinn,[9] (*George D. Blinn*, 56,) b. April 30th, 1863.

6 Josephine Thomson (Joseph Thomson 2/) b. Sept 1. 1870

Joshua, (Hebrew,) Jehovah Savior.

1 Joshua,[5] (*David*, 2,) b. on Long Island, March 25th, 1735, or March
       26th, 1733 ? d. July 6th, 1812.
       M. (1) Anna Paine, May 1st, 1755, at Southold. She b. September
       6th, 1733, d. April 4th, 1781 ; (2) Rhoda Davis, widow of ——
       Emerson.
       Ch. Joshua, Peter, David, Abner, Annie, Joseph, John, Jemima, Ben-
   jamin.
       In 1775, signed engagement to support Congress. In 1776, on census
   list, 331, having in family 2 males, 1 female over 16, and 5 children. He
   removed to Orange Co., N. Y., during or soon after the Revolution. He
   lived near Mt. Hope.

2 Joshua,[6] (*Joshua*, 1,) b. March 6th, 1756, d. ——.
       M. Mary Corwin, T. 3
       Ch. Alsop, Mary, Jemima.   (*Mount Hope, Orange Co., N. Y.*)

3 Joshua Goldsmith,[6] (*William*, 2,) b. February 4th, 1793, d. November
       9th, 1867.
       M. Elizabeth Fordham, daughter of Rev. Lenas Fordham.
       Ch. Mary A., Lemuel F., William, Harriet F.   (*Succasunna, N. J.*)

4 Joshua,[6] (*Jedediah*, 2,) b. ——, 1760–80, d. ——.

---

* Curwen.   † Corwine.

M. (1) Levina Reeve, March 5th, 1803. (*Aquebogue Recs.*) She was
b. 1779, d. December 26th, 1809 ; (2) —— Philips.
Ch. Joshua, Chauncey. (*Southampton, L. I.*)

5 Joshua,[6] (*Thomas*, 1,) b. ——, 1797, d. March 15th, 1867, in Brooklyn,
N. Y.

6 Joshua,[7] (*Joshua*, 2,) b. October 19th, 1785—living.
M. (1) Priscilla Mapes, October 27th, 1807. She b. November 23d,
1788, d. December 4th, 1842 ; (2) Elizabeth Smith. She b. January
2d, 1802.
Ch. Polly M., Selah R., Fanny J., David S., Joshua II., Priscilla E.,
Silas G., Alsop L., Eri C., Benjamin L., Mary E., Naomi A., Sarah E.
(*Mount Hope, N. Y.*)

7 Joshua,[7] (*Abner*, 1,) b. ——, 1784, d. ——, 1860.
M. Parmeley Hawkins.
Ch.

8 Joshua,[7] (*Joshua*, 4,) b. about 1805. (*Southampton, L. I.*)

8½ Joshua Clark,[7] (*Asa*, 3,) b. February 14th, ~~1788,~~ at Aquebogue, L. I.
M. Nancy Little. *1798*
Ch. Charles W., Sarah J., Mary, Caroline, Adelaide.
(There was a Joshua C. Corwin, and Sally his wife, of Hector, Tomp-
kins Co., N. Y., 1825, 1831, 1834, when he deals in land.)

9 Joshua Harvey,[8] (*Joshua*, 6,) b. April 20th, 1815, d. August 29th,
1861.
M. Annesta McCord, March 31st, 1852, in Newark, N. J.
Was a soldier in the Mexican war. Was killed in firing a salute, (at
the opening of the Rebellion,) on Willard's Point. Sixty-Fifth Chasseurs.
(*Newark, N. J.*)

10 Joshua,[8] (*Alsop*, 1,) b. ——, 1810–20. (*Otisville, N. Y.*)

11 Joshua,[8] (*Daniel*, 10,) b. October 30th, 1805, d. February 10th, 1871.
Buried at Upper Aquebogue.

11¼ Joshua,[8] (*Gilbert*, 4,) b. February 29th, 1857, d. May 2d, 1857.

12 Joshua Case,[6] (*Samuel Case*, 20.)
M. ? Deliverance Wells, 1729. (*Moore's Index*, p. 64.)

13 Joshua Skidmore,[8] (*Mary Corwin*, 28,) b. ——, 1810–20.

Josiah, (Hebrew,) whom Jehovah heals.

1 Josiah,[7] (*John*, 13,) b. ——, 1785–95, d. ——, 1831.
(*Brookhaven, L. I.*)

2 Josiah F.,[8] (*Robert*, 1.)

3 Josiah,[8] (*Thomas*, 5,) b. ——, 1856.

4 Josiah Terry,[7] (*Keturah Corwin*, 1,) b. ——, 1780–90.

5 Josiah Smith,[9] (*John C. Smith*, 92,) b. ——, 1845–50.
(*St. Lawrence Co., N. Y.*)

(1) Josiah Wolcot,[3] [*Penelope Curwen*, (1),] b. December 21st, 1690, d. January 4th, 1691.

1 Judson A. Sperry,[8] (*Mary Corwin*, 40,) b. June 27th, 1835.
M. Josephine Van Valkenbergh.

2 Judson Adoniram Woodruff,[8] (*Erastus Woodruff*, 1,) b. ——, 1825–35.

3 Judson L. Smith,[9] (*George Smith*, 74,) b. ——, 1855–65.

(*St. Lawrence Co., N. Y.*)

Julia, (Greek,) soft-haired.

1 JULIA,[7] (*Joseph*, 9,) b. ——, 1814.
M. Ab. Hunsike.
Ch. Helen, Henrietta.

2 JULIA,[7] (*John*, 7,) b. ——, d. young.

3 JULIA,[7] (*John*, 7,) b. ——, 1790–1810.
M. John Jacques.
Ch. Harvey, Abram.

4 JULIA,[7] (*Moses*, 1,) b. September 22d, 1814.
M. Hiram P. Topping, November 23d, 1836.
Ch. Frank.   (*Penn Yan, N. Y.*)

5 JULIA ANN,[7] (*Matthias*, 6,) b. ——, 1790–1800.
M. —— Terry.
Ch. Jeremiah, Mary F., John.

6 JULIA,[7] (*Nathan*, 2.)
M. Matthias Corwin ? 4.

7 JULIA N.,[7] (*Ezra*, 2,) b. August 30th, 1813.
M. Charles Knox, December 19th, 1837. He b. January 13th, 1811.
Ch. Hannah, Samuel C., Fanny M., Adelmer, Mary H., Gertrude, Hialwar H.   (*Chemung Co., N. Y. ; 1866, Virginia.*)

8 JULIA ANN,[7] (*Ichabod*, 1,) b. June 15th, 1810, d. ——, 1855 ?
M. Rev. Thomas J. Price, October, 1831.

9 JULIA ELLA,[8] (*Nathan H.*, 3.) b. May 1st, 1826.
M. Alexander Wilson, September 3d, 1846.
Ch. Olive A., Anna, Frank.   (*Middletown, N. Y.*)

10 JULIA F.,[8] (*Isaac*, 4,) b. September 6th, 1830.
M. Thomas Chattle, (see Adaline, 3,) September 6th, 1859.

11 JULIA A.,[8] (*Hervey*, 1,) b. November 13th, 1842.
M. ——, March, 1864.

11½ JULIA,[8] (*Robert*, 1,) b. ——, 1810–20.

12 JULIA ELEANOR,[8] (*Benjamin*, 14,) b. ——, 1824, d. May 25th, 1845.

13 JULIA FRANCES,[8] (*Silas*, 7.)
M. Hiram Coleman.   (*Michigan.*)

14 JULIA,[9] (*George W.*, 28.)

14¼ JULIA J.,[9] (*John R.*, 52,) b. June 28th, 1859.

15 Julia Horton,[8] (*Eliza Corwin*, 2,) b. ——, 1830–40.
M. William Garlock. *a . Saly , , . ~*

16 Julia Millspaugh,[8] (*John H. Millspaugh*, 73,) b. ——, 1845–55.

17 Julia Ketchum,[8] (*Sophia Corwin*, 2.)

18 Julia Ann Blinn,[8] (*Lucinda Thatcher*, 5,) b. April 29th, 1831.
M. James H. Paris, June 6th, 1853.
Ch. Elizabeth, Jacob, Anna, Linnie, Lucinda, James, Thomas, Walter,
Laura.

19 Julia Davaw,[9] (*Helen Hunsike*, 6.)

20 Julia Nora Derrick,[9] (*Mary M. Blinn*, 137,) b. ——, 1868 ?

21 Julia Ann Tompkins,[9] (*Anna M. Corwin*, 17,) b. March 2d, 1841, d.
November 20th, 1841.

22 Julia L. V. Baker,[9] (*Rosalinda Corwin*, 1,) b. February 11th, 1862.

23 Julia B. Boisseau,[9] (*Bethia Corwin*, 9½,) b. June 10th, 1857.

24 Julia Lauretta Crater,[10] (*Mary Corwin*, 83.)

(A) JULIA,[2] [*Malcolm*, (A),] b. 1830–40, in London, England.

1 JULIETTE,[9] (*David*, 24,) b. ——.

Justina, (Latin,) just.

1 JUSTINA,[9] (*Egbert*, 1,) b. October 1st, 1860, d. September 27th, 1861.

Kate, see Catharine.

(I) * KATE RINGGOLD ROANOKE,[9] [*Amos*, (2),] b. in Panama, at the Con-
sulate, June 7th, 1860.
The name Roanoke was added by the officers of the American flag-
ship of war, Roanoke, on board of which vessel she was christened, by the
chaplain.

2 KATY LOUISA,[9] (*Hubbard*, 1,) b. February 4th, 1854.

Keturah, (Hebrew,) incense.

1 KETURAH,[6] (*Jeremiah*, 1,) b. ——, 1760–70.
M. William Terry.
Ch. Jeremiah, John, Josiah, Albert, Louisa, Esther V., Mary, Aletta
William.

Keziah, (Hebrew,) cassia.

1 KEZIAH,[7] (*Joseph*, 10.)

2 Keziah Duckworth,[7] (*Sarah Corwin*, 9,) b. April 6th, 1809.
M. Henry White.

(1) * KEZIAH,[6] [*Joseph*, (1),] b. March 1st, 1754, in West-Jersey.
Removed to Canada, about 1784.

* Corwine.

(2) * KEZIAH D.,[9] [*John*, (11),] b. ——, 1854.   (*Waverley, O.*)

1 Lacy Corwine Stevenson,[9] [*Sarah A. Corwine*, (5),] b. May 29th, 1850, at Centreville, Ind., d. July 1st, 1851.

1 LA FAYETTE WASHINGTON,[8] (*Daniel*, 19,) b. February 19th, 1824. Unm.   (*Riverhead, L. I.*)

1 LANGFORD,[8] (*Thomas*, 5,) b. ——.

1 Lansing Everett,[7] (*Deborah Corwin*, 1,) b. ——, 1810-20.

Laura, (Latin,) a laurel.

1 LAURA,[7] (*Richard*, 3,) b. ——, 1842.

2 LAURA,[7] (*Nebat*, 1,) b. ——, 1810-20.

3 LAURA,[7] (*Martin L.*, 1,) b. ——, 1791, d. ——, 1794.

4 LAURA,[7] (*James*, 4,) b. ——, 1800 ?
M. William Wells.
Ch. Frank, etc.   (*Mattituck, L. I.*)

5 LAURA BELLE,[8] (*James*, 16,) b. October 28th, 1863.

6 Laura Kate Millspaugh,[8] (*Benjamin F. Millspaugh*, 31,) b. ——, 1850-5.

7 Laura Terry,[8] (*Charlotte Corwin*, 5,) b. ——, 1815-30.

8 Laura Paris,[9] (*Julia A. Blinn*, 18,) b. ——, 1867.

9 Laura A. Ayres,[9] (*Experience A. Corwin*, 2,) b. March 1st, 1855.

10 Laura C. Hood,[9] (*Caroline L. Corwin*, 4,) b. September 16th, 1851.
                    (*Vermont.*)

11 Laura A. Cram,[9] (*Hannah Hackett*, 26,) b. June 20th, 1848, d. July 26th, 1863.   (*Vermont.*)

1 Laurin B. Drake,[8] (*Arvilla Corwin*, 1,) b. ——, 1835-45.

1 Laurinda E. Duckworth,[7] (*Sarah Corwin*, 9,) b. December 15th, 1814. M. —— Ebert, of Dayton, Ohio.

Laury, for Lawrence, (Latin,) crowned with laurel.

1 LAURY,[7] (*James*, 5,) b. September 2d, 1798, d. August 12th, 1799.

Leah, (Hebrew,) wearied.

1 LEAH MARGARET,[8] (*Edward C.*, 5,) b. March 24th, 1832, in New-York City.
M. George S. Corwin, 8.

2 LEAH BERTHA,[9] (*Hubbard*, 2,) b. April 23d, 1863.

3 Leah Ann Brown,[8] (*Daniel C. Brown*, 36.)

Leander, (Greek,) lion-man.

1 Leander Newton Millspaugh,[8] (*Gilbert C. Millspaugh*, 6,) b. January 8th, 1847.

* Corwine.

2 Leander Milnor,[3] (*Harriet A. Millspaugh*, 19,) b. August 30th, 1844.
(*Covington, O.*)

1 Leartus B. Connor,[3] (*Caroline Corwin*, 2,) b. January, 1843.

Lemuel, (Hebrew,) created of God.

1 LEMUEL FORDHAM,[7] (*Joshua G.*, 3,) b. September 14th, 1822.
M. (1) —— Train ; (2) Charlotte Dora Martin.
Ch. Stephen G., Oliver G., Wilber M., George.   (*Succasunna, N. J.*)

2 Lemuel Ellston,[3] (*Esther Corwin*, 3,) b. ——, 1825–35.
M. —— Knight.

1 LENA,[10] (*William H.*, 59.)

Leonard, (German,) brave as a lion.

1 LEONARD,[3] (*James*, 20,) b. July 12th, 1861.

(1) LEONARD,[8] [*William*, (3),] b. ——.
M. ——.   (*Hackettstown, N. J.*)
A Leonard Corwin bought land in Hardwick, Warren Co., N. J., 1852–5.

1 L. Ermina Millspaugh,[8] (*J. M. Millspaugh*, 72,) b. June 5th, 1847.

Le Roy, (French,) the king.

1 Le Roy Fitz-Gerald Pike,[3] (*Dorothy Corwin*, 5,) b.——.
M. Ann Tompkins.

1 LESTER M.,[6] (*William H.*, 38,) b. March 22d, 1861, in Vermont.

2 Lester K. Mitchell,[3] (*Maria Corwin*, 3.)

Letitia, (Latin,) happiness.

1 LETITIA JENNETTE,[8] (*Gabriel*, 3,) b. October 1st, 1844, d. in Elbridge,
N. Y., about 1869.

1 Leveret Conklin,[10] (*Charles J. Conklin*, 47,) b. ——, 1866–70.

Levi, (Hebrew,) union.

1 Levi N. Davis,[9] (*Polly Corwin*, 2,) b. ——, d. young.

1 Levisa Porter,[7] (*Elizabeth Corwin*, 9,) b. ——, 1820–30.

Lewis, (Old High-German,) bold warrior.

1 LEWIS WASHINGTON,[7] (*George*, 2,) b. May 5th, 1814.
M. (1) Lucinda Hammond ; (2) Phebe Ann Chesley.
Ch. John, Ann, Jane, Frances.   (*Seneca Falls, N. Y.*)

2 LEWIS BODEI,[7] (*James*, 5,) b. December 30th, 1807, d. March 6th, 1832.
M. Caroline Haviland, April 3d, 1831.
No Ch.

2½ LEWIS J.,[7] (*Seth*, 1,) b. January 19th, 1818.

M. Roxana G. Fordham.

Ch. Mary S., Nathan F., Jesse H., Joseph B., John.

While master of the bark Luika, of Honolulu, he entered the first foreign cargo at the custom-house, Olympia, Washington Territory, in March, 1854.   *(North Sag Harbor, L. I.)*

3 LEWIS D.,[8] *(Benjamin,* 7,) b. ——, 1825? d.——, aged 8.

4 LEWIS F.,[8] *(Francis F.,* 1,) b. ——.

4¼ LEWIS,[9] *(Daniel,* 28½,) b. ——.

4½ LEWIS ANDREW,[9] *(Andrew,* 4½,) b. November 24th, 1862.

5 Lewis Everett,[7] *(Deborah Corwin,* 1,) b. 1810–20.

6 Lewis Hudson Everett,[7] *(Abigail Corwin,* 2,) b. November 5th, 1803.

7 Lewis Skidmore,[8] *(Mary Corwin,* 28,) b.——, 1810–20.

8 Lewis Little,[7] *(Abigail Corwin,* 3,) b. April 24th, 1805.
   M. Ann Eliza Van Cleft.
   Ch. Lewis.

9 Lewis Little, *(Lewis Little,* 8,) b. ——, 1830–5.

10 Lewis M. Brown,[8] *(Silas C. Brown,* 16.)

11 Lewis Howell,[8] *(Jemima Corwin,* 7.)

12 Lewis Penney,[9] *(Helen M. Corwin,* 3,) b. October 31st, 1854.

13 Lewis Smith,[9] *(Sarah Corwin,* 40.)

14 Lewis Brewster Young,[8] *(Eli H. Young,* 6,) b. December 10th, 1850, in Plymouth, Illinois.

(1) * LEWIS,[8] [*Amos,* (1½),] b. ——.   *(Illinois.)*

Lillian, Lillie, (Latin,) a lily.

1 LILLIAN,[9] *(John,* 56,) b. January 1st, 1859.

2 LILLIE B.,[9] *(William,* 39,) b. ——, 1855–65.
3 Lillian Genevieve Thomson[9] *(Edward Thomson 16) C. Aug 30. 1875*

Lina, same as Salina.

1 LINA,[6] *(John,* 6,) b. April 25th, 1766.
   M. Samuel Philips.
   Ch. George, and two others.

1 Linnie Paris,[9] *(Julia A. Blinn,* 18.)

1 Lloyd Burdette Little,[8] *(George C. Little,* 53,) b. August, 1857.

Lois, (Greek,) good, desirable.

1 LOIS ANN,[9] *(Charles L.,* 19,) b. June 19th, 1849.

1 Loren Smith,[8] *(Sarah Corwin,* 28,) b. ——, 1822, in Tunbridge, Vt.
   M. Helen Stacy, in 1848.
   Ch. Emma L., Adelbert S., Hattie.   *(St. Lawrence Co., N. Y.)*

* Corwine.

1 Louis,[8] (*Gabriel*, 1,) b. ——, 1830–40.   (*Middletown, N. Y.*)

1 Louisa,[7] (*James*, 8,) b. 1800–10.

2 Louisa,[8] (*Harrison*, 1.)

3 Louisa,[8] (*John*, 32,) b. March 5th, 1815, in Vermont.
   M. Abiah Noyes, March 7th, 1837.
     Ch. Sarah L., Henry A. C., Henry A., Warren L., Mary L., Frank A., Emma L., Charles W., Katie A., Willie H., Marcell F., Frederic W., Nellie M.

4 Louisa Jane,[8] (*De Witt C.*, 2,) b. July 3d, 1839.

4½ Louisa,[7] (*Stephen*, 7½,) b. January 11th, 1820, d. May 7th, 1847.
   M. —— Cook.

5 Louisa Augusta,[9] (*Joseph*, 20,) b. February 13th, 1850.

6 Louisa Grace,[9] (*Frederic*, 6,) b. July 22d, 1860.

7 Louisa Grace,[9] (*Charles*, 21,) b. ——, 1868, in Niagara Co., N. Y.

8 Louisa Jemima Duckworth,[7] (*Sarah Corwin*, 9,) b. September 12th 1812.
   M. William Mulford.

9 Louisa Terry,[7] (*Keturah Corwin*, 1,) b. ——, 1780–90.

10 Louisa C. Brown,[8] (*Silas C. Brown*, 16,) b. ——.

11 Louisa Fairchild,[9] (*Eliza C. Corwin*, 3.)

1 Lucas,[5] (*Daniel*, 2,) b. ——, 1710–20.

1 Lucetta,[7] (*Peter*, 2,) b. June 16th, 1801.
   M. George Hulse.
     Ch. Alfred, Parmenas, Mary E., Moses, Angelina, Harriet, Jemima.
                 (*Port Jervis ; now at McGrawville, N. Y.*)

1 Lucilla,[8] (*John*, 30,) b. ——, 1835.
   M. Hiram Tinney.
     Ch. Rosalia, Archibald, Herbert, Ellen.   (*Jerusalem, N. Y.*)

2 Lucilla S.,[8] (*Parker*, 1,) b. January 7th, 1821.
   M. J. W. Treadwell, December 24th, 1859. He d. July, 1866.
   No Ch.   (*Lenawee Co., Mich.*)

1 Lucina,[8] (*James*, 18,) b. April 3d, 1844.

2 Lucinda,[6] (*Daniel*, 3,) b. ——, 1770–80.
   M. John Osborn.

1 Lucinda,[8] (*Richard W.*, 5.)
   M. C. C. Murray.
     Ch. Walter, Frances, Richard W., Harrison, Charles, George, Delphina.   (*Orange Co., N. Y.*)

2 Lucinda,[7] (*Daniel*, 9,) b. ——, 1795–1805.
   M. Gabriel Terry.
   Ch.

3 Lucinda,[7] (*Asa*, 3,) b. May 4th, 1794.
   M. Asa Owen.

4 Lucinda,[7] (*Ichabod*, 1,) b. December 8th, 1800.
   M. Anthony H. Dunlevy, August 20th, 1818.   He was b. at Columbia,
   Ohio, December 21st, 1793.   A lawyer.   He was son of Francis Dunlevy, who was b. in Winchester, Va., January 1st, 1762, and served
   five campaigns in the Revolutionary War.   Francis D. also graduated
   at Carlisle College, Pa.; was a member of the first Constitutional
   Convention of Ohio, and fourteen years presiding judge of the Court
   of Common Pleas, in Southern Ohio.   He died October 3d, 1839.
   His father was a native of Ireland, but originally the family came
   from Spain, being driven to France by Romish persecution, and thence
   to Ireland, by the edict of Nantes, and the massacre of St. Bartholomew's day.
   Ch. Sarah M., Francis, John C., Rebecca J., Eliza, James, William
   Wilberforce, Howard, George, Mary C., Lucy.

5 Lucinda Thatcher,[5] (*Jemima Corwin*, 1,) b. August 19th, 1805.
   M. Jacob Blinn, 1824.   He b. December 28th, 1800, d. September
   13th, 1852.
   Ch. Adam, Amos T., Julia A., Jemima, George D., Jesse C., Mary M.
                                                   (*Frankfort, Ind.*)

6 Lucinda Halsey,[7] (*Elizabeth Corwin*, 2,) b. ——, 1808 ?
   M. —— Cary.

7 Lucinda Duckworth,[7] (*Sarah Corwin*, 9,) b. January 19th, 1807, d.
   July 21st, 1823.

8 Lucinda Thatcher,[8] (*Jesse Thatcher*, 10,) b. April 10th, 1822.

9 Lucinda A. Davis,[9] (*Polly Corwin*, 2,) b. ——, d. young.

10 Lucinda Paris,[9] (*Julia Ann Blinn*, 18,) b. ——, 1860 ?

Lucretia, (Latin,) gain or light.

1 Lucretia,[7] (*Selah*, 1,) b. ——, 1790–1800.
   M. James Shay.

2 Lucretia,[8] (*Anson*, 1,) b. October 9th, 1824, d. November 24th, 1824.

3 Lucretia Rosamond Jane,[8] (*Anson*, 1,) b. February 7th, 1827.
   M. Ira Concklin.
   Ch. Elizabeth J., Charles J., Mary M., Lucretia, Melissa S., Edwin, Ida,
   Sylvester.

4 Lucretia,[9] (*Ezra*, 3,) b. ——, 1834, d. ——, 1860.
   M. James W. Lockwood.
   Ch. Ezra, Eliza, Willie.

5 Lucretia Little,[7] (*Desire Corwin*, 1,) b. September, 1808, d. ——, 1835

6 Lucretia A. Little,[7] (*George C. Little*, 53,) b. October, 1850, d. ——. 1854.

7 Lucretia Weed,[8] (*Charity Corwin*, 1,) b. ——, 1820–30.
M. —— Brown.

8 Lucretia Concklin,[9] (*Lucretia R. J. Corwin*, 3,) b. ——.
M. Payne Fanning.

Lucy, (Latin,) born at break of day.

1 LUCY KARON,[8] (*Jabez*, 3,) b. ——, d. when a young lady.

2 LUCY LEVINA,[9] (*Nathaniel*, 7,) b. August 25th, 1829, d. April 17th, 1863.
M. Ezra Goldsmith, January 7th, 1861.
One Ch.

3 LUCY ELLEN,[9] (*Marcus*, 2,) b. ——, 1868 ?

4 Lucy A.,8 (*Gershom*, 2 ?)
She gave a joint deed with Aaron Corwin, in 1851, in Hannibal, Cayuga Co., N. Y. Possibly only a Corwin by marriage, in which case her name would not belong here. (*See Aaron*, 1.)

4½ LUCY E.,[9] (*James*, 36½,) b. about 1864.

4¾ LUCY A.,[10] (*George*, 48,) b. ——, 1860.

5 Lucy Dunlevy,[x] (*Lucinda Corwin*, 4,) b. July 17th, 1843, d. July 6th, 1856.

6 Lucy Blinn,[9] (*Adam Blinn*, 3,) b. ——, 1857.

7 Lucy M. Cram,[9] (*Hannah Hackett*, 26,) b. November 10th, 1840, d. March 4th, 1866.   (*Vermont.*)

8 LUCY ANN,[9] (*George*, 42,) b. March 4th, 1857, d. March 31st, 1857.

9 Lucy Addie Hackett,[9] (*Samuel B. Hackett*, 32,) b. November 30th, 1855.   (*Vermont.*)

(1) LUCY,[3] [*John*, (1),] b. May 11th, 1670.

1 Lula Florence Howes,[9] (*Theresa Corwin*, 2.)

1 LURA A.,[9] (*Jabin*, 2,) b. June 4th, 1868, at Chelsea, Vt.

Luther, (German,) illustrious warrior.

1 LUTHER S.,[7] (*Jemuel*, 1,) b. January 16th, 1822, d. October 2d, 1823.

1 Luthera P. Smith,[8] (*Philenia Corwin*, 1,) b. April 26th, 1840, d. July 26th, 1864.
M. Horace Durkee, October 5th, 1859.

1 LUTHERIA,[8] (*John*, 25.)

1 LYDIA,[6] (*James*, 2,) bapt. November 6th, 1768, at Southold
M. James Moore.
No Ch.

2 LYDIA,[7] (*John*, 9,) b. November 7th, 1785, d. April 30th, 1860.
    M. John Wells, September 20th, 1802.   (*Aquebogue Records.*)
    Ch. Joseph.

3 LYDIA,[7] (*Nathan*, 2,) b. ——, 1780–90.
    (A Joseph Williamson married a Lydia Corwin, September 24th, 1807.
Perhaps this one.   *Aquebogue Records.*)

3₂ LYDIA,[7] (*Selah*, 1,) b. ——, d. December 23d, 1785.
                                            (*Aquebogue Records.*)

4 LYDIA,[7] (*Selah*, 1,) b. 1790–1800.
    M. —— Turner.

5 LYDIA,[7] (*Hubbard*, 1,) b. June 7th, 1806.

6 LYDIA ANN,[7] (*Asa*, 3,) b. ——, 1818.
    M. Rev. J. M. Harris.   (*Ithaca, N. Y.*)

6½ LYDIA,[8] (*Uriah*, 1,) b. ——.   (*Greenport, L. I.*)

7 LYDIA,[8] (*John*, 32,) b. April 28th, 1817.
    M. Benjamin R. Heald, July, 1848.
    No Ch.

8 LYDIA,[8] (*Thomas*, 5,) b. ——, 185–, d. ——.

8½ LYDIA ANTOINETTE,[8] (*David*, 18,) b. July 16th, 1832.
    M. Peter Clark, October 14th, 1852.   He b. July 3d, 1815.
    Ch. Clinton Corwin, (b. April 6th, 1855,) Henry Porter, (b. December
        11th, 1857,) Evan Peter, (b. April 15th, 1859,) Ella Sarah, (b. Feb-
        ruary 22d, 1861,) Jennie Antoinette, (b. May 12th, 1862,) Minnie
        Hester, (b. June 14th, 1865,) Frank Totten, (b. April 21st, 1867,)
        Birdie, (b. December 31st, 1869.)   (*Brady's Bend, Pa.*)

9 Lydia Thatcher,[7] (*Jemima Corwin*, 1,) b. ——, 1800.   (*Missouri?*)

10 Lydia Hackett,[8] (*Mary Corwin*, 14,) b. January 5th, 1810, d. July
    29th, 1811.

11 Lydia A. Smith,[8] (*Philenia Corwin*, 1,) b. November 3d, 1832.
    M. Hasper G. Cushman, April 3d, 1853.
    Ch. Walter S.

(1) * LYDIA,[6] [*Samuel*, (3),] b. November 24th, 1753.
    M. —— Servis.   (*New-Jersey.*)

(2) * LYDIA,[7] [*Richard*, (2),] b. May 24th, 1798, d. ——, 1854.
    M. (1) Thomas Comly, 1814.   He d. 1815, and a young son in the
        same year; (2) James T. Barkalow.   He d. 1856.
    Ch. William, Samuel, and seven others.   (*Franklin, Warren Co., Ohio.*)

(3) * LYDIA,[6] [*Samuel*, (6½),] b. July 15th, 1813, d. September 26th,
    1834.
    M. John Hitch, September 13th, 1833.
    Ch. Mary.

He now lives at Louisville, Clay Co., Illinois.

* Corwine.

1 Lyman,[8] (*John*, 30,) b. ——, 1828, d. ——, 1847.
   M. Adaline Drake. *No ch.*   (*Elmira, N. Y.*)

2 Lyman H. Swina,[8] (*Bethia Hobart*, 11,) b. September, 1839.

(1) Mae Belle Miller,[9] [*Sarah H. Corwine*, (6),] b. July 6th, 1861.

Mahala, (Hebrew,) sickness.

1 Mahala,[6] (*Daniel*, 3,) b. ——, 1750–60.
   M. James Mills.

2 Mahala,[7] (*John*, 10,) b. July 7th, 1811, d. September 3d, 1854.
   M. John Bertholf.
   Ch. Elizabeth.   (*Howells, N. Y.*)

3 Mahala,[8] (*John*, 32,) b. September 19th, 1836.
   M. Nathaniel H. Austin, April 3d, 1855.
   Ch. Aura J.   (*Vermont.*)

Mahlon, (Hebrew,) sickly.

1 Mahlon,[7] (*Nathaniel*, 5,) b. -——, 1770–90.
   M. (1) ——; (2) Jennette Vail.

1 Major Fullam H.,[8] (*Hazen*, 1,) b. October 18th, 1833.
   M. (1) Amanda S. Abbott, at Lawrenceville, N. Y., October 26th, 1853.
   She d. at Hopkinton, N. Y., June, 1857; (2) Annie P. Wing, at
   Hopkinton, June 18th, 1858.
   Ch. John A.

(A) Malcolm Corwin,[1] (son of John Corwin, of London, England.) Malcolm came to New-York in 1844. He was born in London, 1815, and
   died in New-York, July, 1855. He married Elizabeth Phelan, in
   England, and left two daughters, Julia and Ann Maria. Probably
   he is of a collateral line of the Cumberland County Curwens.

Manasseh, (Hebrew,) forgetfulness.

1 Manasseh Reeves,[7] (*Isaac*, 1,) b. February 7th, 1786.
   M. Catharine Moore.
   Ch. Isaac, Charles, Hannah M., Samuel C., Stephen M., Matilda,
   Manasseh R.

He left Morris Co., N. J., in or before 1812, when his name occurs on
the records, as being of the town of Phelps, Ontario Co., N. Y. He bought
land of John Fuller, and sold land to Jeremiah Andrews, in Ontario Co.,
in 1812. In 1823 and 1827 he is of the town of Hector, now Schuyler
Co., N. Y. Was also once of Newark, N. Y.

2 Manasseh Reeves,[8] (*Manasseh*, 1,) b. ——, 1820–30. (*Lockport, N. Y.*)

1 Manson Blinn,[9] (*George D. Blinn*, 56,) b. February 17th, 1762.

1 Marcell F. Noyes,[9] (*Louisa Corwin*, 3,) b. August 25th, 1857.
                                     (*Vermont.*)

2 Marcell Smith,[9] (*Mary Corwin*, 53,) b. July 18th, 1854.

10

Marcellus, diminutive of Marcus, (Latin,) a hammer.

1 Marcellus Halsey,[8] (*Miller Halsey*, 2,) b. ——.
2. Marcellus ——,[8] (*Hannah Corwin*, 13,) b. ——.

Marcia, feminine of Marcus.

1 MARCIA,[8] (*John*, 32,) b. February 22d, 1833.
   M. Edwin J. Norris, September 17th, 1854.
   Ch. Frank C.
2 MARCIA E.,[9] (*William H.*, 38,) b. November 23d, 1855, d. December
   20th, 1859.   (*Vermont.*)

Marcus, (Latin,) a hammer.

1 MARCUS WHEELER,[8] (*William*, 19,) b. December 23d, 1853.
2 MARCUS HUBBARD,[8] (*Russell*, 1,) b. in Chelsea, Vt., November 28th,
   1839.
   M. Ellen M. Collins, March 1st, 1865.
   Ch. Carl H., Lucy E.
   He graduated at Dartmouth College, 1863. A physician.
                                                      (*Corinth, Vt.*)

Margaret, (Latin,) a pearl.

1 MARGARET,[7] (*Joseph*, 8,) b. August 18th, 1788, d. January 1st, 1845.
   M. Anthony Drake, January 31st, 1818.
   Ch. Nelson, Sylvanus.   (*Flanders, N. J.*)
2 MARGARET,[7] (*Moses*, 1,) b. March 29th, 1820.   (*Middletown, N. Y.*)
3 MARGARET,[7] (*James*, 6,) b. November 2d, 1806, d. ——, 1832, in New-
   York City.
   M. Hugh Hill.
   Ch. Jane L.
4 MARGARET OPHELIA,[7] (*Isaac S.*, 2,) b. August 16th, 1846.
5 MARGARET K.,[8] (*Charles*, 3,) b. August 18th, 1853.
6 MARGARET,[9] (*John D.*, 47.)
7 Margaret Reed,[8] (*Sophia Corwin*, 1,) b. August 8th, 1802.
   M. —— Voorhees.   (*Daggetsville, N. Y.*)
8 Margaret Ketchum,[8] (*Sophia Corwin*, 1.)
9 Margaret Ellston,[8] (*Esther Corwin*, 3,) b. ——, 1825–35.
   M. (1) —— Edwards; (2) ——.
10 Margaret Halstead,[8] (*Cynthia Hobart*, 10,) b. September 17th, 1832.
   M. William Pickard, September 29th, 1850.
   Ch. Eugene.   (*Peru, N. Y.*)
11 Margaret Ann Mongown,[8] (*Elma Corwin*, 1,) b. ——, 1843, d. August
   27th, 1860.
12 Margaret Willinghouse,[8] (*Hannah J. Corwin*, 9,) b. ——.

18 Margaret Deel,[10] (*Elizabeth J. Deel*, 61,) b. ——, 1865–70.

(1) MARGARET,[9] [*Jonathan*, (1),] b. April 15th, 1679, d. November 5th, 1679.

(2) MARGARET,[9] [*Jonathan*, (1),] b. March 30th, 1685, d. February 23d, 1686.

(3) * MARGARET,[6] [*George*, (6),] b. November 28th, 1746, went to Kentucky, 1794, yet living in 1800.

(4) * MARGARET LUCAS,[8] [*Joab*, (1),] b. ——, 1814.
M. J. Clark. (*Covington, Ky.*)

Maria, a Latin form of Mary.

1 MARIA,[6] (*David*, 5.)

2 MARIA,[7] (*Joseph*, 4,) b. April 23d, 1796, d. December 22d, 1827.
M. —— Wheat.

2½ MARIA,[7] (*William*, 4,) b. August 18th, 1818.

3 MARIA,[7] (*Jacob*, 2,) b. ——, 1790?
M. —— Mitchel.
Ch. George C., Uriah, Sarah M., Mary E., Lester K.

4 MARIA,[7] (*Selah*, 1,) b. ——, 1790–1800.
M. John Youngs.

5 MARIA,[7] (*David*, 7½,) b. ——.

6 MARIA,[7] (*James*, 5,) b. May 7th, 1804, d. June 27th, 1805.

7 MARIA,[8] (*John*, 32,) b. ——.

8 MARIA,[8] (*Nathan*, 5,) b. ——, 1825–30.

9 MARIA MILLER,[8] (*Gilbert*, 4,) b. May 28th, 1840.
M. John T. Ward, April 6th, 1858.

10 MARIA,[8] (*Ebenezer*, 3,) b. April 18th, 1823.
M. James H. Burst. (*Florida.*)

11 MARIA LOUISA,[8] (*Thomas*, 4,) b. August 8th, 1834.

12 MARIA,[8] (*Henry*, 10,) b. January 31st, 1818.
M. Albert Williamson. (*Sag Harbor, L. I.*)

18 Maria E. Upright,[7] (*Anna Corwin*, 4,) b. January 2d, 1830.
M. William D., son of Jacob and Julia Van Wagener, March 2d, 1848.
Ch. David T., Anna J.

14 Maria J. North,[8] (*Mehetable Campbell*, —,) b. January 3d, 1836.
M. George Chittenden, July 4th, 1854.

15 Maria Swina,[8] (*Bethia Hobart*, 11,) b. October, 1834.
M. Edward Carts.
Ch. Eunice.

16 Maria Skidmore,[8] (*Mary Corwin*, 28,) b. ——, 1810–20.

* Corwine.

17 Maria L. Smith,[8] (*Philenia Corwin*, 1,) b. October 24th, 1834.
    M. Orville Noyes, March 4th, 1864.
    Ch. Russell, Frederick, Edward E.   (*Vermont.*)
18 Maria Elizabeth Newman,[9] (*Moses B. Newman*, 9,) b. at Sidney, Ohio,
    June 26th, 1855.
19 Maria Louisa Cropper,[9] (*Carrie R. Corwin*, 8,) b. August 20th, 1868.
20 Maria Louisa Tompkins,[9] (*Anna M. Corwin*, 17,) b. May 22d, 1839.
  (1) Maria Barkalow,[8] [*Rachel Corwine*, (2).]
    M. Peter Du Bois.
    Ch. Rachel, etc.   (*Franklin, O.*)

Marietta, diminutive of Mary.

1 MARIETTA,[8] (*Ebenezer*, 3,) b. May 12th, 1829.
    M. James C. Thompson.
1½ MARIETTA,[9] (*Daniel*, 28½,) b. ——.
2 Marietta Woodruff,[7] (*Sibyl Corwin*, 1,) b. ——, 1800 ?
    M. Benjamin Watkins.
    Ch. Abel W., Edward, Sarah J.   (*Newburg, N. Y.*)
3 Marietta M. Armstrong,[8] (*Charlotte Millspaugh*, —,) b. December 25th,
    1847.   (*Booneville, Ind.*)
4 Marietta H. Howell,[8] (*Jemima Corwin*, 7.)
5 Marietta Swina,[8] (*Albert Swina*, 10,) b. ——, 1854.

Marion, a French form of Mary.

1 MARION,[8] (*Polydore B.*, 1,) b. January 21st, 1845, in New-York City.
                                              (*Michigan.*)
2 Marion Byram,[8] (*Mary A. Corwin*, 34,) b. July 3d, 1848, d. November
    11th, 1860.
3 Marion L. Kyte,[9] (*Clarinda Corwin*, 2,) b. January 15th, 1861.
4. *Marion Tompkins,*[10] (*Mary Corwin Clark 176*) *b. Oct 16. 189*

Mark, same as Marcus.

1 Mark Coryell,[9] (*Susan Ann Shannon*, 83.)
1 MARSELUS,[7] (*Daniel*, 6,) b. ——, 1832, d. ——, 1853.
    He went to South-America with a companion from Cincinnati, O., and
both were drowned in the Amazon River.
  (a) *MARSHALL E.,[4] [*George*, (a),] b. ——.   (*Cincinnati, O.*)

Martha, (Syriac,) lady, feminine of Marā, lord.

1 MARTHA,[2] (*Matthias*, 1,) b. ——, 1630–40, yet living in 1698.
    M. (1) Henry Case, November, 1658 ; (2) Thomas Hutchinson, January
    11th, 1665–6, d. ——, 1676–83.

* Curwen.

Ch. Henry, Theophilus ; (2,) Thomas, Matthias, Martha, Samuel, Benjamin ?

Henry Case perhaps lived in Rhode Island, whence his widow probably came with her two children, after his death, (1661–5.) In 1660, a Henry Case had a suit with Theophilus Corwin. In 1681, Martha Hutchinson and her eldest son, Henry Case, give each other releases in respect to her first husband's estate. In 1675, Thomas Hutchinson, of Southold, was rated at £176. (*Doc. Hist. N. Y.* ii. 260.) In 1686, Martha was again a widow, having in her family 3 males and 2 females. In 1698, her name is on census list of Southold, as a widow, with the names of the *Hutcheson* children, and a Hannah Case. (*Doc. Hist.* i. 453.) Respecting Thomas Hutchinson, see *Drake's Boston*, 227, and vol. xx. *N. E. Reg.* 355, and *Moore's Index of Southold*, p. 24. Respecting Henry Case, see *Moore's Index of Southold*, p 11, 64.

2 MARTHA,[7] (*Daniel*, 4,) b.——, 1780–90.
 M. Richard Pierson.
 No Ch.

3 MARTHA,[7] (*Isaac*, 1,) b. October 29th, 1781.
 M. Jonah Hopkins.
 Ch. Nancy.   (*Palmyra, N. Y.*)

4 MARTHA E.,[7] (*Barnabas*, 1,) b. January 10th, 1831.

4½ MARTHA CHRISTIANA,[7] (*Jabez*, 1,) b. January 15th, 1831.
 M. George Whitefield Cooper, January 3d, 1850.
 Ch. Fanny R., George C., Christiana M., Franklin C., Martha F.
          (*Mattituck, L. I.*)

5 MARTHA,[8] (*Moses H.*, 2,) b. ——, 1850? in New-York City.

6 MARTHA,[8] (*John*, 30,) b. ——.

7 MARTHA ALVENIA,[8] (*Franklin E.*, 1,) b. ——, d. in infancy.

7¼ MARTHA EVELINE,[8] (*Ezra*, 2½.)

8 MARTHA M.,[9] (*William H.*, 88,) b. November 4th, 1847.

9 MARTHA,[9] (*Spencer W.*, 2,) b. September 20th, 1862.

9½ MARTHA J.,[9] (*John R.*, 52,) b. March 8th, 1857.

10 Martha Hutchinson,[2] (*Martha Corwin*, 1,) b. about 1670.

11 Martha Beach,[8] (*Eliza Ann Corwin*, 5,) b. December 13th, 1827.

12 Martha J. Flannery,[8] (*Mary Corwin*, 33,) b. at Cleveland, Ohio, October 7th, 1850.
 M. Perry Thatcher, at Cleveland, Ohio, May 12th, 1870. He was born at Pembrook, N. Y., June 14th, 1848. (*Cleveland, O.*)

13 Martha Brown,[8] (*Daniel C. Brown*, 36.)

14 Martha Weed,[8] (*Charity Corwin*, 1,) b. at Rochester, Ohio, March 5th, 1846.

15 Martha Ann Taylor,[8] (*Elizabeth Corwin*, 17.)

16 Martha E. Meeker,[8] (*Harriet F. Corwin*, 4,) b. February 25th, 1860.

17 Martha Blinn,[9] (*Adam Blinn*, 3,) b. November, 1850.

18 Martha Estelle Cooper,[8] (*Martha C. Corwin*, 4½,) b. ——, 1870.

19 Martha Ann Cram,[9] (*Hannah Hackett*, 26,) b. September 25th, 1837.

(1) * MARTHA,[9] [*George*, (11½).]   (*Missouri*.)

(2) * MARTHA THOMAS,[9] [*Samuel*, (12),] b. January 29th, 1863.

Martin, (Latin,) warlike.

1 MARTIN LUTHER,[6] (*James*, 1,) b. on Long Island, December 22d, 1768, d. August 14th, 1845.
M. Rebecca Jane Newman.
Ch. Laura, Charlotte, Horton, Martin.
(*New-York City*, 1795–1803.   *Howells, N. Y.*, 1811.)

2 MARTIN,[6] (*Timothy*, 2,) b. ——, 1786 ? d. ——, 1850.
M. Elizabeth White.
Ch. William, Samuel, Elizabeth, Orange.   (*Newburg, N. Y.*)

3 MARTIN,[7] (*Martin L.*, 1,) b. ——, 1803, d. ——, 1804.

4 MARTIN,[5] (*Horton*, 2,) b. December 26th, 1826, d. September 11th, 1829.

5 MARTIN,[5] (*Horton*, 2,) b. November 14th, 1833.
M. Juliette Palmer, September 16th, 1852.
Ch. Horton, Phebe J., Salmon W., Mary A.
(*Howells, N. Y., East-Saginaw, Mich.*)

6 MARTIN,[8] (*John Eli*, 19,) b. ——, 1845–50.

Mary, (Hebrew,) rebellion.

1 MARY,[3] (*Theophilus*, 1,?) b. ——, 1660–70.
Her name stands alone on census list, 1698, her father probably being dead. (*Doc. Hist. N. Y.* i. 455.)

2 MARY,[3] (*John*, 1,) b. December 15th, 1659, d. probably before 1690. (*See her father's will.*)
M. Jacob or Jabez Mapes.
Ch. Sarah.

3 MARY,[5] (*Matthias*, 3,) b. ——, 1753, d. ——.
M. Joshua Corwin, 2.

4 MARY,[5] (*Daniel*, 2,) b. ——, 1710–15.
M. James Terry, 1734.

4½ MARY,[5] (*Samuel*, 3,) b. ——, 1741, d. ——, 1795.
M. Nathaniel Norton, February 1st, 1776, at Southold.
(*Brookhaven, L. I.*)

5 MARY,[6] (*Theophilus*, 4? or Timothy, 1 ?) b. about 1775.
M. Ezekiel Brockway, of Little Britain, Orange Co., N. Y.
Ch. Betsey, Sally.

* Corwine.

6 MARY,[5] or [6] (*George*, 1,) b. ——, 1770–80.

(A Mary Corwin married Benjamin L'Hommedieu, February 1st, 1781. Another Mary Corwin married Richard Albertson, July 17th, 1783. (*Aquebogue Records.*)

7 MARY,[6] (*Jesse*, 2,) b. ——, 1763.

M. John Matthews, in Pennsylvania, 1780.

Ch. Mary.

Removed to Kentucky, 1784–5, and settled at the mouth of Bear Grass, now Louisville. His wife's health failing, they started to return to Pennsylvania, but she died before reaching Cincinnati. He returned to Pennsylvania, and married again.

8 MARY,[6] (*Silas*, 1,) bapt. at Aquebogue, May 31st, 1767.

M. Captain Jonathan Brown, May 24th, 1792. (*Aquebogue Recs.*)

- *Ch.* Arminda, Daniel C., Polydore, Silas C., Jonathan, Parmenas, Dorastus: • (*Long Island; Orange Co., N. Y., 1800.*)

9 MARY,[6] (*Samuel*, 6,) bapt. August 6th, 1769, at Southold.

10 MARY,[6] (*Thomas*, 1,) b. ——, 1790–1810.

M. Ezra Reeve. (*Long Island.*)

11 MARY,[6] (*John*, 6,) b. ——.

M. (1) David Bishop; (2) Jonathan Rockett.

Ch. Mehetable, Edward, David.

'(*Long Island; Orange Co., N. Y.; Hamburg, Mich.*)

11½ MARY,[6] (*Nathan*, 1½,) bapt. at Aquebogue, September 18th, 1785.

12 MARY,[6] (*David*, 4,) b. ——, d. ——.

M. James H. Reeve.

Ch. Isaac, David, Daniel, Elijah.

13 MARY,[6] (*Joseph*, 1,) b. ——, 1770?

(A Mary Corwin married Ebenezer Wade, 1762, according to Mattituck Records.)

14 MARY,[7] (*Hubbard*, 1,) b. January 29th, 1786, on Long Island, d. November 7th, 1864.

M. Captain Ephraim Hackett, of Tunbridge, Vt., January 5th, 1808. He died May 7th, 1864.

Ch. Hannah, Lydia, James, Samuel B., John S., George, Russell. (*Tunbridge, Vt.*)

15 MARY,[7] (*Richard*, 3,) b. August 31st, 1844, d. September 10th, 1849.

16 MARY,[7] (*Jesse*, 3,) b. ——, 1808?

M. Thomas L. Wood.

Ch. Cornelius, Morrison, Henry, Jesse, William, Mary, Jane. (*Howells, N. Y.*)

17 MARY,[7] (*Joseph*, 10,) b. ——.

18 MARY BAKER,[7] (*Stephen*, 4,) b. ——. (*Pa.*)

19 MARY JANE,[7] (*Daniel*, 5,) b. ——, d. in infancy.

20 MARY,[7] (*Matthias*, 5,) b. December 4th, 1791, d. ——.
   M. Samuel Drake, October 16th, 1814.
   Ch. John, Clayton, Thomas, Matthias, Eliza J., Horace.

21 MARY,[7] (*Ichabod*, 1,) b. May, 1795, d. May, 1835.
   M. John Hart, 1812.

22 MARY,[7] (*Peter*, 2,) b. June 4th, 1810.
   M. George Mullock.
   Ch. Theodore, Albert, Gabriel, Coe, Angeline, Mary.
                                        (*Waverley*, *N. Y.*)

23 MARY E.,[7] (*David*, 10,) b. November 2d, 1853.

24 MARY,[7] (*Daniel*, 4,) b. ——, 1780–90, d. ——.
   M. (1) William Pellett ; (2) Howell Reeves.
   Ch. Harriet, Elizabeth, Gabriel, Howell.

25 MARY,[7] (*Joseph*, 9,) b. March 28th, 1810. *died Feby. 23. 1893.*
   M. William A. Thompson, who was b. May 29th, 1808.
   Ch. Samuel, William, Joseph, Anna M., Emma, Elizabeth, Edward,
     Clara.   (*N. Y. C.*, 122 *W. 23d st.*)

25½ MARY,[7] (*Samuel*, 11,) b. ——, d. ——.
   Unm.

26 MARY,[7] (*Benjamin*, 4,) b. ——, 1770–90.
   M. Joshua Case.   (*Morris Co.*, *N. J.*)

27 MARY ANN,[7] (*Rev. Jacob*, 1,) b. ——, 179–. ?
   M. (1) —— Higgins ; (2) Hurtin.
   Ch. Rockwell ? William S., Charlotte.   (*Yaphank*, *L. I.*)

28 MARY,[7] (*Joshua*, 2,) b. ——, 1790 ?
   M. Richard Skidmore.
   Ch. Harvey, Lewis, Maria, Eliza, Joshua.

29 MARY,[7] (*Rev. Joseph*, 5,) b. about 1795, d. about 1839.
   M. Bethuel Halleck.
   Ch. Bethuel.   (*Jamesport*, *L. I.*)

30 MARY,[7] (*George*, 2,) b. May, 1817.
   M. Henry W. Peters.
   Ch. Matilda, Eleazar, Horace H., William H., Mary E., Sarah E.,
     George W.   (*Elmira*, *N. Y.*)

31 MARY,[7] (*Jason*, 1,) b. January 26th, 1817, d. April 28th, 1817.

32 MARY HORTON,[7] (*David*, 7,) b. January, 1813. *1823 —*

33 MARY,[7] (*Moses*, 1,) b. November 22d, 1824.
   M. John Flannery. He was b. in Tipperary, Ireland, August 12th,
     1818.
   Ch. Charles E., John, John L., Daniel, Martha, Mary A., Joanna A.
   Was a merchant tailor.   (*Cleveland*, *O.*)
   He enlisted in the 16th Michigan Regiment, at Bay City, April 15th,

1863 ; was taken prisoner at Petersburg, Va., June, 1863 ; died in the rebel prison at Richmond, Va., October, 1863.

34 MARY ANN,[7] (*Joshua G.*, 3,) b. December 23d, 1819.
    M. Francis T. Byram, January 23d, 1845. He was born November 11th, 1821.
    Ch. Marion E., Harriet O., Anna L., Etta D., George G., Mary E., Charles M.

34½ MARY,[7] (*Amaziah*, 3,) b. ——, 1825–30.

35 MARY,[7] (*Jacob*, 2,) b. ——, 1790–1800.
    M. —— King.
    Ch. Sarah M., William H.

36 MARY FRANCES,[7] (*Matthias*, 6,) b. ——, 1790–1800.
    M. Daniel Corwin, 23 ?

37 MARY,[7] (*Jeremiah*, 2,) b. March 31st, 1790.
    M. (1) —— Young ; (2) Abner Corwin, No. 2.
    Ch. Frank, Mary.

38 MARY,[7] (*Henry*, 5,) b. ——, 1827 ?

39 MARY,[7] (*Jedediah*, 3,) b. ——, 1780–90.
    M. Erastus Wells.

40 MARY,[7] (*Ezra*, 2,) b. February 12th, 1805.
    M. Ambrose Sperry, August 7th, 1824.
    Ch. Stephen D., Elias C., Alburtis B., Judso A., Helen M.
                                    (*Cortlandt, N. Y.*)

41 MARY,[7] (*Abel*, 1,) b. ——, 1790–95.

42 MARY ANN,[7] (*Richard*, 2,) b. ——, 1810–20.
    Unm.

43 MARY,[7] (*John*, 14,) b. February 14th, 1806 ?

44 MARY ELIZABETH,[7] (*Silas*, 5,) b. September 22d, 1832.
    M. John B. Cox, 1854.
    Ch. Edgar, Francis, Dan. Irwin, George, Imogene, Idomey.
                     (*Mount Kisko, Westchester Co., N. Y.*)

45 MARY A.,[7] (*Barnabas*, 2,) b. September 28th, 1833.

46 MARY,[7] (*James*, 7,) b. ——, 1810–30.

47 MARY,[8] (*David*, 10,) b. August 26th, 1833.
    M. John Wilcox.
    Ch. William, Harry Houston.   (*New-York City.*)

48 MARY C.,[8] (*Nathan H.*, 3,) b. February 13th, 1828.
    M. Rev. P. Manning Stryker, October 8th, 1846.
    Ch. Elizabeth, James, Belle.

Mr. Stryker was son of Judge James Stryker, of Buffalo, N. Y. He was born December 26th, 1820, d. August 19th, 1862. He entered Hobart College, 1837, but was obliged to leave on account of ill health. He was ordained to the ministry October, 1846, by Bishop De Lancey. His first charge was East-Bloomfield, N. Y.   He afterward taught and preached at Pottsville, Pa. ; while here, his health failed.

49 MARY ELIZABETH,[8] (*Joshua,* 6,) b. May 9th, 1828.
    M. Charles Mapes.
    Ch. Aletta, Frances, Harriet.   (*Middletown, N. Y.*)

50 MARY,[8] (*Abram,* 1,) b. ——, 1820–30.

51 MARY,[8] (*Archibald,* 1,) b. ——, 1840–50.

52 MARY FRANCES,[8] (*Peter,* 4,) b. ——, 1835–40.

53 MARY,[8] (*John,* 32,) b. November 3d, 1824, d. December 7th, 1865.
    M. Nelson W. Smith, May 31st, 1849.   He d. August 31st, 1864.
    Ch. Marcell, Carlos.   (*Vermont.*)

54 MARY E.,[8] (*William,* 11,) b. February 6th, 1836.
    M. Isaiah H. Mapes.   He was b. ——, 1828, d. March 7th, 1859.
    Ch. Harrison.

55 MARY ANTOINETTA,[8] (*Epenetus II.,* 1,) b. May, 1836 ?
    M. D. S. De Vinne, son of Rev. Daniel De Vinne, a Methodist.
                                   (*New-York City.*)

56 MARY ANN,[8] (*Shubal,* 1,) b. September 16th, 1817.
    M. David R. Terry.   He b. ——, 1815.
    No ch.

57 MARY,[8] (*Richard S.,* 7,) b. ——, 1815–30.

58 MARY,[8] (*George E.,* 13,) b. ——, 1830–40.
    M. William Rorick.

59 MARY JANE,[8] (*Hervey,* 1,) b. January 24th, 1848, d. October 12th, 1855.

60 MARY CATHARINE,[8] (*Nathaniel,* 6,) b. June 30th, 1835, d. July 30th, 1835.

61 MARY,[8] (*Adam,* 1,) b. ——, 1845.

62 MARY FRANCES,[8] (*John,* 21,) b. February 22d, 1859.

63 MARY ANN,[8] (*Henry W.,* 8,) b. May 17th, 1823, d. July 31st, 1825.

64 MARY ELIZABETH,[8] (*Samuel II.,* 14,) b. September 15th, 1842.
    M. Samuel Vanton, September 22d, 1870.   (*Otisville, N. Y.*)

65 MARY,[8] (*Peter,* 3,) b. ——, 1810–20.

66 MARY E.,[8] (*William,* 19,) b. April 2d, 1852.

67 MARY EMMA,[8] (see Mary Frances Corwin, 36.)

68 MARY EMMA,[8] (*Harrison,* 1.)

69 MARY,[8] (*Nathan,* 5,) b. ——, 1825–30.
    M. Franklin Lane.

70 MARY,[8] (*Jeremiah,* 4.)

70½ MARY IDUELLA,[8] (*Albert,* 2,) b. April 17th, 1851.

71 MARY AUGUSTA,[8] (*Alsop II.,* 2.)

72 MARY ESTHER,[8] (*Samuel B.,* 16,) b. July 4th, 1841.

M. Orville Benedict. He b. February 14th, 1841.
One ch. (*Janesville, Wis.*)

73 MARY B.,[8] (*Benjamin,* 14,) b. about 1805 ?
M. Epenetus Corwin, 1, about 1835 ?

74 MARY ELIZABETH,[8] (*Timothy,* 6.)

75 MARY,[8] (*Joshua C.,* 8½,) b. ——.
M. —— Barlow.

76 MARY,[8] (*William,* 29,) b. ——.
M. Chauncy Clark.
They lived in the old homestead of their grandfather, Eli, 3.
(See Oliver B., 2½.) (*Wallkill, Orange Co., N. Y.*)

77 MARY ELIZABETH,[8] (*Silas,* 6,) b. ——, d. in infancy.

78 MARY O.,[8] (*Silas,* 5,) b. ——.

79 MARY,[8] (*John,* 30,) b. ——.
M. Elisha Ingraham.
Ch. (*Jerusalem, N. Y.*)

80 MARY GARETTE,[8] (*Joseph,* 16,) b. February 14th, 1850, d. September 9th, 1851.

80½ MARY ELIZA,[8] (*Rev. Ira,* 4,) b. June 10th, 1851, at Marietta, O., d. October 3d, 1862, at South-Bend, Ind.

81 MARY JANE,[8] (*John M.,* 46,) b. March 18th, 1823.

81½ MARY JANE,[8] (*David,* 12,) b. March 11th, 1830.
M. Clark Crandell.
Six ch. (*Sag Harbor, L. I.*)

82 MARY,[8] (*Benjamin,* 15,) b. ——.
M. —— Gallagher. (*Lebanon, O.*)

82¼ MARY E.,[8] (*Silas,* 7,) b. ——, d. young.

82⅓ MARY E.,[9] (*William II.,* 38,) b. January 17th, 1838, in Vermont.

82¾ MARY,[9] (*Barnabas,* 6,) b. May 2d, 1851.

83 MARY,[9] (*Nicholas,* 1,) b. ——, 1841 ?
M. Serin Crater.
Ch. Mary E., Henry A., William E., George N., Julia L., Ella J., Frank.
(*Near Chester, N. J.*)

84 MARY,[9] (*Amos,* 1.)

85 MARY AUGUSTA,[9] (*Halsey,* 1.)

86 MARY ADELAIDE,[9] (*Alva,* 1,) b. November 23d, 1829.
M. Seamen Shirley, July 14th, 1848.
Seven ch. (*Brooklyn, N. Y.*)

87 MARY EMMA,[9] (*Charles L.,* 19,) b. April 3d, 1847.

88 MARY MINNETTA,[9] (*Henry W.,* 19,) August 9th, 1851.

89 MARY ELIZABETH,[9] (*Hewlett C.*, 1,) b. July 27th, 1847.
  M. Harrison Reeve, February 11th, 1865.   (*Riverhead, L. I.*)

89½ MARY D.,[9] (*Daniel*, 24,) b. ——.

90 MARY F.,[9] (*De Forest*, 1,) b. September 20th, 1859.

91 MARY ELIZABETH,[9] (*John D.*, 47.)

92 MARY,[9] (*Peter F.*, 7,) b. August 14th, 1859.

93 MARY LANGSTAFF,[9] (*Joseph*, 20,) b. March 2d, 1855.

94 MARY,[9] (*James II.*, 34,) b. at Bergen, N. J., December 28th, 1856.

95 MARY ADA,[9] (*Martin*, 5,) b. February 14th, 1861.

96 MARY ELIZABETH,[9] (*Silas G.*, 12,) b. April 4th, 1855.

96¼ MARY E.,[9] (*George W.*, 23½) b. July 30th, 1847, d. September 4th, 1849.

96½ MARY ELIZABETH,[9] (*John V.*, 57,) b. March 8th, 1852.

97 MARY II.,[10] (*Bartholomew*, 1,) b. September 11th, 1859.

97¼ Mary Osman,[4] (*Sarah Corwin*, 3,) b. ——, 1690–1700.

97½ Mary Case,[4] (*Henry Case*, 1,) b. ——, 1697, d. April, 1777.
  M. —— Terry.   (*Moore's Index of Southold*, p. 65.)

97¾ Mary Case,[5] (*Samuel Case*, 29.)
  M. William Reeve, May 7th, 1744. (*Moore's Index of Southold*, p. 64.)

98 Mary Case,[6] (*Benjamin Case*, 27.)
  M. Abner Wells, December 10th, 1758.

98½ Mary Ann Millspaugh,[7] (*Cynthia Corwin*, 1,) b. June 15th, 1811, d. March 22d, 1843.

99 Mary J. Millspaugh,[7] (*Dorothy Corwin*, 1,) b. January 10th, 1816.
  M. (1) William Y. Smith, March 13th, 1839; he died October 9th, 1839; (2) Rev. J. M. Boyd, November 18th, 1840.
  Ch. Samuel Millspaugh.   (*Circleville, O.*)

100 Mary Sophia Woodhull,[7] (*Hannah Corwin*, 2,) b. October 13th, 1807, d. ——, 18—.
  M. Daniel Wilson.   (*New-York and Brooklyn.*)

101 Mary Halsey,[7] (*Elizabeth Corwin*, 2,) b. ——, 1810?
  M. —— Hicks.

102 Mary Davis,[7] (*Elizabeth Corwin*, 4,) b. ——, 1780–90.

103 Mary Matthews,[7] (*Mary Corwin*, 5,) b. about 1780–85.
  M. John Prawl.   (*Washington, Pa.*)

104 Mary Thatcher,[7] (*Jemima Corwin*, 1,) b. ——, 1790? in Warren Co., Ohio.   (*Missouri.*)

105 Mary Philips,[7] (*Hannah Corwin*, 2.)
  M. James Gilaspy.
  Three Ch.

106 Mary W. Duckworth,[7] (*Sarah Corwin*, 9,) b. February 19th, 1811.
  M. Luther Babbitt.

107 Mary Terry,[7] (*Keturah Corwin*, 1,) b. ——, 1780-90.

108 Mary Frances Terry,[8] (*Julia A. Corwin*, 5.)

109 Mary Ann Moffat,[8] (*Sarah Corwin*, 11,) b. ——, 1810-20.

110 Mary Horton,[8] (*Eliza Corwin*, 2,) b. ——, 1830-40.
    M. —— Hunsike.
    Ch. Mary, Millard.

111 Mary A. Odell,[8] (*Bethia Corwin*, 4.)

112 Mary Miller,[8] (*Sarah J. Corwin*, 32,) b. ——, 1825-35.

113 Mary E. Byram,[8] (*Mary A. Corwin*, 34,) b. February 10th, 1859.

114 Mary Ellston,[8] (*Esther Corwin*, 3,) b. ——, 1825-35.   (*West.*)

115 Mary Emma Peters,[8] (*Mary Corwin*, 30,) b. ——, 1835-45.

116 Mary Catharine Drake,[8] (*Arvilla Corwin*, 1,) b. ——, 1835-45.

117 Mary Jane Willinghouse,[8] (*Hannah J. Corwin*, 9.)
    M. George Rickett.   (*Newburg, N. Y.*)

118 Mary Woodruff,[8] (*Erastus Woodruff*, 1,) b. ——, 1825-35.

119 Mary Halstead,[8] (*Cynthia Hobart*, 10,) b. October 6th, 1820.
    M. —— Goodhue.   (*Illinois.*)

120 Mary Swina,[8] (*Bethia Hobart*, 11,) b. July, 1849.

121 Mary M. North,[8] (*Sarah Hobart*, 65,) b. December, 1846.

122 Mary P. Campbell,[8] (*Daniel Campbell*, 34,) b. November 18th, 1850,
    d. December 18th, 1859.

123 Mary Jane Millspaugh,[8] (*Gideon H. Millspaugh*, 1,) b. October 22d,
    1840, d. March 18th, 1841.

124 Mary C. Millspaugh,[8] (*Benjamin F. Millspaugh*, 31,) b. October 14th,
    1841.

125 Mary Ann Millspaugh,[8] (*William H. Millspaugh*, 71,) b. April 20th,
    1847.

126 Mary Smith,[8] (*Sarah Corwin*, 28,) b. ——, 1820, in Tunbridge, Vt.
    M. Daniel Leach, 1842.
    Ch. Frances H., Ellen.   (*St. Lawrence Co., N. Y.*)

127 Mary Letitia Millspaugh,[8] (*John H. Millspaugh*, 73,) b. ——, 1850.

128 Mary Terry,[8] (*Charlotte Corwin*, 5,) b. ——, 1815-20.
    M. Victor Drake.
    Ch. Franklin.   (*Newton, N. J.; Goshen, N. Y.*)

129 Mary Ann Flannery,[8] (*Mary Corwin*, 33,) b. at Cleveland, O., January
    21st, 1849.   (*Cleveland, O.*)

130 Mary Baldwin,[8] (*Jane Corwin*, 1,) b. ——.
    M. Virgil Smith.   (*Goshen, N. Y.*)

131 Mary Galloway,[8] (*Amanda Corwin*, 2,) b. ——.
    M. Henry Potter, of Yorktown, Westchester Co., N. Y.

132 Mary Morris,[8] (*Rhoda Corwin*, 1.) b. ——.
   M. Robert Wickersham.   (*Ohio.*)

133 Mary Eliza Dill,[8] (*Phebe Corwin*, 6.)
   M. Charles Kimball.   (*Minneapolis, Minn.*)

134 Mary Osborn,[6] (*Amelia Corwin*, 2,) b. June 12th, 1828.
          (*Teacher, Quincy, Ill.*)

135 Mary Elizabeth Howell,[8] (*Elizabeth Corwin*, 20.)

136 Mary Charlotte Young,[8] (*Eli H. Young*, 6,) b. July 10th, 1854, in Plymouth, Ill.

137 Mary Margaret Blinn,[8] (*Lucinda Thatcher*, 5,) b. September 23d, 1845.
   M. Harrison Derrick, November, 1863.
   Ch. William, Charles, Julia N.

138 Mary Sophia Brown,[6] (*Elizabeth Woodhull*, 45,) b. January 27th, 1827.
   M. Philip J. Crater, 1849.   (*Raritan, N. J.*)

139 Mary Craig Dunlevy,[8] (*Lucinda Corwin*, 4,) b. May 5th, 1840, d. September 15th, 1840.

140 Mary Wood,[8] (*Mary Corwin*, 16.)

141 Mary Mullock,[8] (*Mary Corwin*, 22.)

141½ Mary F. Colgrove,[8] (*Bela H. Colgrove*, 1,) b. May 3d, 1836.
   M. James H. Jewett, June 7th, 1854.
   Ch. Bela H., Nellie, Alice, Mary, Grace.   (*Buffalo, N. Y.*)

141¾ Mary Call,[8] (*Bethia Corwin*, 8,) b. October 21st, 1830.
   M. Nehemiah Hand, April 27th, 1867.
   One child.   (*East-Setauket, L. I.*)

142 Mary Elizabeth Howell,[8] (*Susan Corwin*, 1,) b. ——, 1825.

143 Mary Reed,[8] (*Sophia Corwin*, 1,) b. October 19th, 1798, d. December 8th, 1798.

144 Mary Corwin Reed,[8] (*Sophia Corwin*, 1,) b. July 4th, 1815.
   M. —— Lawrence.

145 Mary E. Mitchel,[8] (*Maria Corwin*, 3.)

146 Mary A. Smith,[8] (*Philena Corwin*, 1,) b. February 3d, 1829.
   M. John A. Reynolds, April 3d, 1853.
   Ch. Addie M., Nellie M.   (*Vermont.*)

147 Mary Girard,[8] (*Jerusha Corwin*, 2.)

148 Mary Young,[8] (*Mary Corwin*, 37.)

149 Mary Ann Pike,[8] (*Dorothy T. Corwin*, 5.)

150 Mary H. Knox,[8] (*Julia Corwin*, 7,) b. July 16th, 1848.

151 Mary Meyers,[8] (*Arminda Brown*, 2.)

152 Mary Brown,[8] (*Daniel C. Brown*, 36.)

153 Mary C. Brown,[8] (*Silas C. Brown*, 16.)

154 Mary Brown,[8] (*Parmenas Brown*, 1.)

155 Mary E. Brown,[8] (*Dorastus Brown*, 1.)

156 Mary Ann Howell,[8] (*Jemima Corwin*, 7.)
   M. John Talmadge.

157 Mary Elizabeth Hulse,[8] (*Lucetta Corwin*, 1.)
   M. William Wilson.

158 Mary Velnette Wells,[8] (*Puah Corwin*, 1.)
   M. James Scribner.

159 Mary Elizabeth Peters,[9] (*Eleazer Peters*, 3.)

160 Mary Davaw,[9] (*Helen Hunsike*, 6.)

161 Mary Eleanor Ketchum,[9] (*Nancy A. Corwin*, 3,) b. August 4th, 1841.
   M. Ebenezer Milice.   (*Indiana.*)

162 Mary Ann Tuthill,[9] (*Ann Corwin*, 22.)

163 Mary Denton Concklin,[9] (*Sarah M. Corwin*, 34,) b. September 11th,
   1833.

164 Mary Ada Penney,[9] (*Helen M. Corwin*, 3,) b. October 26th, 1855.

165 Mary D. Campbell,[9] (*Daniel E. Campbell*, 39,) b. August, 1861.

166 Mary J. Padden,[9] (*William A. Padden*, 94.)

167 Mary Hunsike,[9] (*Mary Horton*, 110.)

168 Mary Matilda Concklin,[9] (*Lucretia R. J. Corwin*, 3,) b. ——.
   M. Gilbert Squires.
   Ch. Eli.

168¼ Mary Belle Colgrove,[9] (*Rev. Clinton Colgrove*, 1,) b. May 26th, 1867.

169 Mary Chase,[9] (*Phebe Corwin*, 8½.)

169¼ Mary Jewett,[9] (*Mary J. Colgrove*, 141¼,) b. about 1861.

170 Mary Ann Moffat,[9] (*Sarah Corwin*, 49.)

171 Mary L. Noyes,[9] (*Louisa Corwin*, 3,) b. June 17th, 1844, d. June 20th,
   1864.   (*Vermont.*)

172 Mary E. Chamberlin,[9] (*Jennette Corwin*, 1,) b. April 18th, 1848.
                                                           (*Vermont.*)

172¼ Mary Sears,[9] (*Eliza N. Colgrove*, 26¼,) b. about 1852.

173 Mary H. Cram,[9] (*Hannah Hackett*, 26,) b. September 15th, 1833.
                                                           (*Vermont.*)

174 Mary E. Boisseau,[9] (*Bethia Corwin*, 9½,) b. June 10th, 1857.

175 Mary Elizabeth Crater,[10] (*Mary Corwin*, 83.)
   176 Mary Corwin Clark (Anna M. Thomson, 33.) b. June 29. 1874.
   (1) *MARY ANN,[8] [John, (4),] b.——.   (*Baltimore, Md.*)

   (2) *MARY ANN,[8] [Samuel, (6½),] b. June 1st, 1811, d. September 30th,
      1834.

   (3) *MARY,[8] [George, (7¼),] b. ——, 1810–20.   (*Indiana.*)

   (4) *MARY SPINGLER,[9] [Samuel, (12),] b. and d. October 22d, 1868.

                              * Corwine.

*176. Mary Corwin Clark,[9] (Anna Mary Thomson 33.)
born June 29. 1874, married May 30. 1894, Robert Tompkins
of Dingding, N.Y.*
Ch. Maurice Tompkins b. July 7. 1895
Marion Tompkins, b. Oct 16. 1896.

(5) Mary Lindall,[4] [*Elizabeth Curwen*, (3),] b. December 14th, 1705.

(6) Mary Jennette Stevenson,[9] [*Sarah Corwine*, (5),] b. February 21st, 1855, at Danville, Ky.

(7) Mary Hitch,[9] [*Lydia Corwine*, (2).]
 M. Dr. J. B. Roy.   (*Omega*, *O.*)

(8) Mary Sharp,[9] [*Sarah Corwine*, (9).]   (*Kingston*, *O.*)

(9) * MARY W.,[9] [*John*, (11),] b. ——, 1854. . (*Waverley*, *O.*)

(a) † MARY,[4] [*George*, (a),] b. ——.

(I) * MARY,[2] [SAMUEL, (I),] b. about 1660–80, perhaps.
 M. Thomas Smith, May 9th, 1701.
 Ch. Thomas, etc.

Mary's name is written Curwin, Corwine, Corwin, and Curran. Her father-in-law was also a Thomas Smith, of Boston. His will, (1688,) mentions his wife Rebecca, and five children, namely, Annie, Thomas, John, Elizabeth, and Rebecca. They were all living in 1706. Thomas, their son, (who married Mary Corwin,) was Captain of the Ancient and Honorable Artillery Company. His mother's name was Rebecca Glover, daughter of Habakkuk Glover and Hannah Eliot, of Boston. Hannah was the only daughter of Rev. John Eliot, the famous missionary to the Indians. (See *Glover Memorials*, 105, 114, and *Mass. Hist. and Gen. Rec.*)

Matilda, (Old High-German,) mighty battle-maid ; heroine.

1 MATILDA,[8] (*John Eli*, 19,) b. ——, 1850 ?

2 MATILDA,[8] (*Manasseh R.*, 15,) b. ——, 1805–15.
 M. —— Garlock.

3 Matilda Peters,[8] (*Mary Corwin*, 30,) b. ——, 1830–40.
 M. Asa French.
 Ch. Norman B.
 4 . *Matilda A. Call.* (B[mc] E. Ball. 90½) b 7h.y 23. 1862. d Aug 19.

Matthew, same as Matthias.

1 MATTHEW H.,[7] (*John*, 10,) b. September 1st, 1805, d. December 12th, 1805.

Matthias, (Hebrew,) gift of Jehovah.

1 MATTHIAS CORWIN,[1] (*the first settler of the name in America*,) was probably born in England, between 1590–1600, d. Sept. 1–12th, 1658.
 M. Margaret (Morton ?) (See *Appendix G.*)
 Ch. John, Martha, Theophilus.

His name also appears written sometimes *Curwin*, and even Currin, for a probable explanation of which, as well as the origin of the family, see *Introduction*. His name appears on the Commoner's Record at Ipswich, Mass., in 1634, as Currin, when he receives a second grant of land in that place.

   * Corwine.   † Curwen.

Ipswich was settled under the lead of Rev. Nathaniel Ward. When the tide of emigration turned from the older settlements, toward the beautiful Connecticut valleys, he joined the companies which went thither, probably by selection of the citizens at Ipswich, (Felt's *Ipswich*, p. 55,) and his name, as Curwin, accordingly appears among the first founders of New-Haven. (*Hollister*, i. 506.) But about this time Rev. John Youngs with several families, arrived at New-Haven, and this company bought the east end of Long Island of the Indians, and Mr. Youngs led his families thither, and settled at Southold, in October, 1640. Matthias Corwin joined this company. (Trumbull's *Connecticut*, i. 119.) The record at Ipswich notes that he emigrated thence to Long Island. He received a lot of land for a house, directly opposite the present Congregational church in Southold. The new lecture-room of that church now stands on the very plot. Here he lived for eighteen years, till his death, having been in America, altogether, about twenty-five years. He died between August 31st and September 15th, 1658. On December 11th, 1656, he was appointed, together with William Wells, Lieutenant Budd, Barnabas Horton, and William Purrier, to order town affairs. They were to meet in the "meeting-house," and were to be fined twenty shillings for non-attendance, unless they could show a good reason. (*Southold Records.*)

On December 5th, 1655, the property of Matthias Curwin is described as follows:

(1.) House lot, 4 acres, with lot of Barnabas Horton on the west, and that of Thomas Z. Keryhes ? on the east.

(2.) 18 acres of woodland, toward the north sea, (L. I. Sound,) bounded north by lands of Henry Whitnie, formerly, (now Pastor Young's,) south by lands of Thomas Moore.

(3.) 21 acres of woodland, toward the north sea, behind, bounded by Mr. Wells's (formerly Thomas Peakin's) lands east, and Mr. Hubbard's west.

(4.) 6 acres of woodland in the neck, near Tom's Creek Head; lands of Philemon Dickinson, north, and John Hubbard s, south.

(7.) 2 acres of meadow at the head of Tom's Creek.

(8.) 15 acres of woodland, adjoining the meadow of John Tucker and John Herbert; lands of Purrier, west, and of Henry Whitnie, east.

(9.) 3 acres of meadow, by Tom's Creek Pond; the meadow of Pastor Young's, north, and of Lieutenant Budd, south.

(10.) 1½ acres of arable land, in the old field, between lands of Joseph Horton, south, and John Tucker, north-west.

(11.) 2 acres of meadow, at Goose Neck, adjoining Pine Neck; the meadow of Henry and John Scudder, (now Mr. Elton's,) lying north.

(12.) 1 acre of meadow, on north-east side of Hog Neck, Mr. Wells's meadow, (formerly Edward Ketchum's,) lying north-east.

(13.) 3 acres of woodland in the Calves Neck.

(14.) 4½ acres of meadow, in Oyster Pond meadows, near to the south

11

end of the hedge, about the tobacco houses. (His son John exchanged this with Gideon Youngs for a meadow lot, containing six acres, having lands of Charles Glover on the east, and the meadow of Thomas Moore, Sr., (formerly John Tuthill's,) on the west.)

(15.) A third lot of meadow at Accoboack; Jer. Vail's meadow, lying north, and Abr. Whitehares, south-west.

(16.) A third lot of meadow at Corchacke, lying on the west side of the old fields, and extending from the head of the meadow to the sea, containing 3 acres; Joseph Horton's meadow lying west.

(17.) A fourth lot of meadow, on south side of Pechaconnicke River, containing 8 acres; Thomas Mapes's meadow, (formerly John Paine's,) west, and John Elton's, north.

(18.) 200 acres of woodland at Corchack; John Elton's land, east, and William Wells, west.

(19.) 40 acres of woodland at Corchacke; Richard Benjamin's land, north, and John Budd's meadow, south, and lying partly inclosed in said woodland.

(*Southold Records, Lib. B*, p. 11, *on new Book*, *Lib. B*, p. 127.)

His will, as recorded in the Southold town records, is yet preserved. The following extracts from the New-Haven records are found, referring to this will:

"A writing was presented, for the last will and testament of Matthias Curwin, but returned for legal probate. An inventory of the estate of said Matthias Curwin was presented, taken 15th of the seventh month, (1658,) by William Purrier and Charles Glover, amounting to £313 8d. That the appraisement was just, was witnessed upon by the above appraisers before

"BARNABAS HORTON."

(*Extract from Hoadley's Recs. New-Haven*, vol. ii. 35.)

On p. 400, he adds, "The last will and testament of Matthias Curwin, late of Southold, was presented, subscribed by John Underhill, and deposed by Barnabas Horton, at the Court, March 5th, 1660. Before William Wells, John Youngs."

The original will may possibly yet exist in the New-Haven archives. The following is a copy:

"SOUTHOLD, August 31, 1658.

"The last will & testament of me, Mathias Curwin, beinge in perfect memorie, do, in the name of our Lord God, & all men, comit my soule to God, & my bodie to the dust from whence it came. My will is, that my wife, Margarett Curwin, & my sonn, John Curwin, shall be my true & lawfull executors to administer upon my present estate as followeth: Item: I give to my Daughter, Martha, twentie pounds sterlinge, and over and above such goods my wife shall buy for her in the bay (?) It.: I give unto my sonn, Theophilus, 20 lbs. sterlinge, to bee payed to them when they shall leave their mother, accordinge to the law and custome of this Colonie. It.:

The remainder of my estate I give to my wife and son, John, equally betwixt them untill my wife marrie. But in case shee shall marrie, then my son, John, is to pay unto her 40 lbs. sterlinge uppon the day of her marriage. And this, my will and testiment, to stand in full force & power after my decease.

"In presence of John Underhill, Barnabas Horton.

"MATHIAS CURWIN."
(*Southold Town Records, Lib. B, p. 95.*)

The following is the inventory of his estate :
An inventory of the estate of the said Mathias Curwin, lately deceased.

	£	s.	d.
Impris. Houses and lands with th' appertenes, ...	50	00	00
It. Neate cattle, ...	97	00	00
It. In horse flesh, ...	20	00	00
It. Sheepe kinde, ...	23	16	00
It. In goats & swyne, ...	14	00	00
It. Cart, plow, with all furniture, ...	05	17	00
It. Working tools, with other things, ...	06	14	00
It. Corne & hay, ...	29	00	00
It. Bedsteads & beddinge, ...	19	17	00
It. The man's warcinge & tools, & lynnen, ...	09	16	00
It. Arme's wooll & some cheese, ...	16	13	06
It. Pewter, brasse, & other trumpery, ...	15	18	00

Apprisers: BARNABAS HORTON,
WILLIAM PURRIER, } Sworn.
CHARLES GLOVER.

(Immediately preceding this will on the town book, p. 92, is copied, "September 15th, 1658," the inventory of the estate of Joseph Youngs, mariner ; and following it, p. 97, Thomas Cooper's will, "15th September, 1658," and p. 98, his inventory, " January 20th, 1658-59.")

2 MATTHIAS,[3] (*John,* 1,) b. ——, 1676, d. March 9th, 1769.

M. Mary ——, in or before 1708. A Mary Corwin, wife of Matthias, died December 8th, 1725.

Ch. Matthias ? Jesse ?

In 1695 ? received a deed of land from his father. 1705, he gave a deed to Margaret Corey or Cooper. 1708, he gave a deed to his brother John. In same year, sold homestead to his brother John, containing 22 acres, for £76. 1714, sold 10 acres of land to Samuel Sweezey, for £23. His name on census list for 1698. (*Doc. Hist. N. Y.* i. 450.)

(*Southold, L. I.*)

3 MATTHIAS,[4] (*Matthias,* 2,) b. ——, d. October, 1782.

M. (1) Mary ——? (2) Elizabeth Benjamin, in 1750, who d. June 4th, 1783.

Ch. Jeremiah, Matthias, Mary, Rev. Jacob, Gershon, Gilbert, Naomi.

In 1775, a Matthias Corwin was in Captain Lupton's Company, and signed engagement to support Congress. In 1776, on census list, 393, having in his family one male over fifty, one male and one female over sixteen, and one boy.

4 MATTHIAS,[5] (*Matthias*, 3,) b. ——, 1732, d. June 10th, 1801.
    M. (1) Naomi Davis, 1754; (2) Elizabeth ——, who was b. ——, 1728, d. June 1st, 1783.
    Ch. Elizabeth.
On census list, 444, in 1776, (or 1766 ?) having in family one male, one female over sixteen, and one child.

5 MATTHIAS,[6] (JESSE, 2,) b. February 19th, 1761, in Morris Co., N. J., d. ——, 1829.
    M. (1) Patience Halleck, in Fayette Co., Pa., April 8th, 1782. She was b. ——, 1761, d. ——, 1818; (2) Mrs. Elizabeth Corbley, January 2d, 1820, by whom there were no children.
    Ch. Elizabeth, Benjamin, Matthias, Mary, Hon. Thomas, Jesse, Rhoda, Phebe, Amelia.
He lived successively in Morris Co., N. J., Fayette Co., Pa., Kentucky, and Ohio. He moved in 1785 from Pennsylvania to Kentucky, settling first in Mason Co., on a stream called Sharon, and afterward moved to Bourbon Co., in the same State. In 1798, he removed to the north-western territory, (now Ohio,) on account of bad land-titles in Kentucky, and settled near where the town of Lebanon, Warren Co., now stands. He was elected a Justice of the Peace, in Kentucky, and was eleven times chosen as a member of the Ohio Legislature, being sometimes speaker of the Assembly. He was also Associate Judge of the Court of Common Pleas. He was universally esteemed for his goodness of heart, his kindness to all, but especially to the poor and afflicted. He was a devoted and consistent Christian, (of the Baptist order,) which led him at length to desire respite from the turmoil of public life. He was above medium size, very stout, dark skin, black hair and eyes, and in moderately comfortable circumstances.

6 MATTHIAS,[6] (*Jeremiah*, 1,) b. about 1777, d. about 1850.
    M. Julia A. Corwin, No. 6, January 8th, 1801. (*Aquebogue Records.*)
    Ch. Mary F., Julia A., Sophia, Harrison, Charles H., Matthias, Nathan, Jeremiah, Gilbert.

7 MATTHIAS,[7] (*Matthias*, 5,) b. September 29th, 1789, d. October 8th, 1822.
    M. Minerva Brown.
    Ch. Emeline, Franklin.
He was a captain under General Harrison, in 1812, and served with distinction; was clerk of the Court of Common Pleas, in Warren Co., Ohio. Was a lawyer. (*Ohio.*)

8 MATTHIAS,[7] (*Matthias*, 6,) b. ——, 1800–1810.
    M.

Ch. Harriet, Charles M., Adelaide, Elizabeth.   (*New-London, Ct.*)

9 MATTHIAS,[7] (*Ichabod,* 1,) b. February 7th, 1818, d. January 15th, 1862.
    M. Lavinia A. Williamson, August 20th, 1839.

10 MATTHIAS,[7] (*Joseph,* 10.)

11 MATTHIAS,[8] (*Charles,* 3,) b. December 31st, 1848.
    M. Charry M. Paine, October, 1870.

12 MATTHIAS,[8] (*Nathan,* 5,) b. ——, 1822–30.

13 Matthias Hutchinson,[3] (Lieutenant,) (*Martha Corwin,* 1,) b. ——, 1668,
    d. January 16th, 1724.
    M. Mary ——. She b. —— –, 1674, d. February, 1721.
                             (*See Moore's Index of Southold, p. 94.*)

14 Matthias Drake,[8] (*Mary Corwin,* 20,) b. ——, d. ——.   (*Ohio.*)

15 Matthias Osborn,[8] (*Amelia Corwin,* 2,) b. December 12th, 1834.
    M. Jane Colton.
    Ch. Amelia.   (*Mobile, Ala.*)

(1) * McCOLLISTER,[9] [*George,* (11½,)] b. ——.   (*Missouri.*)

1 - Maurice Tomblein (Mary Corwin Clark 176) b. July 7, 180

    Mehetable, (Hebrew,) benefited of God.

1 MEHETABLE,[3] (*Theophilus,* 1 ?) b. probably 1660–70.
    Name occurs in census list of Southold, 1698, standing alone, her father being probably dead.

2 MEHETABLE,[5] (*John,* 4.)

2¼ MEHETABLE,[6] (*Nathan,* 1½,) bapt. at Aquebogue, September 20th, 1772.

2½ MEHETABLE,[6] (*David,* 4,) b. ——, d. young.

3 MEHETABLE,[6] (*Phineas,* 1,) b. May 29th, 1777, d. May 4th, 1858.
    M. Samuel Hobart.
    Ch. Fanny, Cynthia, Peter, Bethia, Sarah, Charlotte, Emily, Desire.

4 MEHETABLE,[6] (*James,* 1,) b. ——, 1773 ? d. ——, 1794.
    M. James Boak.   (*Brooklyn, N. Y.*)

5 MEHETABLE,[7] (*George,* 2,) b. ——, 1798.
    M. Thomas Shannon.
    Ch. Nehemiah, Fanny, Sarah A., George, Noahdiah, Parshall, Eunice.
                                (*Weston, N. Y.*)

6 MEHETABLE,[7] (*James,* 8,) b. ——, 1800–10.

7 MEHETABLE,[7] (*Benjamin,* 8,) b. ——, 1790–1800.
    Unm.

8 MEHETABLE,[8] (*Richard S.,* 7,) b. ——, 1815–25.

10 Mehetable Campbell,[7] (*Sarah Corwin,* 6,) b. November 6th, 1805.
    M. Alonzo North, October 12th, 1836.
    Ch. Sarah, Chloe, Joseph, Erastus, Maria J., Mehetable.
                                (*Elbridge, N. Y.*)

* Corwine.

11 Mehetable Little,[7] (*Desire Corwin*, 1,) b. June, 1810, d. January, 1826.
12 Mehetable Taylor,[7] (*Fanny Corwin*, 1,) b. September, 1828.
M. George Fowler, September, 1855.
Ch. Fanny.
13 Mehetable Corwin Warner,[7] (*Elizabeth Corwin*, 10,) b. ——.
M. Epenetus Purdy.
14 Mehetable Bishop,[7] (*Mary Corwin*, 11,) b. ——, 1800–10.
M. Jonas Winters.
Ch. Mehetable.   (*Beaverdam, L. I.*)
15 Mehetable North,[8] (*Mehetable Campbell*, 10,) b. January 26th, 1840.
16 Mehetable Winters,[8] (*Mehetable Bishop ?* 14.)
M. Joseph C. Beers.
17 Mehetable Skult,[8] (*Fanny Hobart*, 9,) b. ——, 1820–30.
M. George Haskins.   (*New-York City.*)
18 Mehetable North,[8] (*Sarah Hobart*, 65,) b. June, 1837.
M. William Smith.
Ch. Franklin, and two others.
(1) * MEHETABLE,[6] [*George*, (5),] b. January 23d, 1741, d. April 4th, 1813.
M. Richard Ward, of Salem, November 8th, 1764; was b. April 5th, 1741, d. November 24th, 1824.
Ch. George C., Samuel C., Sarah, Mehetable, Elizabeth, Richard, Martha, Daniel.

He, with his father, Joshua Ward, Esq., ardently espoused the popular cause in the Revolution. He was a member of the "Committee of Safety and Protection" during the entire struggle. He was a subaltern officer in Colonel Pickering's Regiment, marched on the Lexington alarm, and likewise on the day of the battle of Bunker Hill, though on neither occasion in season to participate in the conflict. In June, 1776, was commissioned as a captain, but family affairs compelled his retirement in 1777. He was for a long time at the head of the town government, and a member of the State Legislature.

For the Ward genealogy, see *Historical Collections* of the Essex Institute, Mass., Vol. V., No. 5, October, 1863. The descendants of Samuel Curwen Ward, only, are contained in this volume.

Melissa, (Greek,) a bee.

1 Melissa H. Hobart,[8] (*Peter Hobart*, 9,) b. July 24th, 1840.
M. Merritt Chatfield, 1861.
Ch. Sarah L.   (*Beloit, Wis.*)
2 Melissa Sarepta Concklin,[9] (*Lucretia R. J. Corwin*, 3,) b. ——.
Unm.
1 Melvin J. Davis,[9] (*Rachel Corwin*, 4,) b. ——, 1845.

* Curwen.

M. Adaline E. Belknap, 1867.
Ch. Ella.

1 MERCY,[8] (*Bestir,* 1,) d. 6 years old.
1 MERINDA,[8] (*Peter,* 3,) b. ——.

Michael, (Hebrew,) who is like God.

1 MICHAL,[5] (*Daniel,* 2,) b. ——, 1710–30.
M. Mr. —— Sweazy?
(*This name, Michal, is feminine.*)

1 Miles A. Davis,[9] (*Rachel Corwin,* 4,) b. March 5th, 1843.
Printer and writer for the press. (*Branchport, N. Y.*)

1 MILLA FRANCES,[8] (*Gilbert,* 4,) b. July 28th, 1833.
M. Daniel A. Griffin, January 12th, 1859.

1 Millard Hunsike,[9] (*Mary Horton,* 110.)

1 Miller Halsey,[7] (*Elizabeth Corwin,* 2,) b. ——, 1814?
M.
Ch. Cary, Marcellus. (*Drakesville, N. J.*)

1 MILTON,[7] (*Joseph,* 10.)
2 MILTON,[7] (*Phineas,* 3,) b. December 8th, 1827, d. October 5th, 1853.
3 Milton Goble,[5] (*Elizabeth H. Corwin,* 19.)
M. Catharine Lasher.
4 Milton C. Connor,[5] (*Caroline Corwin,* 2,) b. September 6th, 1853.
5 Milton Higby,[8] (*Harriet Corwin,* 2,) b. ——, 1862.
6 Milton Padden,[5] (*Anna Campbell,* 29,) b. ——, 1822–30, d. April, 1843.

1 MINERVA,[7] (*David,* 8.)
1 MINISSA,[8] (*Cortland E.,* 1,) b. ——.

Minnie, see Mary.

1 MINNIE,[9] (*George,* 40,) b. ——, 1858.
2 Minnie A. Whitney,[10] (*Ann M. Hackett,* 43,) b. August 9th, 1867.
(*Vermont.*)

Miranda, (Latin,) admirable.

1 MIRANDA,[8] (*John,* 30,) b. ——, 1813.
M. Elijah Dean.
No ch. (*Newfield, Tompkins Co., N. Y.*)

Miriam, same as Mary.

1 MIRIAM,[8] (*Alsop H.,* 2.)
2 MIRIAM GRANT,[10] (*Henry H.,* 26,) b. about 1869.

1 MONROE,[8] (*William H. H.,* 18,) b. May 8th, 1843, d. April 17th, 1852.
1 MORGIANA,[8] (*Harrison,* 1.)

1 Morrison Mercein,[9] (*George H.*, 24,) b. March 21st, 1858.

2 Morrison Wood,[8] (*Mary Corwin*, 16.)

1 Mortimer,[10] (*George*, 48,) b. ——, 1859. *- ᵗʸ ᵗʸ ᵗʸ α ᵗʸ ᵗʸ*

Moses, (Hebrew,) drawn out of the water.

1 Moses,[6] (*James*, 1,) b. at Middletown, N. Y., January 16th, 1785, d. March 29th, 1866.

  M. Martha Stuart, September 28th, 1809. She b. June 24th, 1791.

  Ch. Fanny J., Hiram, Julia, Charles E., Elizabeth, Margaret, George S., Mary, Joanna, Moses H., Samuel D., James, Daniel, William H.

(Middletown, N. Y.)

2 Moses Horton,[7] (*Moses*, 1,) b. October 22d, 1826, at Middletown, N. Y., d. ——, 1854, in New-York City.

  M. Mary Ann Parker.

  Ch. Martha, Charles, Theresa. (New-York City.)

3 Moses,[7] (*Samuel*, 12,) b. January 5th, 1793, d. April 13th, 1869.

  M. Vilura Morley. She was b. April 12th, 1797, d. February 19th, 1868.

  No ch.

He settled in Madison Co., Indiana, in 1823, (where he afterward lived,) about two and a half miles south-west of Pendleton. A farmer. Though he had no children, he raised six or seven orphan children, and left to them, at his death, about $10,000.

4 Moses B.,[7] (*Ichabod*, 1,) b. in Bourbon Co., Ky., January 5th, 1790.

  M. Margaret Fox, 1811.

  Ch. Ichabod, John A.

He spent his boyhood on a farm in Ohio, receiving, however, a good education. He studied law, and was admitted to the bar in 1812. In 1838–9, he was a member of the Ohio Legislature, and was a representative in Congress, 1849–55, serving as a member of the committee on the Post-Office Department. In his last canvass, his own son, John A., was the opposing candidate; the campaign was very amusing, perfectly friendly and courteous, but highly exciting. (Urbana, O.)

5 Moses Young,[8] (*George L.*, 14,) b. May 28th, 1838.

6 Moses,[8] (*William H. H.*, 18,) b. and d. 1841.

6½ Lieut. Moses Case,[5] (*Samuel Case*, 29,) b. ——, 1723, d. ——, 1814.

(Moore's Index of Southold, p. 64.)

7 Moses Everett Young,[8] (*Eli H. Young*, 6,) b. April 9th, 1856, in Plymouth, Illinois.

8 Moses Hulse,[8] (*Lucetta Corwin*, 1.)

9 Moses B. Newman,[8] (*Eliza Corwin*, 6,) b. at Lebanon, O., March 9th, 1816.

M. Harriet A. Sheppard, at Sidney, O., December 10th, 1851. She was b. at Frome, Eng., December 31st, 1829.

Ch. Eliza C., Maria E., Frank P., Cornelia L., Edmund P., Robert W.

A lawyer of Wyandotte, Kansas. His wife's father was Richard G. Sheppard, who was born at Frome, England, October 21st, 1791. He married Rosamond Rawlings, December 11th, 1813, and died at Syracuse, N. Y., October 24th, 1834. Rosamond Rawlings was born at Frome, August 29th, 1792, and died in St. Lawrence Co., N. Y., May 5th, 1865.

1 MYRON,[7] (——,) b. ——.

He was of Camillus, N. Y. His will is dated February 22d, 1853. His wife, Ruth Ann, was sole executrix.

1½ MYRON,[8] (*Ira*, 4¼,) b. June 6th, 1847.

Enlisted in Federal Army, in 1864, Co. I, 53d Illinois. Was wounded in the right arm. Returned home in 1865. (*La Salle, Ill.*)

2 Myron Swina,[8] (*Bethia Hobart*, 9,) b. May, 1847.

3 Myron E. Whitney,[10] (*Emma E. Hackett*, 19,) b. January 7th, 1868.

(*Vermont.*)

Naboth, (Hebrew,) fruit.

1 NABOTH DAVIS,[7] (*David*, 4,) b. February 13th, 1787, d. April 13th, 1813.

Nancy, a familiar form of Anne or Hannah.

1 * NANCY,[5 or 6] (*George*, 1,) b. ——, 1770–80.

Samuel Philips, of Brookhaven, married a Nancy Corwin, February 18th, 1789. (*Aquebogue Records.*)

1½ NANCY HANNAH,[6] (*Edward*, 2,) b. ——, 1758, bapt. at Aquebogue, under the name of Hannah, July 29th, 1764, d. May 11th, 1820, at Coventry, R. I.

M. John Colgrove, (a widower,) of Coventry, R. I. He died about 1816.

Ch. Bela H.

She and her husband are both buried at Sterling, Ct. He had several small children when he married his second wife. She removed to Connecticut, from Long Island, probably about 1763–5. Her only son remembers her allusions, (a half-century ago,) to Southold, Potopog, and Oyster Ponds. She became at an early age a member of the Congregational church of Rev. Samuel Nott, of Franklin, Ct., where she was married as above, and removed to Rhode Island. Mr. Colgrove's youngest sister, Phebe, married Phineas Corwin, 1½.

1¾ NANCY,[7] (*Edward*, 3,) b. January 27th, 1811.

M. O. Gallup.

Twelve ch.

2 NANCY,[7] (*Isaac*, 1,) b. October 6th, 1797.

M. Jacob Rieger.

Ch. Isaac. (*Chester, N. J.*)

* Curran.

2¼ NANCY,[7] (*Asa*, 2,) bapt. at Aquebogue June 16th, 1796.

3 NANCY ANN,[8] (*Benjamin*, 14,) b. January, 1818.
  M. (1) Charles Ketchum; (2) Azariah Julian.
  Ch. Mary E., Charles H., Frank, Jenny.   (*Warsaw, Ind.*)

4 NANCY JANE,[8] (*Jesse*, 4,) b. ——, 1815–25.

5 NANCY,[8] (——,) b. ——, 1831, d. January 30th, 1859.
  M. Isaac H. Rieger. He b. November 30th, 1833, d. October 26th,
    1865.
  (A Nancy Corwin died at Aquebogue, May, 1846.)

5½ NANCY PARMELA,[8] (*David*, 12,) b. December 20th, 1819, d. March 13th,
    1856.
  M. Anson Hewlett.
  No ch.   (*Connecticut.*)

5¾ NANCY,[8] (*Phineas*, 1¾.)

6 Nancy Woodhull,[7] (*Hannah Corwin*, 2,) b. May 19th, 1804, d. ——.
  M. (1) —— Chipps; (2) —— Prudens.   (*Near Baskingridge, N. J.*)

7 Nancy Terry,[8] (*Charlotte Corwin*, 5,) b. ——, 1815–30.
  M. Victor Drake.

8 Nancy Hopkins,[8] (*Martha Corwin*, 3,) b. ——, 1815–25, d. age 8.

Naomi, (Hebrew,) my pleasantness.

1 NAOMI,[6] (*Matthias*, 3,) b. ——, d. ——, 1833.
  M. David Corwin, 6.

2 NAOMI,[7] (*Joseph*, 4,) b. July 24th, 1803, d. November 20th, 1814.

3 NAOMI ANN,[8] (*Joshua*, 6,) b. June 3th, 1831.
  M. Robert Tompkins.
  Ch. Egbert.   (*Newark, N. J.*)

(1) * NAOMI,[6] [*Joseph*, (1),] b. April 16th, 1752, in West-Jersey. Removed
    to Canada, about 1784.

Nathan, (Hebrew,) a gift.

1 NATHAN,[5] (*Daniel*, 2,) b. probably after 1726, d. before 1783.
  M. Phebe, daughter of Isaac Howell. She died a widow June 1st,
    1783, (or April, 1784.)
  Ch.

In 1775, he signed engagement to support Congress. In 1776, his name
appears in census list, 308, having in his family one male, one female over
sixteen, and five children.

1½ NATHAN,[5 or 6] (——,) b. ——, d. ——.
  M. Mary Williamson, August 5th, 1771. Both of Aquebogue.
  Ch. Mehetable, Phebe, Mary, James.   (*Aquebogue Records.*)

2 NATHAN,[6] (*Jedediah*, 2,) b. ——, 1758, d. June 21st, 1830.

* Corwine.

M. Lydia Young, June 4th, 1781. She was b. February, 1758, d. August 16th, 1852. (*Aquebogue Records.*)

Ch. Julia, Lydia, Charry, Finey, Sarah.

He united with the church of Aquebogue, October 10th, 1790.

3 NATHAN HOWELL,[7] (*Eli*, 3,) b. February 19th, 1796, d. October 7th, 1857.

M. (1) Olive Belknap, daughter of David and Mary Case, February 18th, 1819. She was b. October, 1797, d. October 11th, 1839; (2) Phebe Bannister, October 6th, 1841. She was b. July 28th, 1814.

Ch. Samuel W., Julia E., Mary A., Adelaide, Emma. (*Widow now lives in Newark.*) (*Middletown, N. Y.*)

4 NATHAN,[7] (*John Calvin*, 11,) b. ——, 1790–1800.

5 NATHAN,[7] (*Matthias*, 6,) b. October 17th, 1801.

M. Mehetable Tuthill.

Ch. Matthias, Wallace, Hannibal, Mary, Jane, Amanda, Rosabella, Maria. (*A merchant; Riverhead, L. I.*)

6 NATHAN,[7] (*James*, 9.)

7 NATHAN,[7] (*Stephen*, 2,) b. ——, 1790–1800.

M. Sophronia ——. (*Chautauqua Co., N. Y.*)

8 NATHAN,[7]? (——,) b. about 1779, d. May 15th, 1850.

There was a Nathan Corwin (according to record of wills in Auburn, Cayuga Co., N. Y.) who died at the above date. He was of the town of Metz. He appointed Washington Bogardus, and Polly Coleman, his sister, of Blooming Grove, Orange Co., N. Y., his executors. He left no widow. Derick Ireland, of Huntington, L. I., was a nephew.

9 Nathan H. Moffat,[8] (*Sarah Corwin*, 31.)

Nathaniel, (Hebrew,) a gift of God.

1 NATHANIEL,[5] (Samuel, 3 ?) b. ——, 1740–50.

M. Deborah Hutchinson, January 1st, 1767, at Southold, L. I.

Ch. Cynthia, Vincent, Pitt, Watts, Stephen.

(*From Southold Church Records.*)

In 1775, he signed engagement to support Congress. In 1776, his name is in census list, 58, having in his family one male, one female over sixteen, and three children.

1½ NATHANIEL,[5]? (——,) b. ——, d. October 20th, 1783.

2 NATHANIEL,[6] (*William*, 2,) b. September 26th, 1783, d. February 24th, 1849.

M. (1) Elizabeth, daughter of Barnabas and Elizabeth Horton. She was b. November, 1786, d. May 5th, 1806. (Barnabas Horton was b. September 27th, 1759, d. December 6th, 1800. His wife Elizabeth was b. January 3d, 1760, d. January 26th, 1831;) (2) —— Monroe; (3) Adaline Pickle; (4) Sarah Bell.

Ch. William H. H., Elizabeth. (*Succasunna, N. J.*)

3 NATHANIEL,[7] (*Joseph*, 8,) b. ——, 1785, ? d. ——, 1860.
M. Betsy Biles.
Ch. William, George, Drake.   (*Warren Co., N. J.*)

In Warren Co., N. J., wills, there is a will of a Nathaniel Corwin approved in 1861, having wife Elizabeth, and children or children-in-law, (or relatives,) namely, Joseph, John B., George B., William B., Anthony Drake, Mary, Dennis, and Anna P. Howell, or Honnell. George B. and William B. were executors. (Probably No. 3.)

4 NATHANIEL,[6] (*John*, 6,) b. August 20th, 1759, d. young.

5 NATHANIEL,[6] (*Daniel*, 3,) b. ——, 1760–70, d. ——, 1858.
M. —— Culver.
Ch. Ardon, Mahlon, Gamaliel, Daniel, James. See will.
                                      (*Southampton, L. I.*)

6 NATHANIEL,[7] (*Eli II.*, 2,) b. December 29th, 1809, d. February 14th, 1839.
M. Jane E. Felter, November 19th, 1832.
Ch. Abel D., Mary C.

7 NATHANIEL,[7] (*Abel*, 1,) b. at Baiting Hollow, L. I., December 11th, 1792, d. June 16th, 1868.
M. Mary Ann Lemmaa, September 29th, 1818. She was daughter of John and Thankful (Jeffrey) Lemmaa.
Ch. Cecilia A., John M., Mary J., Elizabeth S., Seth M., Lucy L., Walter E., Frances E., Amelia H., Chatham.

He was one of the principal founders of Greenport, L. I. He named the place, built the first wharf, and had the first whaling-ship which sailed from that place. He was studious and gentlemanly, a true philanthropist, and one of nature's noblemen—a self-made man. He had a clear head, a pure heart, untiring energy, yet modest as a child. More than once he refused to allow his name to run for some of the highest offices in the State. Merchant, New-York City, 1817–29. He afterward became an agent for whaling-ships sailing from Greenport, L. I.

8 NATHANIEL,[8] (*Richard*, 4.)

9 NATHANIEL,[7] (——,) b. about 1800.
M.
Ch. Jehiel.
(*Possibly this No. 9 is the same as some of the preceding.*)

10 Nathaniel Corwin Wells,[6 or 7] (*Cynthia Corwin*, 3,) b. ——, 1795, d. ——, 1831.
M. Patience Clark. She afterward married (2) Rev. Mr. Grier, and (3) Rev. Mr. Clark ; having
Ch. William H. Wells, of Cold Spring, N. Y., and John Clark.

11 Nathaniel Chase,[8] (*Bethia Corwin*, 5.)

12 Nathaniel C. Reed,[8] (*Sophia Corwin*, 1,) b. February 13th, 1814.

1 N. BENJAMIN,[8] (*John*, 25.)

Nebat, (Hebrew,) a view.

1 NEBAT,⁶ (*David*, 4,) b. ——, 1783 ? d. ——.
M. Mary Howard.
Ch. De Witt C., Howell ? Laura, Sarah J.

Nehemiah, (Hebrew,) comfort of Jehovah.

1 Nehemiah Shannon,⁸ (*Mehetable Corwin*, 5,) b. ——, 1820–30.
(*Hector, N. Y.*)

Nellie, see Ellen.

1 NELLIE E.,⁹ (*William H.*, 38,) b. October 19th, 1859, d. August 26th, 1864.

2 Nellie M. Reynolds,⁹ (*Mary A. Smith*, 146,) b. January 15th, 1864.
(*Vermont.*)

3 Nellie F. Hood,⁹ (*Caroline L. Corwin*, 4,) b. October 23d, 1856.
(*Vermont.*)

4 Nellie M. Noyes,⁹ (*Louisa Corwin*, 3,) b. September 5th, 1862, d. August 21st, 1869.
(*Vermont.*)

5 Nellie C. Boisseau,⁹ (*Bethia Corwin*, 9½,) b. June 30th, 1868, d. September 28th, 1869.

6 Nellie Jewett,⁹ (*Mary F. Colgrove*, 141¼,) b. about 1857.

1 NELSON HARVEY,⁸ (*Noah*, 2,) b. October 13th, 1847.
M. Hattie Heddin, October 27th, 1870.

2 Nelson Drake,⁸ (*Margaret Corwin*, 1,) b. July, 1823.
(*Keeper of State Prison, Trenton, N. J.; Home, Flanders, N. J.*)
There was a Nelson T. Corwin in New-York City, 1837–39, 1844, 1862, and a Nelson G. Corwin in 1864–5. (See *Directory.*)

3 Nelson Green ——,⁸ (*Charry Corwin*, 1.)

1 Nettie Blinn,⁹ (*George D. Blinn*, 56,) b. December 20th, 1867.
2 *Nettie Thomason (William Thomason 77) b. April 30. 1864*
Nicholas, (Greek,) victory of the people.

1 NICHOLAS,⁸ (*Peter*, 3,) b. ——, 1805, d. ——, 1861.
M. (1) Elizabeth Howell ; (2) Julia Howell ; (3) —— Fansheur.
Ch. Mary, William H., Sarah A., Henry H.
(*Chester, N. J.*)

1 Nichols Squires,⁸ (*Patty Corwin*, 1.)

Noah, (Hebrew,) rest; comfort.

1 Noah Smith,⁸ (*Sarah Corwin*, 24.)

2 NOAH,⁷ (*Stephen*, 2,) b. near Scotch Plains, N. J., January 15th, 1796.
M. (1) Eunice Madison, April 6th, 1820. She b. October 13th, 1797, d. April 28th, 1839 ; (2) Eliza Decker, August 23d, 1839. She b. July 16th, 1816, d. February 4th, 1851 ; (3) Anna Price, August 28th, 1851.

Ch. Isaac, George, Stephen, Caroline, Andrew, Abigail, Nelson.
He lived in Barrington, 1826, and Jerusalem, 1836, Yates Co., N. Y.

3 NOAH,[8] (*John*, 30,) b. ——, 1816.
M. Eliza Jane Buck, of Schuyler Co., N. Y., (or Tioga Co., Pa. ?)
Ch. John, Delilah E., Gideon.

(*Near Caton Centre, Steuben Co., N. Y.*)

Noadiah, (Hebrew,) rest from Jehovah.

1 NOADIAH,[7] (*George*, 2,) b. ——, 1812, d. ——, 1833.
2 Noadiah Shannon,[8] (*Mehetable Corwin*, 5,) b. ——, 1820–30.

Norman, (German,) a northman.

1 Norman B. French, (*Matilda Peters*, 3.)

Obadiah, (Hebrew,) servant of Jehovah.

1 OBADIAH,[6] (——,) b. ——, d. August 9th, 1746.

Olive, (Latin,) an olive.

1 OLIVE,[7] (*Edward*, 3,) b. July 2d, 1803, d. October 2d, 1826.
M. L. Wolcott.
Two sons.
2 Olive Augusta Wilson,[9] (*Julia E. Corwin*, 9.)

Oliver, (Latin,) an olive-tree.

1 OLIVER P.,[7] (*Jason*, 1,) b. September 9th, 1815, d. November 14th, 1815.
2 OLIVER HAVENS,[8] (*Epenetus II.*, 1,) b. ——, 1830–40.
M. Eliza J. Tosick, (daughter of William and Elizabeth Tosick,) March 24th, 1870.
3 OLIVER C.,[8] (*John*, 22,) b. at Mattituck, L. I., August 2d, 1830.
M. Mary Ann, daughter of J. B. and Mary Jane Stevens, at Charlestown, Mass., January 24th, 1853.
No ch.
Was originally a seaman, now a farmer; served in the federal army in the rebellion, as an officer on a transport; was present at the taking of Hill's Head, (near Charleston,) and was with General Banks at New-Orleans; was on board the steamer Curlew when it was sunk in Chesapeake Bay. (*Topham, Vt.*)
4 OLIVER GOLDSMITH,[8] (*Lemuel F.*, 1,) b. August 31st, 1851.
(*Succasunna, N. J.*)
5 OLIVER,[8] (*Henry*, 10,) b. July 25th, 1800, d. September 7th, 1851.
M. Eliza Winship.
Ch. Daniel. (*New-York City.*)

6 OLIVER B.,[8] (*William O. ?* 29,) b. ——.
 M. Mary Clark. She d. 1843, leaving a will.   (*Wallkill, N. Y.*)
7 OLIVER B.,[8] (——,) b. ——.
 Possibly the same as No. 6. He sold land about 1835–40, in Chemung
Co., N. Y. (Deeds, vol. xxxv. p. 359.)
8 OLIVER COREY,[9] (*Joseph*, 19,) b. September, 1848.
 M. Betsy Robinson.
 No ch.   (*Riverhead, L. I.*)
8½ OLIVER,[9] (*Alca*, 1,) b. September 7th, 1839.
 M. Sarah Concklin, December 31st, 1862.
 Ch. Halsey, Henry F.
9 Oliver Millspaugh,[8] (*Gilbert C. Millspaugh*, 6,) b. July 20th, 1831.
10 Oliver Terry,[8] (*Charlotte Corwin*, 5,) b. ——, 1815–30.
11 Oliver Chase, (*Bethia Corwin*, 5.)

Olivia, same as Olive.

1 OLIVIA,[6] (*Eli*, 1,) b. July 17th, 1796, d. September 15th, 1818.
     (*Middletown, N. Y.*)
2 OLIVIA,[7] (*Daniel*, 7,) b. July 3d, 1819.
 M. Thomas E. De Kay, April 19th, 1839.   (*Hamptonburg, N. Y.*)
4 Olivia Connor,[8] (*Caroline Corwin*, 2,) b. August 30th, 1840.
5 Olivia Pike,[8] (*Dorothy T. Corwin*, 5.)
1 OMER R.,[9] (*David*, 24,) b. ——.

Ophelia, (Greek,) a serpent.

1 OPHELIA,[8] (*Richard S.*, 7,) b. ——, 1820–30.
2 OPHELIA,[9] (*Andrew*, 2,) b. January 20th, 1858.
1 Ora Blinn,[9] (*Jesse C. Blinn*, 13,) b. October 19th, 1867.
1 ORANGE,[6] (*Timothy*, 2,) b. ——, 1790–1800, d. young.
2 ORANGE,[7] (*Martin*, 2,) b. ——, 1805–15, d. ——.   (*Albany, N. Y.*)
1 ORIET,[9] (*Daniel*, 28½,) b. ——, d. young.
1 ORIN,[8] (*Henry* 11,) b. October 16th, 1846.   (*Lake City, Minn.*)
2 ORIN EMERSON,[9] (*Frederick*, 6,) b. April 19th, 1850.
 Now in Union Christian College at Merom, Sullivan Co., Ind.

Orlando, same as Rowland, (Teutonic,) fame of the land.

1 Orlando Taylor,[8] (*John C. Taylor*, 70,) b. August, 1845.
2 Orlando L. Crego,[8] (*Clarissa Taylor*, 4,) b. March, 1860.
3 Orlando B. Whitney,[10] (*Ann M. Hackett*, 43,) b. March 28th, 1870.
     (*Vermont.*)
1 ORSAMUS,[7] (*Ezra*, 2,) b. July 16th, 1798, d. October 8th, 1814.

2 Orsamus Pike,[8] (*Dorothy T. Corwin*, 5.)
   M. Wealthy Allen.

1 ORTHENAL,[7] (*Eli*, 3,) b. December 22d, 1804, d. January 20th, 1805.

2 Orthenal Moffat,[8] (*Sarah Corwin*, 31.)

1 ORTON,[8] (*Rev. Jason*, 4,) b. August 26th, 1831, d. April 15th, 1844.

Oscar, (Celtic,) bounding warrior.

1 Oscar II. Little,[7] (*Desire Corwin*, 1,) b. October, 1825.

1 PARKER,[7] (*John*, 13,) b. in Baiting Hollow, L. I., February 22d, 1796;
   yet living.
   M. (1) Betsy, daughter of Daniel Tuttle,* of Speonk, L. I., and Sarah
   Parshal,† of Aquebogue, July 12th, 1817. Daniel Tuttle died March
   8th, 1845, aged 87; his wife died September 15th, 1852, aged 87.
   Betsy was born September 6th, 1795, and died November 6th, 1855,
   in Niagara Co., N. Y.; (2) Mrs. Rebecca Campbell, of Youngstown,
   Niagara Co., N. Y., in 1856.
   Ch. Frederic II., Lucilla S., Rosalinda, Clarinda, John, Charles,
   George J.
   Settled in Moriches, L. I., 1819. Removed to Niagara Co., N. Y., 1834.
                                                    (*Youngstown, N. Y.*)

2 PARKER EUGENE,[9] (*Frederic*, 6,) b. October 20th, 1852.

3 PARKER,[9] (*Charles*, 20¼,) b. ——, 1863, in Niagara Co., N. Y.

4 Parker C. Baker,[10] (*Edna Kyte*, 2,) b. January 29th, 1871.

1 PARMENAS HORTON,[7] (*Silas*, 2,) b. January 18th, 1815.
   M. Caroline Roe, February 6th, 1839.

2 Parmenas Brown,[7] (*Mary Corwin*, 8,) b. ——, 1800 ?
   M. (1) Susan Holley; (2) Charity Holley.
   Ch. Mary, Cornelia, Susan, Egbert, Daniel.

3 Parmenas Hulse,[8] (*Lucetta Corwin*, 1.)

1 PARNELL WICKHAM,[7] (*John*, 9,) b. May 5th, 1794, d. October 5th, 1795.

2 PARNELL W.,[7] (*John*, 9,) b. November 16th, 1796, yet living.
   M. Asaph Youngs.   (*Near Riverhead, L. I.*)

1 Parshall Shannon,[8] (*Mehetable Corwin*, 5,) b. ——, 1820–30.

1 PATIA ELIZA,[8] (*George E.*, 13,) b. ——, 1830–40.

Patience, (Latin,) suffering, endurance, contentment.

1 PATIENCE MARIA,[7] (*Jabez*, 1,) b. March 21st, 1824.
   M. Rev. James Trowbridge Hamlin, November 26th, 1857.
   Ch. Sarah Roselle.

---

* Son of John Tuttle, of Wading River, L. I., and Elizabeth Wells.
† Daughter of James Parshal and Elizabeth Sweezy, of Aquebogue, L. I.

Rev. Mr. H. studied in Burr Seminary, Vt., and settled at Mattituck in 1846, where he yet remains.

2 PATIENCE,[7] (*Richard*, 1,) b. ——, 1790–1800.
Unm.

3 PATIENCE FRANCES,[8] (*Jabez*, 3.) ●
   M. —— Tuthill.

4 Patience Dill,[8] (*Phebe Corwin*, 6,) b. ——.
   M. —— Helmick.   (*New-Paris, O.*)

5 Patience Osborn,[8] (*Amelia Corwin*, 2,) b. July 10th, 1832.
   M. —— Orr.   (*Ferris, Ill.*)

6 Patience C. Wells,[8] (*Puah Corwin*, 1.)
   M. J. G. Mead.   (*Jamesport, L. I.*)

1 PATTY,[7] (*Henry*, 6,) b. ——, 1770–90.
   M. —— Squires.
   Ch. Nichols, Alvan, Seth, Harriet, Bethia.

Paulina, (Latin,) little.

1 Paulina Clark,[8] (*Sarah J. Millspaugh*, 66,) b. November 19th, 1838.

2 PAULINE,[6] (*James*, 15,) b. about 1856.

Pelatiah, (Hebrew,) whom Jehovah delivers.

1 PELATIAH,[5] (*Daniel*, 2,) b. ——, 1710–15.
   M. Bazelah Osman, 1735.

Penelope, (Greek,) a weaver.

(1) * PENELOPE,[2] [*George*, (1),] b. December 7th, (or August 9th,) 1670, d December 28th, 1690.
 · *M.* Hon. Josiah Wolcott, February 19th, 1684–5, of Salem, Mass.
   Ch. Elizabeth, Josiah.

1 Pennina Osman,[4] (*Sarah Corwin*, 1,) b. ——, 1690–1700.

1 Permelia Chase,[9] (*Phebe Corwin*, 8½.)

1 PERRY,[9] (*Amos*, 1,) b. ——, d. in the federal army during the Rebellion.

2 Perry F. Chappel,[6] (*Charlotte Hobart*, 9,) b. November, 1854, d. April, 1863.

Peter, (Greek,) a stone.

1 PETER,[6] (*Joshua*, 1,) b. December 6th, 1757, d. September 20th, 1779.

2 PETER,[6] (*Silas*, 1,) b. 1762, bapt. at Aquebogue, July 22d, 1764, d. September 30th, 1850.
   M. Jemima Young, April 1st, 1787. (*Aquebogue Records.*) She b. June 30th, 1766, d. June 25th, 1850.

                        * Curwen.

Ch. Elizabeth H., Israel Y., Jemima, Azubah, Lucetta, Mary.

He left Aquebogue before 1793, as on June 9th, 1793, his son, Israel Y., was baptized at Aquebogue, and the record says that Peter was then of Goshen, N. Y. Possibly he subsequently lived in Morris Co., N. J.

(There was a Peter (or Prance) on census list, 242, of Southold, having in family one male, one female, and five children. There seems to be no Peter here to whom these facts would apply.)

3 PETER,[7] (*Joseph*, 8,) b. ——, 1781? d. ——, 1835?
    M. Sarah, daughter of Nicholas Emmons.
    Ch. Nicholas, Joseph, Mary, Merinda, Ann E., Ellen.

4 PETER,[7] (*John*, 7,) b. ——, d. ——, 1841.
    M. Lydia Jane Harding.
    Ch. Mary Frances, Elizabeth, (Isaac?)

5 PETER,[7] (*Selah*, 1,) b. ——, 1800, d. September 15th, 1825.
                                              (*Riverhead, L. I.*)

6 PETER,[8] (*William*, 16,) b. ——, 1825–30.   (*Milwaukee, Wis.*)

7 PETER FRANCIS,[8] (*William*, 16,) b. September 11th, 1833.
    M. Jane Ann Kent.
    Ch. Mary.   (*Spring Valley, N. J.*)

7½ PETER,[9] (*Selah*, 2,) b. ——, d. September, 1825. (*Aquebogue Records.*)

8 Peter Millspaugh,[8] (*Gilbert C. Millspaugh*, 6,) b. March 18th, 1842.

9 Peter Hobart,[7] (*Mehetable Corwin*, 3,) b. February 8th, 1806.
    M. Sarah Coo, July, 1827.
    Ch. Susannah, Electra M., Melissa H.

10 Peter Brown,[8] (*Elizabeth Woodhull*, 45,) b. April 3d, 1832, d. ——, 1834.

(1) Peter Bellos,[7] [*Alice Corwine*, (1).]

Phebe, (Greek,) pure, radiant.

1 PHEBE,[8] (*Theophilus*, 1?) b. ——.
    A Phebe Corwin is mentioned in a will of A. Whittier. (*Moore.*)

2 PHEBE,[5] (*Samuel*, 3,) b. ——, d. ——.
    M. John Tuthill, October 27th, 1763, at Cutchogue.
        (*Or* 1769, *according to Moore. Southold Church Records.*)

2½ PHEBE,[6] (*Daniel*, 3,) b. ——, 1750–60.
    A Phebe Corwin married Thomas Jennings, February 1st, 1778. Perhaps this one. (*Aquebogue Records.*)

2¾ PHEBE,[6] (*Nathan*, 1½,) bapt. at Aquebogue, May 2d, 1773.

3 PHEBE,[6] (*Samuel*, 6,) bapt. April 26th, 1767, at Southold, L. I.

4 PHEBE,[6] (*Stephen*, 1,) b. ——, d. ——.
    M. Richard, Harris.   (*Ohio.*)

5 PHEBE,[7] (*Samuel*, 11,) b. ——, 1790–1800.
    Unm.

6 Phebe,[7] (*Matthias,* 5,) b. August 9th, 1802.
M. Francis Dill.
Ch. Mary E., Joseph, Patience. '

6½ Phebe,[8] (*Phineas,* 1¾.)

7 Phebe,[9] (*Hewlett C.,* 1,) b. November 20th, 1836.
M. John Henry Corwin, 67.

8 Phebe,[8] (*John,* 30,) b. ——, 1818.
M. Godfrey Chase.
Ch. Permelia A., Cordelia L., Almeda, Mary, Fidelia J. (*Milo, N. Y.*)

9 Phebe Jane,[9] (*Martin,* 5,) b. January 2d, 1858, d. November 9th, 1861.

10 Phebe A. Padden,[8] (*Anna Campbell,* 5,) b. September 4th, 1821.
M. John Lankton, September 27th, 1843.
Ch. Charles M., Anna M., Zada L., Isabel H., Sarah E.
(*Elbridge, N. Y.*)

(A Phebe Corwin is buried at Jamesport, L. I., b. ——, 1772, d. September 29th, 1822. Perhaps No. 5.)

(1) Phebe,[7] [*John,* (3),] b. September 28th, 1770, d. February 5th, 1844, at Pennington, N. J.
M. Richard Allen, March 17th, 1806, of Mine Brook, Somerset Co., N. J. He died March 13th, 1810. One child, who died young. A letter to her from Ruth Corwine, 1800, is preserved.

(2) Phebe,[9] [*Gideon R.,* (1),] b. August 27th, 1821.
M. Isaac Farley, of Titusville, N. J.
Two ch. (*Titusville, N. J.*)

Phila, or Phile, (Greek,) beloved.

1 Phila,[6] (*John,* 6,) b. June 22d, 1770, died young.

2 Phila,[7] (*Abel,* 1,) b. ——, 1787 ?

3 Phila,[8] (*Abel,* 3.)

1 Philenia,[7] (*Hubbard,* 1,) b. May 26th, 1797.
M. Stephen Smith, May 26th, 1828. He d. April 6th, 1859.
Ch. Mary, Diantha, Lydia, Maria, Russell, George, Luthera.

2 Philenia,[8] (*Jabin,* 1,) b. April 25th, 1825, at Tunbridge, Vt.
M. Benjamin Franklin Durkee, September 16th, 1841. He was born July 9th, 1815.
Ch. Alida M., Sarah A., Sophronia. (*Randolph, Orange Co., Vt.*)

1 Philetus,[7] (*Edward,* 3,) b. September 7th, 1814, d. July 7th, 1857.
M. C. Gallup.
Eight ch.

2 Philetus W. Howell,[8] (*Sophia Woodruff,* 5,) b. August 8th, 1839.
M. Victoria Vaugh. (*Newburg, N. Y.*)

Philip, (Greek,) a lover of horses.

1 PHILIP,[7] (*Asa,* 3,) b. April 1st, 1802.
　M. Margaret Labar.
　Ch. Sarah J., Ransom, William.

2 PHILIP FORDHAM,[8] (*William*, 19,) b. July 29th, 1861.

Philistia, (Hebrew,) sojourner's land.

1 PHILISTIA,[8] (*Polydore B.*, 1,) b. March 8th, 1855, d. ——, 1861.

Phineas, (Hebrew,) mouth of brass.

1 PHINEAS,[5] (*David*, 2,) b. September 11th, 1749, d. December 24th,
　1828.
　M. Mehetable Parshall, who was b. October 27th, 1750, and d. April
　4th, 1837.
　Ch. George, Phineas, James, Sibyl, Mehetable, Sarah, Bethia, Desire,
　Fanny.
　A Phineas Corwin bought land in Orange Co., N. Y., 1803.

1¼ PHINEAS,[5] (*Edward*, 2,) b. in Plainfield, Ct., or in Coventry, R. I.,
　about 1753, d. July 12th, 1831.
　M. Phebe Colgrove, of Coventry, R. I.　She was born about 1749, and
　died at Bridgewater, N. Y., 1839.　(See Nancy, 1½.)
　Ch. Phineas, Stephen, Horton, Ira, Jason, and sisters.
　He lived at Norwich, Ct., and afterward joined the church in Franklin,
Ct., January 17th, 1796, by letter from First, church in Lisbon, New-
London Co., Ct.　On January 20th, 1807, he was made deacon.　(See Dr.
Ashbel Woodward's *History of Franklin.*)　In March, 1816, he removed
to Cazenovia, N. Y., where he died, honored and respected as a faithful
and consistent member of the Congregational church in that place.　His
wife was a Baptist.　He and a sister, Elizabeth, were baptized at Mattituck, L. I., February 22d, 1756.　His parents were probably then on a
visit to Long Island, from their new Connecticut home ; perhaps on account of the death of John, 3.

1¾ PHINEAS,[6] (*Phineas*, 1½,) b. ——, 1780-90, in Lisbon, Ct.
　M. Margaretta Gardner.
　Ch. Phineas F., Phebe, Eveline, Nancy, Elizabeth, Cordelia, Ange-
　nette, Calista J.　(Another letter gives Antoinette, instead of Ange-
　nette, Emeline for Eveline, and adds another, Avanda.)
　(The wife of a Phineas Corwin died at Camillus, N. Y., 1837.　There
lived a Phineas at Elbridge, N. Y., in 1835.)

2 PHINEAS,[6] (*David*, 4,) b. ——, 1784? d. young.

3 PHINEAS,[6] (*Phineas*, 1,) b. August 22d, 1788? d. ——, 1851.
　M. (1) Charity Curtis ; (2) Rebecca Moore.
　Ch. Charity, Gabriel, Hervey, Caroline, Harriet, Alfred, Sarah, Milton.
　　　　　　　　　　　　　　　　　　　(*Camillus, N. Y.*, 1830-4.)

4 Phineas,[7] (*David*, 6,) b. ——, 1790–1800.　　　　(*Ohio.*)

4½ Phineas Fernandez,[7] (*Phineas*, 1¾.)　　　　(*Pompey, N. Y.*)

5 Phineas,[7] (*George*, 1,) b. ——, 1804 ? d. young.

6 Phineas Campbell,[7] (*Sarah Corwin*, 6,) b. November 21st, 1801.

7 Phineas Padden,[8] (*Anna Campbell*, 29,) b. ——, 1825 ? d. October, 1857.
M. Dranthe Shepherd.
Ch. Willard M.

Pithom, (Egyptian,) the narrow place.

1 Pithom Halsey,[7] (*Elizabeth Corwin*, 2,) b. ——, 1806 ?　　(*Wisconsin.*)

1 Pitt,[6] (*Nathaniel*, 1,) bapt. December 20th, 1771, at Southold.
Polly, see Mary.

1 Polly,[7] (*Stephen*, 2,) b. ——, 1790–1800.

2 Polly,[8] (*John*, 30,) b. ——, 1820.
M. Jesse H. Davis, 1838.
Ch. Joel, Imogene, Ann L., Lucinda, Levi N.　　(*Jerusalem, N. Y.*)

3 Polly Amanda,[7] (*William*, 4,) b. ——, 1827 ? d. 1828 ?

4 Polly Maria,[8] (*Joshua*, 6,) b. August 2d, 1808.
M. J. C. Harding.
Ch. Celinda, Jemima, Ira, Fanny, Elmira.　　(*Middletown, N. Y.*)

Polydore, (Greek,) many gifts.

1 Polydore Brown,[7] (*Ezra*, 2,) b. March 30th, 1801.
M. (1) Sarah Thompson, February 12th, 1824. She b. May 11th, 1803 ;
(2) Phebe Candell, May 14th, 1842. She b. April 23d, 1819; (3)
Catharine Palmater, September 18th, 1847. She b. October 6th, 1819.
Ch. Almerin, De Forest, Ezra ; (2) Theresa, Sarah R., Isabella, Marion, Harriet, Dudley G., Philistia.
He was baptized at Aquebogue, and said to be son of Ezra Corwin, from the Highlands.　　(*Cortlandt, N. Y.*)

2 Polydore Brown,[7] (*Mary Corwin*, 8,) b. ——, 1790–1800.

Priscilla, (Latin,) somewhat old.

1 Priscilla,[8] (*Joshua*, 6,) b. ——, d. at 14 years of age.

2 Priscilla Campbell,[7] (*Sarah Corwin*, 6,) b. September 1st, 1814.
M. William Evans, September, 1857.　　(*Elbridge, N. Y.*)

3 Priscilla Brown,[8] (*Elizabeth Woodhull*, 45,) b. February 16th, 1832.
M. Philip Latourette.　　(*Lesser Cross-Roads, Somerset Co., N. J.*)

4 Priscilla Gordon,[9] (*Fanny J. Corwin*, 4.)

Prudence, (Latin,) far-seeing.

1 PRUDENCE,[6] (*James*, 2,) bapt. April 13th, 1766, at Southold, d. ——, 1829.
M. —— Halsey.
Ch. Prudence, Eliza A., James C., and a daughter.
(*Southampton*, *L. I.*)

2 PRUDENCE,[7] (*Joseph*, 4,) b. September 5th, 1800.

3 Prudence Halsey,[7] (*Prudence Corwin*, 1,) b. ——, 1790–1800.

Puah, (Arabic,) mouth.

1 PUAH TUTHILL,[7] (*Jabez*, 1,) b. December 27th, 1805.
M. Charles, son of Obadiah Wells, November 30th, 1826.
Ch. Patience C., Mary V., Charles E., Rosabella, Puah F., Isaiah C.
(*Jamesport*, *L. I.*)

2 Puah Frances Wells,[6] (*Puah T. Corwin*, 1.)
M. Albert Valentine.

1 QUINCY,[8] (*Robert G.*, 2,) b. December 3d, 1841.
Graduated at Miami University, taking the first honor. Was a private for three months, under President Lincoln's first call for troops. Now a lawyer. (*Dayton, O.*)

(1) * QUINTON, [*Richard M.*, (6),] b. ——.
A lawyer. (*Cincinnati, O.*)

Rachel, (Hebrew,) a ewe.

1 RACHEL,[5] (*Simeon*, 1,) bapt. at Mattituck, November 7th, 1755.

2 RACHEL,[6] (*Stephen*, 1,) b. ——.
M. John Pack. (*Ohio.*)

2¼ RACHEL,[6] (*Jonathan*, 1,) bapt. at Aquebogue, December 9th, and d. December 10th, 1772. (*Aquebogue Records.*)

2½ RACHEL MESSENGER,[7] (*Stephen*, 4.) (*Richland Co., O.*)

3 RACHEL,[7] (*Selah*, 1,) b. ——, 1785–1800.
M. Ebenezer Terrill, August 30th, 1801. (*Aquebogue Records.*)

4 RACHEL,[8] (*John*, 30,) b. ——, 1815.
M. Rev. Joseph N. Davis, 1836, a local preacher in the Methodist Episcopal Church.
Ch. Edgar E., Miles A., Melvin J., George D., Harriet J.
(*Jerusalem, N. Y.*)

5 RACHEL,[8] (*Timothy*, 6.)

6 RACHEL JANE,[8] (*John B.*, 28.)

7 Rachel Squires,[8] (*Elizabeth Corwin*, 21.)

8 Rachel Smith,[8] (*Sarah Corwin*, 24.)

(1) * RACHEL,[6] [*Samuel*, (3),] b. February 27th, 1752.
M. Jacob Williamson.

* Corwine.

Ch: Richard, Samuel, William, Isaac, John, Amos.

(2) * RACHEL,[6] [*Richard*, (2),] b. November 20th, 1783, in New-Jersey, d. ——, 1864.

M. Derick Barkelow, of New-Jersey, November 2d, 1802. He died in 1847.

Ch. Arthur, Maria, Sarah, and two others.

They moved from Mason Co., Ky., to Warren Co., Ohio, soon after their marriage.   *(Portsmouth, Scioto Co., O.)*

(3) * RACHEL ANN,[8] [*George*, (9),] b. ——, 1833.

M. .Peter Q. Hoagland.   *(Mechanic ; Somerset Co., N. J.)*

(4) Rachel Barkalow,[8] [*Maria Corwine*, (1),] b. ——.

M. Jacob Johnson.   *(Mississippi.)*

(5) * RACHEL M.,[9] [*John*, (11),] b. ——, 1857.   *(Waverley, O.)*

1 RANSOM,[8] *(Philip,* 1.)

Rebecca, (Hebrew,) an ensnarer, or one of fascinating beauty.

1 REBECCA,[3] *(John,* 1,) b. ——, 1660–70.

M. Abram Osman before 1690.

Ch. Joseph, John, Rebecca ? Sarah ? Abraham ? Dorothy ?

     (Doc. Hist. N. Y. i. 452. *Moore's Index of Southold,* 105, 106.)

Probably Ab. Osman married a Martha, who died 1706, and Abram and Dorothy, and perhaps Sarah, may be the children of this second wife.

2 REBECCA,[7] *(Jonathan,* 2,) b. ——, ~~1790–1800~~.

M. Jeremiah Randall.

3 REBECCA JANE,[6] *(Horton,* 2,) b. April 18th, 1828.

M. James L. Ketcham, January 25th, 1853. He was born in Mount Hope, Orange Co., N. Y., March 4th, 1819.

Ch. Salmon C., Elizabeth, James L.   *(East-Saginaw, Mich.)*

4 Rebecca Osman,[4] *(Rebecca Corwin,* 1,) b. about 1700.

5 Rebecca Bowers,[8] *(Elmira Corwin,* 1,) b. April 13th, 1836.

           (Plattekill, Ulster Co., N. Y.)

6 Rebecca Jane Dunlevy,[8] *(Lucinda Corwin,* 4,) b. November 24th, 1824.

M. W. J. Drake.   *(Chicago, Ill.)*

(1) * REBECCA,[6] [*Samuel*, (3),] b. about 1757.

M. Levi Mitter.   *(Amwell, N. J.)*

(2) * REBECCA,[7] [*Samuel*, (4),] b. October 18th, 1793.

M. Allen Edwards.   *(Pennsylvania.)*

(3) * REBECCA,[8] [*Gideon*, (1),] b. August 11th, 1816.

Unm.   *(Amwell, N. J.)*

(4) Rebecca Snyder,[7] *(Esther Corwine,* (1).]

          * Corwine.

M. Isaac Simonson.

(5) Rebecca Bellos,[7] [*Alice Corwine*, (1).]

(6) * REBECCA,[7] [*John*, (3),] b. about 1782, d. young.

1 Reed Case,[5] (*Samuel Case*, 29.)
    M. —— Clark, about 1730. (*Moore's Index.*)

Reuben, (Hebrew,) behold! a son.

1 REUBEN,[8] (*Jabin*, 1,) b. July 21st, 1820, d. ——, 1821.

Rhoda, (Greek,) a rose.

½ RHODA,[6] (——,) b. ——, 1750–60.
    M. Richard Howell, March 18th, 1779. (*Aquebogue Records.*)

1 RHODA,[7] (*Matthias*, 5,) b. November 14th, 1799. –
    M. Isaiah Morris, 1840.
    Ch. Mary, Thomas.   (*Wilmington, O.*)

2 RHODY,[6] (*Jedediah*, 2,) b. ——.
    M. Moses Philips.

Richard, (Old High-German,) rich-hearted.

1 RICHARD,[6] (*Jedediah*, 2,) b. ——.
    M. Elizabeth Marthers, November 9th, 1775. (*Aquebogue Records.*)
    Ch. Richard, Alsop H., William, Patience, Hannah.

In 1775, signed engagement to support Congress. He was also a Revolutionary soldier, being present at the battle of Yorktown, and at the surrender of Cornwallis. Washington once tested his fidelity as a guard by attempting to pass him in the night, but he would not allow him to pass, and afterward received commendations for his fidelity. He was then known and received his pension under the name of Currin. (*See index.*)

2 RICHARD,[6] (*Timothy*, 2,) b. ——, 1790, d. ——, 1857.
    M. Rachel Bloomer.
    Ch. Timothy, Eliza, Mary A., William B., Hannah, John B.
                              (*Balmville, N. Y.*)

3 RICHARD T.,[6] (*David*, 5,) b. November 14th, 1810.
    M. Eunice, daughter of Ira and Laura Lyon, September 25th, 1835.
    Ch. Eliza J., George, Laura, Mary, John, Charles.

3⅓ RICHARD,[6] (*Jonathan*, 1,) b. ——, d. October 7th, 1777.
                              (*Aquebogue Records.*)

4 RICHARD,[7] (*Richard*, 1,) b. ——, 1785–95, d. ——.
    M. (1) Dency Hawkins.
    Ch. Richard, Fanny, Nathaniel, William.   (*Brookhaven, L. I.*)

5 RICHARD WARREN,[7] (*Asa*, 3,) b. April 2d, 1791, at Aquebogue.
    M. Elizabeth Knapp. She b. February 3d, 1799.
    Ch. Lucinda, Walter S., Elizabeth.   (*Narrowsburg, N. Y.*)

6 RICHARD,[7] (*Joseph*, 10.)

* Corwine.

(A Richard Corwin married Hannah Halleck, December 31st, 1789.

(*Aquebogue Records.*)

7 RICHARD SWEEZEY,[7] (*Rev. Joseph*, 5,) b. about 1807, d. February, 1868.
   M. Mehetable Griffin.
   Ch. Egbert C., Ann A., Bartlett, Mehetable, Mary, Ophelia, Eliza T.,
      Augustine, William.   (*Riverhead, L. I.*)

8 RICHARD,[8] (*Alsop*, 1,) b. ——, 1810-20.   (*Luzerne Co., Pa.*)

9 RICHARD,[8] (*Richard*, 4.)

10 RICHARD S.,[9] (*Walter S.*, 2.)

11 Richard W. Murray,[9] (*Lucinda Corwin*, 1.)

  (1) *RICHARD,[6] [*Bartholomew*, (1),] b. March 26th, 1720, d. young.

  (2) *RICHARD,[6] [*George*, (6),] b. in Amwell, N. J., December 29th, 1748,
     d. in Mason Co., Ky., November 19th, 1813.
    M. Sarah Snyder, who was b. April 3d, 1760, d. in Maysville, Ky.,
      September 14th, 1820.
    Ch. Richard, George, Samuel, Amos, Rachel, Lydia, Sarah, Abby,
      Elizabeth, Jacob, John.

The Hunterdon Co., N. J., Records show him as selling land in Amwell
township in 1804, which he had received by will from his father in 1781.
In 1788, he removed to Washington, Bourbon Co., Ky., but the next year,
the county of Mason was set off, including his residence. He was a soldier
in the Revolutionary war, being in the battle of Trenton and other engagements. A letter of his, written April 13th, 1806, (yet preserved,) to his
brother John, in New-Jersey, breathes a deep spirit of piety, and refers
with pleasure to the pious spirit of John's letter to him. Richard mourns
the low state of religion in Kentucky. He refers to the anti-slavery cry
already heard in Kentucky, and to the distraction caused thereby in the
Baptist Church, of which he was a member. Says he, " Some cry out for
liberty, and others, Pharaoh-like, refuse to let them go." Another letter
to the same, in 1804, is also preserved. The funeral service of Richard,
preached by William Grinstead, pastor of the Baptist Church of Maysville,
Ky., was printed at Lexington, Ky., 1813, but contains no historical matter.

  (3) *RICHARD,[7] [*Amos*, (1),] b. ——, 1790-95, d. ——, 1815.
    He volunteered in the war of 1812, repaired to the northern frontier of
Ohio, where he contracted a severe cold, which settled upon his lungs, and
from the effects of which he soon after died.

  (4) *RICHARD,[7] [*Samuel*, (4),] b. February 27th, 1803, d. April, 1829.
    M. Hannah Soothoof.
    Ch. Hannah C., Samuel.   (*Amwell, N. J.*)

  (5) *REV. RICHARD,[7] [Richard, (2),] b. in Mason Co., Ky., August 29th,
     (or 19th,) 1789, d. in Louisville, Ky., February 12th, 1843.

* Corwine.

M. Sarah Hitt, August 20th, (or 5th,) 1824, in Bourbon Co., Ky. She
was b. 1792, d. at Shelbyville, Ky., March 23d, 1862.

Ch. Rev. Jesse D. H., Sarah A.

He was licensed, and joined the Methodist itineracy, July 17th, 1817,
and continued therein till 1843. He was admitted into conference at
Zanesville, Ohio, in the fall of 1817. As the bounds of the conference ex-
tended then beyond the confines of Ohio, he was appointed to preach in
Kentucky, and was presiding elder of the Louisville district when he
died. He was a prominent minister, eminent for his piety and good judg-
ment. In 1824, while attending conference in Baltimore, he met his cousin
John Corwine, and made inquiries about the family, and sent a letter to
Gideon Corwine, in New-Jersey, soliciting a transcript of the family record.
His letter is still preserved.

(5½) * RICHARD,[8] (*George,* 7¾,) b. ——.   (*Indiana.*)

(6) * RICHARD MORTIMER,[8] [*Joab,* (1),] b. in Mason Co., Ky., April 12th,
1812.

M. Mary Eliza Quinton, February, 1842, in Cincinnati. She d. ——,
1861 ; (2) Julia W., daughter of Col. Sloo, of Indiana, in 1862. She
d. in a few months; (3) Jessie H., daughter of Austin W. Morris, in
1864, of Indianapolis, Ind.

Ch. Quinton, John, Richard.

He first learned the printer's business with his father, and afterward
studied law, moving to Cincinnati, Ohio. He became co-editor with his
father and brother, of the Cincinnati *Daily Herald.* He moved to Indiana
about 1833, and was admitted to the bar in 1834. In 1836, he removed to
Mississippi, and took an active part in politics, as a Whig, canvassing a
part of that State in 1840 for General Harrison, who was then a candidate
for the presidency. He was afterward clerk of the ——, the only Whig
ever holding that office in that State. In 1842, he removed to Cincinnati,
and founded a law partnership with Judge Oliver M. Spencer, to which
firm was united, in 1853, Caleb B. Smith, afterward Secretary of the Inte-
rior under President Lincoln. Subsequently, Mr. R. B. Hays, (now
Governor of Ohio,) was a member of the same firm. Mr. Corwine was a
major of cavalry, and was on General Fremont's staff, in the early part of
the rebellion. He subsequently served on Major H. G. Wright's staff, and
on Major-General Burnside's. In 1864, he resigned, and returned to the
practice of law, being United States District-Attorney for the Southern
District of Ohio. He was also Judge Advocate-General, during his military
service, (except during the last four months,) of Department 14, of the Department of the West 22, of the Mountain Department 30, and of the Department of the Ohio. Was Provost Marshal-General of the Mountain
Department of Kentucky, and commanded the river defenses of Cincinnati
during the siege in 1862. Now engaged in the practice of law in the Su-

* Corwine.

preme Court of the United States, at Washington, D. C., and in the Court of Claims, and before the Mexican North-American Commission. He published, *Digest Cases in High Chancery in England and America*, etc., in 1845, at Cincinnati, Ohio. 8vo. Allibone, in his *Dictionary of Authors*, says, quoting from Marvin's *Legal Bibliography*, "The material of this volume seems to have been carefully collected, and is very well arranged."

(7) *RICHARD,⁹ [*Richard M.*, (6).]

(8) *RICHARD,⁸ [*Samuel*, (6½),] b. March 20th, 1827, d. July 5th, 1828.

(9) *RICHARD,⁸ [*Amos*, (1½),] b. ——.   (*Broadwell, Ill.*)

(10) Richard Taylor Stevenson,⁹ [*Sarah Corwin*, (5),] b. September 24th, 1858, at Taylorsville, Ky., now in Transylvania University, at Lexington, Ky.

Robert, (Old High-German,) bright in fame.

1 ROBERT,⁷ (*John*, 13,) b. ——, 1792, bapt. at Aquebogue, March 12th, 1794, d. ——, 1858.

M. Margaret Hallock.

Ch. Julia, Sarepta, Hannah M., John R., Josiah F.
(*Baiting Hollow, and Riverhead, L. I.*)

2 ROBERT GRIFFIN,⁷ (*Ichabod*, 1,) b. July 28th, 1815.

M. Eliza Bruin, January 29th, 1839.

Ch. David B., Quincy, Robert I., Susan B., Sallie G., Thomas, Eliza.

Studied in Denison University, Ohio, and was admitted to the bar in 1837.   /   (*Dayton, Ohio.*)

3 ROBERT WALSH,⁸ (*Howard*, 1,) b. March 13th, 1857.

4 ROBERT LOWELL.⁸ (*Joseph*, 16,) b. ——. 1870.

5 ROBERT L.,⁸ (*Robert G.*, 2,) b. October 10th, 1844, d. May 6th, 1849.

6 Robert Duckworth,⁷ (*Sarah Corwin*, 9,) b. January 19th, 1817.

7 Robert Willard Newman,⁹ (*Moses B. Newman*, 9,) b. at Wyandotte, Kansas, April 13th, 1870.

1 ROCKWELL B.,⁸ ? (——.)

M. Lydia ——.

His name appears on the Cayuga Co., N. Y., records, at Auburn, as a resident of Auburn, 1845–51, buying and selling land at Genoa, in that county.

2 Rockwell, (Smith?) (*Sarah Corwin*, 10,) b. ——, 1800–10.

1 Rollin Pike,⁸ (*Dorothy T. Corwin*, 5.)

Rosa, (Latin,) a rose.

1 Rosa Jefford,¹⁰ (*Stella L. Baker*, 1,) b. September 3d, 1867.

Rosalia, (Latin,) little blooming rose.

1 Rosalia Tinney,⁹ (*Lucelia Corwin*, 1.)

* Corwine.

Rosabella, (Latin,) a fair rose.

1 ROSABELLA,[8] (*Nathan*, 5,) b. ——, 1825–30.
  M. Charles Hallett.

2 Rosabella Wells,[8] (*Puah Corwin*, 1,) d. in infancy.

Rosalinda, (Latin,) beautiful as a rose.

1 ROSALINDA,[8] (*Parker*, 1,) b. August 9th, 1823.
  M. Warren, son of David and Cynthia (Shaddoc) Baker, November 17th, 1842. David Baker died in 1833, and was of New-Jersey.
  Ch. Herbert, Stella L., Elma E., Julia L. V.   (*Niagara Co., N. Y.*)

Rosamond, (Teutonic,) famous protection.

1 Rosamond,[8] (*Samuel B.*, 16.)

Rose, (Latin,) a rose.

1 ROSE,[8] (*Joseph W.*, 12,) b. August 25th, 1847, d. September 15th, 1847.

2 ROSE ADEL,[9] (*George W.*, 23,) b. April 16th, 1853.

1 ROSETTA,[7] (*Silas*, 2,) b. May 29th, 1809, d. July 7th, 1847.
                                             (*New-Milford, Pa.*)

1½ ROSETTA,[7] (*Stephen*, 7½,) b. October 22d, 1831, d. June 3d, 1845, at Sardinia, N. Y.

2 ROSETTA,[7] (*Daniel*, 8,) b. March 5th, 1799.
  M. William Evans, January 18th, 1815. He b. July 22d, 1796.
  Ch. William.   (*Minisink, N. Y.*)

3 ROSETTA ANN,[8] (*Isaac S.*, 5,) b. September 14th, 1839.

4 ROSETTA E.,[8] (*Ebenezer*, 3,) b. June 10th, 1825.
  M. Sylvester Meade.

5 ROSETTA EVANS,[8] (*Silas*, 7,) b. ——, d. ——.
  M.
  No Ch.

6 ROSETTA MOORE,[8] (*Jabez*, 3.)
  M. G. W. Young.

7 ROSETTA II.,[9] (*Daniel*, 24,) b. May 29th, 1842, d. August 24th, 1865.

8 Rosetta Meyers,[8] (*Arminda Brown*, 2.)

9 Rosetta Smith,[8] (*Sarah Corwin*, 24.)

1 ROSINA JANE,[8] (*Hazen*, 1,) b. November 19th, 1830.
  M. Israel Putnam, at Stockholm, N. Y., January 4th, 1853.

1 Royal M. Smith,[9] (*George Smith*, 74,) b. ——, 1855–65.
                                    (*St. Lawrence Co., N. Y.*)

Ruhama, (Hebrew,) having found mercy.

1 RUHAMA,[7] (*Benjamin*, 4,) b. ——, 1789, d. ——, 1857.
  Unm.

1 RUSSELL,[7] (*Hubbard*, 1,) b. January 29th, 1802.
   M. (1) Mary P. Hood, of Chelsea, Vt., 1830. She died of consumption,
October 1st, 1854, aged 48. (2) Lucy A. Peck, October 21st, 1855.
She was born in Guildhall, Vt., October 17th, 1820.
   Ch. John A. R., Marcus H.
   In 1811, he went to Tunbridge to live with his sister Mary, where he
remained till 1823. Then at Lowell, Mass., till 1828. In 1830, was per-
manently lamed by a fall.   (*Fruit Merchant, Chelsea, Vt.*)

2 Russell C. Hackett,[8] (Mary Corwin, 14,) b. October 24th, 1823, d. June
24th, 1846.
   M. Mary Ann Richardson, ——, 1846.
   Ch. Russell C.

3 Russell H. Smith,[8] (*Philenia Corwin*, 1,) b. April 12th, 1836, d. ——,
1836.   (*Vermont.*)

4 Russell Noyes,[9] (*Maria L. Smith*, 17,) b. August 10th, 1866.
     (*Vermont.*)

5 Russell C. Hackett,[9] (*Russell C. Hackett*, 2,) b. ——, 1846.

6 Russell Colgrove,[9] (*James B. Colgrove*, 45½,) b. ——, 1865 ?

Ruth, (Hebrew,) a female friend.

1 RUTH,[6] (*Daniel*, 3,) b. ——, 1750-60.
   M. Jonas Winters, of West-Hampton, July 3d, 1789. (*Aquebogue Recs.*)

2 RUTH Ross,[7] (*Stephen*, 4,) b. ——.   (*Coshocton Co., Ohio.*)

3 RUTH MARIA,[7] (*Stephen*, 7⅓,) b. January 22d, 1816, d. February 1st,
1829, at Dunkirk, N. Y.

4 Ruth A. Porter,[7] (*Elizabeth Corwin*, 8,) b. ——, 1820-30.

(1) RUTH,[6] [*George*, (6),] b. July 23d, 1754.
   M. George Corwine, (7½.)
   A Ruth Corwine was a witness to a deed of Amos, in Hunterdon Co.
Records, (N. J.,) in 1792, and Ruth was executrix of George Corwine, in
same county, in 1788. A letter of hers from Mason Co., Ky., to Phebe
Corwine, her niece, in Amwell, New-Jersey, is still preserved. The letter
shows deep piety.

Salah, (Hebrew,) a dart.

1 Salah Kyte,[9] (*Clarinda Corwin*, 2.)   (*Niagara Co., N. Y.*)

Salem, (Hebrew,) peace.

1 SALEM,[7] (——.)
   Ch. Henry E., Eunice, Ann Eliza, Sarah J.
     (*Green Point, Fire Place, L. I.*)

Salmon, (Hebrew,) shady.

1 SALMON,[8] (*Horton*, 2,) b. November 20th, 1829, d. September 3d, 1855.

2 SALMON W.,[9] (*Martin*, 5,) b. January 10th, 1860.

3 Salmon Corwin Ketcham,⁹ (*Rebecca J. Corwin*, 3,) b. December 1st, 1855, d. September 5th, 1856.

Samuel, (Hebrew,) asked of God.

1 SAMUEL,³ (*John*, 1,) b. not before 1677, d. December 28th, 1705 ?
M. Anne ——.
Ch. Jedediah ? Samuel ?
In 1698, on census list, (*Doc. Hist.* i. 450,) his name under the name of his father. Hence probably not married yet in 1698.

2 SAMUEL,⁴ (*Theophilus*, 2,) b. ——, 1710 ? d. January, 1762.
M. Experience Corwin, April 13th, 1732.
Ch. Benjamin, Henry, Sarah, David, Samuel, Sarah, Samuel, Asa.
(*A weaver. Southold, L. I.*)
In 1737, his name occurs on list of freeholders. (*Doc. Hist. N. Y.* iv. 132.) In 1740, he is mentioned in the will of his brother John. His own will is in New-York City. (Liber 23, p. 389.)

3 SAMUEL,⁴ (*Samuel*, 1 ?) b. ——, 1706, d. March 18th, 1784.
M. (1) ——; (2) Phebe ——, who d. September 21st, 1783.
Ch. Phebe, Samuel, Stephen, James, Mary B., Nathaniel, Ezra.
In 1775, he signed the engagement to support Congress. In 1776, his name occurs on census list, 56, having in his family one male over 50, one female over 16, and 6 children. According to Aquebogue Records, a widow, Phebe Corwin, died May 4th, 1783.

4 SAMUEL,⁵ (*Samuel*, 2,) b. November 9th, 1741, d. December 19th, 1741.

5 SAMUEL,⁵ (*Samuel*, 2,) b. May 14th, 1743, d. January, 1769.

6 SAMUEL,⁵ (*Samuel*, 3,) b. ——, d. about 1857.
M. Hannah Concklin, February 17th, 1757, at Cutchogue, L. I.
Ch. Susannah, Phebe, Mary, David, Esther, Samuel, Betsy, Eunice.
(*Southold, L. I.*)
In 1775, he signed engagement to support Congress. One of his daughters married an Ichabod Case.

6½ SAMUEL,⁵ ? (——.)
His name appears as a private in Pawling's Co. (8th Company, 2d Regiment) of New-York Revolutionary soldiers. His name appears spelled Curwine and Curvin. He received 600 acres, (being lot 15 in town 3, Cato, Cayuga Co., N. Y.,) on July 8th, 1790. In 1794-7, the records of that county show him as selling land. From 1795-7, he is said to be of Ulster Co., N. Y. He was a Revolutionary soldier for three years. He received the land above mentioned in pay for services.

7 SAMUEL,⁵ ? (——,) b. ——, d. February 6th, 1744–5. (*According to Moore.*)

8 SAMUEL,⁶ (*Amaziah*, 1,) b. ——, 1794.
Went to sea, and never returned.

9 SAMUEL,⁶ (*Timothy*, 2,) b. January 9th, 1788.

M. Rachel, daughter of Samuel and Sibyl (Scudder) Burr, of Massachusetts, on October 14th, 1809. She was b. March 3d, 1788, and d. March 14th, 1849.

Ch. Jemima, Elizabeth, Celia, Sibyl. (*Middle Hope, N. Y.*

10 SAMUEL,⁶ (*David,* 3,) b. ——, 1770 ?

M. ——.

No Ch. (*A carpenter. New-York City.*)

11 SAMUEL,⁶ (*Samuel,* 6,) baptized at Southold, June 16th, 1776, d. ——, 1853.

M. —— Cheeseborough, perhaps, in Connecticut.

Ch. Henry, Phebe, Denniston, Mary, Samuel. (*Peconic, L. I.*)

12 SAMUEL,⁶ (*Stephen,* 1,) b. about 1760–65.

M. Rachel ? ——.

Ch. Moses ? Stephen ?

He moved to Kentucky, (or Ohio,) from Central New-York, where he afterward married, and lived till his death. There was a Samuel Corwin who bought land of Mr. Sage, in Ontario Co., N. Y., in 1846. A Samuel C. Corwin bought land in same county, in 1848. (Lib. 85, p. 410.)

13 SAMUEL,⁷ (*John,* 7,) b. ——, d. ——, 1824.

Unm.

14 SAMUEL HARLEM,⁷ (*John,* 10,) b. January 12th, 1809.

M. Adaline Sweezey.

Ch. Gilbert H., Mary E. (*Howells, N. Y.*)

15 SAMUEL DALSEN,⁷ (*Moses,* 1,) b. April 1st, 1828, d. November 29th, 1851.

M. Mary Howell, March 20th, 1850.

No Ch. (*Middletown, N. Y. ; Jersey City, N. J.*)

He left for California in 1851, but died on the passage, and was buried at St. Diego, on the coast of Mexico.

15½ SAMUEL,⁷ (*Samuel,* 11,) b. ——, d. ——.

Unm. (*Peconic, L. I.*)

16 SAMUEL B.,⁷ (*Ezra,* 2,) b. February 9th, 1808.

M. Abby A. Lewis.

Ch. Mary E., Gillespie B., Emma, Rosamund, Frank. (*Wisconsin.*)

17 SAMUEL,⁷ (*Martin,* 2,) b. ——, 1805–15. (*California.*)

18 SAMUEL WICKHAM,⁷ (*Jesse,* 3,) b. ——, 1814 ? d. ——, 1860, in Cincinnati, O.

M. Amelia ——.

Ch. Two daughters. (*Cincinnati, O.*)

19 SAMUEL WICKHAM,⁷ (*Eli,* 3,) b. December 13th, 1784, d. ——, 1804, of yellow-fever.

20 SAMUEL W.,⁸ (*Isaac,* 4,) b. March 20th, 1826.

M. Kate Hunt, February 18th, 1851.

Erie Railroad conductor. (*Port Jervis, N. Y.*)

21 SAMUEL,[8] (*John,* 16,) b. June 19th, 1826.
    M. —— Westfall.   (*Milton, N. Y.*)

22 SAMUEL,[8] (*Adam,* 1,) b. ——, 1838.

23 SAMUEL BUEL,[8] (*Hudson,* 1,) b. February 12th, 1827.
    M. Mary Brown, June 4th, 1851.
    Ch. Charles H.   (*Greenport, L. I.*)

24 SAMUEL CALDWELL,[8] (*Manasseh R.,* 1,) b. ——, d. October 14th, 1866,
    in Phelps, Ontario Co., N. Y.
    M. Elizabeth J. ——.
    Ch. Samuel A., Edwin, George, Olin, Emma J., Irving A.
    Charles G. Corwin, of Tyre, Seneca Co., N. Y., was appointed guardian
for these children.

25 SAMUEL WICKHAM,[8] (*Nathan H.,* 3,) b. May 13th, 1824.
    M. Frances Broadwell, of Middletown, N. Y.
    Ch. Charles, Carrie, William.
    Graduated at the College of New-Jersey, Princeton, in 1843.   Studied
law, and was licensed to practice, but never did.   (*Newark, N. J.*)

26 SAMUEL ALBERT,[9] (*Hewlett C.,* 1,) b. May 28th, 1840.
    M. Mary Hubbard, February 10th, 1865.
    Ch. Emeline, Frederic.   (*Riverhead, L. I.*)

27 Col. Samuel Hutchinson,[2] (*Martha Corwin,* 1,) b. ——, 1670? d. Janu-
    ary 9th, 1737.
    M. Elizabeth ——.   She b. ——, 1681, d. ——, 1751.
    Ch. Samuel, Benjamin ?
    (*See Moore's Index of Southold, 95, and Doc. Hist. N. Y. i. 453, where
these names occur on census of* 1698, *except Benjamin.*)

28 Samuel Hutchinson,[4] (*Samuel Hutchinson,* 3.)
    M. perhaps, Hannah, daughter of Daniel Tuthill, 1753.

29 Samuel Case,[4] (*Henry Case,* 1,) b. ——, 1687, d. May 10th, 1755.
    M. Zeruiah ——, about 1712.
    Ch. William, Samuel, Joshua, Israel? Moses, Elizabeth, Zeruiah,
    Bethia, Mary, Reed.
    Will in N. Y. Lib. 19, p. 274.   (*See Moore's Index, Southold,* p. 64.)

30 Samuel Case,[5] (*Samuel Case,* 29,) probably d. 1772, on Shelter Island.
                                                  (*Moore's Index,* 64.)

30½ Samuel Corwin White,[6]? (——.)
    M. Lydia Brown, April 17th, 1767.   (*Aquebogue Records.*)

31 Samuel Moffat,[8] (*Sarah Corwin,* 31.)

32 Samuel Barnet Hackett,[9] (*Mary Corwin,* 14,) b. December 10th, 1814.
    M. Sarah Ann Williams, July 16th, 1840.
    Ch. George H., Ann M., Emma A., Lucy A., Cora E.
                                                  (*Tunbridge, Vt.*)

33 Samuel Jennings,[9] (*Sarah Corwin,* 4,) b. ——, 1780–90.

34 Samuel Moffat,⁶ (*Sarah Corwin*, 11,) b. ——, 1810–20.
   M. —— Rockfellow.

35 Samuel Bowers,⁶ (*Elmina Corwin*, 1,) b. May 30th, 1842.
   Unm.

36 Samuel Thompson,⁶ (*Mary Corwin*, 25,) b. February 22d, 1835. *died 7 a... 6.1891.*

37 Samuel Milnor,⁸ (*Harriet A. Millspaugh*, 19,) b. June 4th, 1842.
                                   (*Covington, O.*)

38 Samuel Millspaugh Smith,⁹ (or Boyd,) (*Mary J. Millspaugh*, 99,) b. ——,
   1840–50.   (*Circleville, O.*)

39 Samuel Wells Millspaugh,⁸ (*John II. Millspaugh*, 78,) b. ——, 1840–50.

40 Samuel C. Knox,⁸ (*Julia Corwin*, 7,) b. May 8th, 1842.

(1) * SAMUEL,² [*John*, (1),] b. August 12th, 1674.

(2) * SAMUEL,⁴ [*Rev. George*, (4),] b. in Salem, Mass., December 17th,
   1715, d. April 9th, 1802.

   M. Abigail, daughter of Hon. Daniel and Rebecca (Chambers) Russell,
[and granddaughter of Hon. James and Abigail (Curwen) Hathorne
Russell,] May, 1750. She was, therefore, a great-grand-daughter of
George (1). She died March 31st, 1793.
No Ch.

The following sketch is culled from the Introduction to his *Journal and
Letters*, last edition :

He graduated at Harvard College in 1735, and then studied for the
ministry; but his health proving inadequate, he relinquished his design,
when disappointment in an affair of the heart induced him to travel in
England and on the continent. On his return he became a successful
merchant, though his business was injured by French cruisers from Cape
Breton. Accordingly, New-England undertook an expedition against
Louisburg in the winter of 1744–5, and, under General Pepperell, captured
that French city, which was considered a great feat for the colony. Mr.
Curwen was a captain in this expedition. He was impost officer for
Essex Co., 1759–74, his commission being renewed every three years; he
was judge of admiralty at the time of his departure for England. He
was also a member of a club instituted for improvement in philosophy and
literature in his native town, in which originated the Social Library, in
1760, and the Philosophical Library, which were afterward united, and became the foundation of the Athenæum, in 1810.

On the departure of Governor Hutchinson for England, June 1st, 1774,
he signed the Salem address to that officer, expressing entire approbation
of his public conduct. The governor had already received addresses from
a hundred and twenty merchants of Boston; from all the gentlemen of the
law, with few exceptions; from the magistrates of Middlesex and Plymouth, and the principal gentlemen of Salem and Marblehead, all expressing entire approbation of his public conduct, and their affectionate wishes

for his prosperity. Many of these "*addressers*" were compelled to make public recantation in the newspapers. This Mr. Curwen refused to do, saying that conscience would not allow it; and being unwilling to live under the character of reproach which the fury of party might throw upon him, he resolved to withdraw from the impending storm. He went first to Philadelphia, and then took refuge in England, where he arrived, July, 1775, and remained there as an American loyalist during the Revolution. He returned to Salem September, 1784. During his exile he kept a journal, and wrote many letters, giving many remarks on the men and measures of the day. A selection of these was published in 1842, by George Atkinson Ward, (a great-grandson of Samuel's brother,) and has passed through several editions. It contains more information of the American loyalists than any other work. Many of the biographical notices in the earlier editions are omitted in the last, (1864,) to make room for other notices. Charles Dickens, in his *Household Words*, for May and June, 1853, devoted twenty-three closely-printed columns to a notice of this work. He describes Mr. Curwen as "a small, thin, precise-looking man, in a dress of grave, square cut, with a large bush-wig, sharp features, long nose and chin, a keen, restless eye, a step as active and firm as though it carried sixteen instead of sixty winters, and a complexion certainly not tanned by an English sun."

Dickens further says, " When Judge Curwen fled from the rebellious colonies, he was sixty years old; when he went back to the triumphant young republic, he was sixty-nine; and of the eventful years which formed the interval . . . . . he left a curious record in a diary, ... in which those past days, with all their pains and pleasures, their hopes and their misgivings, still live for us with a vivid and singular reality. For the record was honest and genuine, as in the main the diarist himself was. He does not indeed appear to have been of the heroic stuff of martyrs. If the liberty of opinion he craved had been conceded to him, it would probably have involved nothing graver than the liberty to change his opinion; for he was clearly a man impressible by events, and would probably have saved himself a very long voyage and very great inconvenience, if he could only have held his tongue, till after the first blows were struck in the war of his fellow-citizens for independence. . . . . He was a man of fair learning, and more than average accomplishment; not at all intolerant of opinions at issue with his own; in religion, a dissenter of the class still most prevalent in New-England; in his tastes, scholarly and refined, not ill-read in general literature, prone to social enjoyments, a reasonably good critic of what he saw—altogether an excellent example of the class of men out of whom the fathers and founders of that great republic sprang; and a companion not less pleasant than instructive to pass a few hours with, as I hope the reader will find."

Dickens uses Curwen's *Journal* as the very best picture extant of England, its cities, towns, institutions, people, manners, customs, of not quite a century ago.

Mr. Curwen had been in the commission of the peace for thirty years, and at the time of his departure for England was a judge of admiralty, in which office he was immediately succeeded by the patriotic Timothy Pickering. He returned home in the autumn of 1784, much to the satisfaction of his friends, and was never molested for his political course.

The *Christian Examiner* for January, 1843, has an able review of his *Journal and Letters*, giving copious extracts. The *North-American Review*, January, 1843, has also a lengthy review of the same work, having giving a partial review before. The *Southern and Western Literary Messenger and Review*, for January, 1847, has a reply by Mr. Ward, the editor of Curwen's *Journal*, to a former review in this magazine. Duyckinck also, in his *Cyclopedia of American Literature*, gives the *Journal* a favorable notice. It was first published in New-York and Boston, simultaneously, in 1842. The fourth edition, with a newly-written historical introduction, was published by Little, Brown & Co., Boston, 1864, and it is again out of print.

(3) *SAMUEL,[6] [*Bartholomew*, (1),] b. March 10th, 1728, d. ——, 1776. *M.*

Ch. Rachel, Lydia, Samuel, Amy, Rebecca, Abigail, George.

(*Amwell, N. J.*)

(4) *SAMUEL,[6] [*Samuel*, (3),] b. December 28th, 1762, d. May 20th, 1837. *M.* Rachel Newman, December 26th, 1786. She b. December 26th, 1763, d. June 12th, 1839.

Ch. Sarah, Abigail, William, Hannah, Alice, Rebecca, Richard, Samuel D., George. (*Amwell, Hunterdon Co., N. J.*).

(5) Samuel Curwen Ward,[6] [*Mehetable Curwen*, (1),] b. in Salem, Mass., June 29th, 1767, d. November 26th, 1817.

M. (1) Jane, daughter of Hon. Nathaniel and Priscilla (Sparhawk) Ropes, October 31st, 1790. She was b. January 22d, 1767, and d. January 18th, 1808 ; (2) Malvina Tabitha, daughter of Benjamin Stacy and Tabitha Glover, December 17th, 1807. She was b. September 18th, 1784—living.

Ch. George A., Samuel C., Charles, Jane S., Eliza W., Malvina G., William R. L., Henry O. (*See Ward Genealogy, in Essex Institute,* Vol. 5, No. 5.)

(6) SAMUEL CURWEN [*Ward*,][7] (*Samuel C. Ward*, 5,) b. November 26th, 1795, d. July 3d, 1831, near Bellona Arsenal, Virginia.

M. Priscilla, daughter of James and Eunice (Carlton) Barr, March 22d, 1818. She was b. in Salem, March 31st, 1788, and died November 27th, 1863.

Ch. James B., Samuel R., George R.

The surname, Curwen, of the Salem family, would have become extinct in Massachusetts with the death of Samuel Curwen, (2,) by failure of,

* Curwen.

male heirs ; but at the request of the latter, the Legislature of Massachusetts passed an act permitting Samuel Curwen Ward (6) to take the name of Samuel Curwen, (when seven years of age,) by which also he became heir of Samuel Curwen, (2.)

(6½) * SAMUEL,⁷ [*Richard*, (2),] b. in New-Jersey, July 31st, 1786, d. October 7th, 1865.

M. Mary, daughter of Samuel Wilson, April 8th, 1808. (She was sister to his brother George's wife.) She died July 7th, 1870.

Ch. Abby, Mary Ann, Lydia, Sarah, George, Elizabeth, John W., Jerusha, Richard, Clarissa.

He was taken by his father to Washington, Kentucky, in 1788.   In 1808, he removed to Pike Co., Ohio. He occupied a high position in society. He was commissioner of the county for several terms, and was held in high esteem. He took a deep interest in the political affairs of the county, but was conservative and generous in his views.   He was the unyielding enemy to persecution for opinion's sake, and held the right of free speech to be the great safeguard of our liberties. Upon one occasion, when political prejudice was at its highest pitch, some speakers were denied admission to the school-house, or to any place wherein to hold a meeting ; the residents of the little village being all of the opposite party, and the majority being extremely radical.   The speakers repaired to the residence of Mr. Corwine, to consult with him. He was then old and feeble ; but he sprang up with the vigor of youth, proceeded to the village, opened a church which he and his neighbors had erected upon generous religious principles for all true religious congregations, and cautioned the speakers that their speeches should accord with the place, and permitted the meeting to proceed.   All classes from the excitement had assembled, and the speakers, true to their friend's advice, discussed public affairs in that liberal spirit.   His firmness and liberality upon that occasion was an example in that community the effects of which are apparent to the present time.

He was a Baptist, and was in the war of 1812.

(7) †SAMUEL ROPES,⁸ [*Samuel*, (6),] b. December 28th, 1820, d. November 11th, 1870.

M. Mary Smith, daughter of Jonathan and Betsy (Barr) Holman, May 15th, 1846.

Ch. Henry, Samuel H., Caroline R., Charles F., Betsy H.

(*Salem, Mass.*)

(8) † SAMUEL ENDICOTT,⁹ [*James B.*, (1),] b. November 8th, 1848, d. July 6th, 1849.

(9) † SAMUEL HOLMAN,⁹ [*Samuel*, (7),] b. August 5th, 1849.

10) * SAMUEL DAVIS,⁷ [*Samuel*, (4),] b. March 10th, 1808.

(*Central N. Y.*)

* Corwine.   † Curwen.

(11) *SAMUEL,⁸ [*Richard*, (4),] b. ——, 18 .

(12) *SAMUEL LUCAS,⁸ [*Joab*, (1),] b. July 4th, 1818, d. May 28th, 1871.
M. Belle Thomas, July, 1854.
Ch. Anna F., Amos B., Samuel L., Martha T., Joab II., Mary S.
(*Loveland, O.*)

He was originally a printer. He was associated with his brother Amos
for several years in the publication of the *Yazoo Banner*, at Benton, Mississippi, before and after the presidential campaigns of 1840 and 1844. About
1849, he was again connected with his brother Amos and E. D. Mansfield
in the publication of the old Cincinnati *Chronicle*, then published on the
corner of Third and Hammond streets. In 1851, he joined his brother,
who was then Consul at Panama. From 1858–60, he was Clerk of the
Cincinnati City Council. Subsequently, for a number of years, he was
Secretary of the La Fayette Insurance Company, and was at the time of his
death U. S. Marshal under Mr. Thrall.

There were few citizens who were better known, or who had a larger
and more attached circle of friends and acquaintances. In his nature and
disposition he was preëminently social, and was the life of the society in
which he moved. He possessed great information of men and things, and
was exemplary and honest in all his dealings and intercourse with society.
He was an excellent representative of a gentleman of the old school.

(12) Samuel Barkalow,⁸ [*Lydia Corwine*, (2),] b. ——.   (*Franklin, O.*)

(13) †SAMUEL LUCAS,⁹ [*Samuel L.*, (12),] b. July 4th, 1860.

(14) Samuel Corwine Miller,⁹ [*Sarah H. Corwine*, (6),] b. August 1st, 1848,
in Cincinnati, O.

(15) Samuel Sharpe,⁹ [*Sarah Corwine*, (9).]   (*Kingston, O.*)

(I.) SAMUEL CORWINE, (or Curwin,†) was settled in Essex and old Norfolk,
Mass., in 1652. (*New-England Hist. Reg.* 6, 249.) His wife's name
was Elizabeth ——. He was probably of some collateral English
Curwen branch. (*See also New-Eng. Hist. Coll.* 7, 18, 333, where he
is mentioned in 1650.) He died November 16th, 1698. Possibly
he was the father of Mary Corwin, who married, May 9th, 1701
Thomas Smith, of Boston, afterward captain of the fort at Saco.

Sarah, (Hebrew,) a princess.

1 SARAH,³ (*John,* 1,) b. about 1660.
M. Jacob Osman, before 1690. He was b. about 1670, d. ——, 1758.
Ch. Mary, Sarah, Eliza, Hester, Pinnina, Hannah, Peter, Deborah, Al-
modan, Elizabeth.
His will in Liber 21, p. 108, N. Y. C. (*See also Doc. Hist. N. Y.*, i. 450,
and *Moore's Index of Southold*, p. 105.)

* Corwine.   † Farmer, in his First Settlers N. E., gives this orthography.

2 SARAH,[2] (*John*, 2,?) b. ——, d. October 29th, 1938.
  M. Peter Simons, ? who died November 3d, 1720?
  Names occurs on census list of 1698, directly under that of John and
Sarah Corwin, as Sarah Corwin, Jr.   (*Doc. Hist. N. Y.*, i. 450.)   Moore
thinks she was daughter of Theophilus, 2.

3 SARAH,[5] (*John*, 3,) b. about 1739.
  (A Sarah Corwin married John Penney, October 26th, 1786.)
                                              (*Aquebogue Records.*)

4 SARAH,[6] (*Timothy*, 1,) bapt. at Mattituck, January 18th, 1756.
  M. James Jennings.
  Ch. James, Samuel, Daniel, Harmony, Sarah.   (*Southampton, L. I.*)

4¼ SARAH, (——,) daughter of Widow Hannah Corwin, d. January 31st,
  1767. (*Aquebogue Records.*)

4½ SARAH,[5] (*Samuel*, 2,) b. January 10th, 1747, d. April 29th, 1747.

4¾ SARAH,[5] (*Samuel*, 2,) b. November 10th, 1736, d. September 21st, 1738.

5 SARAH,[6] (*William*, 2,) b. January 13th, 1771; (bapt. at Mattituck,
  March 1st, 1771.)
  M. Jabez Kelsey.   (*Chester, N. J.*)

5¼ SARAH,[6] (*Joseph*, 1,) b. ——, 1772 ?   (*Near Hartford, Ct.*)

6 SARAH,[6] (*Phineas*, 1,) b. July 29th, 1779, d. July 19th, 1863.
  M. Daniel Campbell, March 5th, 1801.
  Ch. Phineas, Anna, Mehetable, Daniel, Sibyl, Priscilla, William, Jonah.

6¼ SARAH,[6] (*Edward*, 2?) bapt. at Aquebogue, July 22d, 1764.

7 SARAH,[6] (*Benjamin*, 2,) b. ——.   (*Morris Co., N. J.*)

8 SARAH MARIA,[6] (*David*, 5,) b. November 19th, 1807, d. February 11th,
  1844.

9 SARAH,[6] (*Jesse*, 2,) b. March 8th, 1780, d. ——, 1850, in Ohio.
  M. George Duckworth, 1804.   He was b. in England, and died in
  1849.
  Ch. Lucinda, Keziah, Mary W., Louisa J., Laurinda E., Robert, Jesse
  C., Sarah, Hannah A. M.
  She removed from Kentucky to Champaign Co., O., in 1797, and subse-
quently to Warren Co., O.

10 SARAH,[6] (*Rev. Jacob*, 1,) b. ——, 1790 ?
  M. John Smith.
  Ch. Rockwell ?

11 SARAH,[6] (*Jedediah*, 2,) b. ——, 1760–80.

12 SARAH,[6] (*John*, 6,) b. December 12th, 1777, died young.

13 SARAH J.,[7] (*Elisha*, 1,) b. ——, 1810–20. *d.* ᴀ · · ᴀ·
  M. Ferdinand Crassons.
  Ch. Ferdinand, Eliza.   (*Jersey City, N. J.*)

14 SARAH,[7] (*Joseph*, 4,) b. May 30th, 1798.

15 SARAH,[7] (*Phineas*, 3,) b. November 6th, 1824, d. June 28th, 1855.

16 SARAH,[7] (*David*, 7,) b. January 6th, 1824.
M. Alvern Smith, February 25th, 1844.

17 SARAH,[7] (*John*, 9,) b. August 26th, 1777, d. September 18th, 1847.
M. Isaiah Benjamin, December 29th, 1793. (*Aquebogue Records.*)

17½ SARAH J.,[7] (*Seth*, 1,) b. May 22d, 1831.
M. John Crook..
Ch. Sarah Jane.

18 SARAH,[7] (*James*, 4,) b. ——, 1801 ?
M. —— Crawford.

19 SARAH,[7] (*Isaac*, 1,) b. December 29th, 1791, d. April 4th, 1841.
Unm.   (*Chester, N. J.*)

20 SARAH,[7] (*William*, 5,) b. ——, 1785–90, d. ——, 1836.
M. Thomas De Boyce.
Ch. Cortlandt, Elizabeth, Thomas, Charlotte.

21 SARAH,[7] (*John Calvin*, 11,) b. ——, 1807 ?

22 SARAH,[7] (*Nathan*, 2,) bapt. at Aquebogue, May 7th, 1809. ? ——

23 SARAH CHARLOTTE,[7] (*John*, 13,) b. ——, 1800 ?
M. Daniel H. Skidmore ?

24 SARAH,[7] (*Henry*, 6,) b. December 21st, 1783.
M. —— Smith.
Ch. Noah, Alfred, Rosetta, Rachel, Sarahetta, Eliza.   (*Sayville, L. I.*)

25 SARAH,[7] (*William*, 7,) b. March 8th, 1839.
M. Daniel Lee, son of John D. and Sarah Hornbeck, January 21st, 1868.

26 SARAH,[7] (*Benjamin*, 4,) b. ——, 1782, d. ——, 1865.
Unm.

27 SARAH JANE,[7] (*Nebat*, 1,) b. ——.

28 SARAH,[7] (*Hubbard*, 1,) b. April 7th, 1795.
M. Josiah Smith, of Tunbridge, Vt., in 1817. He d. in 1863, aged 67 years.
Ch. John C., Mary, Loren, George, Hannah.
In 1833, the family removed from Vermont to Franklin, Franklin Co..
N. Y. In 1838, removed to Hopkinton, St. Lawrence Co., N. Y.

29 SARAH,[7] (*Daniel*, 5,) b. ——, d. young.

30 SARAH GRIFFIN,[7] (*Ichabod*, 1,) b. October, 1813.
M. Rev. Muncier Jones, October, 1835.

30½ SARAH E.,[7] (*David*, 10,) b. July 26th, 1850.
M. James C., son of James and Sarah Vine, December 25th, 1868.
Ch. Frederick.

31 SARAH,[7] (*Eli*, 3,) b. December 21st, 1790, d. ——, 1850.
M. John Moffat.

Ch. Samuel W., Sarah, Nathan H., William O., Orthenal, Wickham.

(*Goshen, N. Y.*)

32 SARAH JANE,[7] (*Joseph,* 9,) b. ——, 1812? *d J any 20 . 1892*
   M. George Miller.   *1890 .*
   Ch. Sylvanus K., Mary, Charles A.

33 SARAH ANN,[8] (*Abel,* 3.)

34 SARAH MARIA,[8] (*Benjamin,* 14,) b. September 15th, 1814.
   M. Israel Concklin, April 26th, 1832.
   Ch. Mary D., Charles W., William H., Theresa M., Jane, Sarah, Israel.

(*Riverhead, L. I.*)

35 SARAH C.,[8] (*Daniel,* 19,) b. August 21st, 1814.
   M. Gilbert Corwin, No. 4.

36 SARAH JANE,[8] (*Philip,* 1.)

37 SARAH JANE,[8] (*Joshua C.,* 8¼.)
   M. —— Skinner.

38 SARAH A.,[8] (*David,* 16,) b. April 9th, 1827.
   M. Rev. Gabriel S. Corwin, No. 5.

39 SARAH ELIZABETH,[8] (*Joshua,* 6,) b. May 2d, 1834, d. October 26th, 1834.

40 SARAH A.,[8] (*William,* 11,) b. March 18th, 1842.
   M. James L. Smith, of Williamsburg, L. I.
   Ch. Lewis, Alice D.

41 SARAH ALICE,[8] (*William H.,* 12,) b. December 30th, 1842.

42 SARAH,[8] (*Stephen,* 3,) b. ——, 1810–20.

43 SARAH,[8] (*Benjamin,* 7,) b. ——, 1820–30.
   M. Daniel Gordon.
   Ch. Charles L., Benjamin, George W., Clara B., Ella.

44 SARAH JANE,[8] (*Ebenezer,* 3,) b. November 4th, 1821.
   M. Orrin H. Kingman.

45 SARAH ROSALIA,[8] (*Polydore B.,* 1,) b. July 23d, 1840.
   M. Le Roy Wheeler.
   Ch. Charles.   (*Cortland, N. Y.*)

46 SARAH,[8] (*Tuthill,* 1,) b. ——, 1823.
   M. William Ackerman.   (*New-York City.*)

47 SARAH G.,[8] (*Robert G.,* 2,) b. March 18th, 1852.

47½ SARAH ELIZABETH,[8] (*David,* 18,) b. April 12th, 1836.
   M. Evan Evans, March 27th, 1855. He was b. May 4th, 1827.
   Ch. Charles Daniel, (b. February 24th, 1856,) Hester Kate, (b. March
      17th, 1858,) Sarah Caroline, (b. May 28th, 1860,) Evan Corwin, (b.
      November 13th, 1863,) Richard Jennings, (b. October 15th, 1866.)

(*Brady's Bend, Pa.*)

48 SARAH FRANCES,[8] (*De Witt C.,* 2,) b. August 24th, 1846.

49 SARAH,⁸ (——.)
   M. —— Moffat.
   Ch. Mary Ann.

50 SARAH,⁸ (*John*, 30,) b. ——, 1811.
   M. (1) John Rouse; (2) Jacob Herrick; (3) Aaron Selbeck.
   No Ch.   (*Elkhorn, Wis.*)

51 SARAH JANE,⁸ (*Salem*, —.)
   M. Nelson Akerly.   (*Green Point, L. I.*)

52 SARAH,⁸ (*Jesse*, 7.)

53 SARAH REBECCA,⁸ (*Isaac*, 4,) b. April 24th, 1828.
   M. Leander Baird, November 3d, 1854. He d. January 20th, 1855.

54 SARAH ANN,⁸ (*Hector*, 1,) b. ——, 1830–40.

55 SARAH E.,⁹ (*Hewlett C.*, 1,) b. May, 1833, d. August 9th, 1835.

56 SARAH,⁹ (*Nicholas*, 1,) b. October 12th, 1836.
   M. George Gray, December 29th, 1857.
   Ch. Elbert H., Theodore W., Annie E., Frederic, George.
                          (*Chester, N. J.*)

57 SARAH FRANCES,⁸ (*Silas G.*, 12,) b. March 21st, 1848. ——

58 SARAH,⁹ (*George*, 40,) b. ——, 1862. ——   (*Michigan.*)

58¼ Sarah Osman,⁴ (*Sarah Corwin*, 1,) b. ——, 1690–1700.

58½ Sarah Mapes,⁴ (*Mary Corwin*, 1,) b. before 1690, when she is referred to in the will of John, 1.

59 Sarah Jennings,⁶ (*Sarah Corwin*, 4,) b. ——, 1780–90.

60 Sarah Brockway,⁶ (*Mary Corwin*, 5,) b. about 1800, d. ——, 1867, in Newburg, N. Y.
   M. —— Davis.

61 Sarah Duckworth,⁷ (*Sarah Corwin*, 9,) b. April 30th, 18—.

62 Sarah Ann Thatcher,⁷ (*Jemima Corwin*, 1,) b. in Warren Co., O., 1794, d. in Vigo Co., Ind.

63 Sarah Philips,⁷ (*Hannah Corwin*, 2.)
   M. Samuel Taylor. *Four ch.*

64 Sarah Everett,⁷ (*Abigail Corwin*, 2,) b. March 25th, 1811.

65 Sarah Hobart,⁷ (*Mehetable Corwin*, 2,) b. August 11th, 1811.
   M. Linus North, October, 1833.
   Ch. Hobart, Mehetable, Ellen, Mary, Eliza.   (*Elbridge, N. Y.*)

66 Sarah Jane Millspaugh,⁷ (*Cynthia Corwin*, 1,) b. October 9th, 1813.
   M. Shelby W. Clark, October 24th, 1833.
   Ch. Harvey, Cortland W., Paulina? Celeste, Virginia.
                      (*Mount Washington, O.*)

67 Sarah Woodhull,⁷ (*Hannah Corwin*, 2,) b. November 1st, 1798?
   M. —— Hatfield.   (*Middletown Point, N. J.*)

68 Sarah Meyers,⁸ (*Arminda Brown*, 2.)

69 Sarah Jane Howell,[8] (*Jemima Corwin*, 7.)

70 Sarah Lippincott,[8] (*Elizabeth Corwin*, 18.)

71 Sarah Girard,[8] (*Jerusha Corwin*, 2.)

72 Sarah Horton,[8] (*Eliza Corwin*, 2,) b. ——, 1830–40.

73 Sarah Alice Connor,[8] (*Caroline Corwin*, 2,) b. March 10th, 1851.

74 Sarah Jane Woodruff,[8] (*Silas Woodruff*, 15,) b. ——, 1825–35.

75 Sarah Padden,[8] (*Anna Campbell*, 29,) b. ——, 1827 ?

76 Sarah Belle Armstrong,[8] (*Charlotte Millspaugh*, 10,) b. October 23d, 1846.   (*Booneville, Ind.*)

77 Sarah Vance Beach,[8] (*Eliza Ann Corwin*, 5,) b. March 8th, 1836.
   M. William L. Windsor, June 6th, 1861.
   Ch. Henry.

78 Sarah Jane Moffat,[8] (*Sarah Corwin*, 11,) b. ——, 1810–20.

79 Sarah Maria Dunlevy,[8] (*Lucinda Corwin*, 4,) b. August 22d, 1819.
   M. S. Suydam.   (*Toledo, O.*)

80 Sarah Brown,[8] (*Elizabeth Woodhull*, 45,) b. April 9th, 1835.

81 Sarah Roselle Hamlin,[8] (*Patience M. Corwin*, 1,) b. March 16th, 1860.

82 Sarah Moffat,[8] (*Sarah Corwin*, 31.)

83 Sarah Ann Shannon,[8] (*Mehetable Corwin*, 11,) b. ——, 1820–30.
   M. Harvey Coryell.
   Ch. Mark.   (*Havana, N. Y.*)

84 Sarah Eliza Peters,[8] (*Mary Corwin*, 30,) b. ——, 1835–45.

85 Sarah Reed,[8] (*Sophia Corwin*, 1,) b. January 3d, 1810.

86 Sarah Frances Higby,[8] (*Harriet Corwin*, 2,) b. ——, 1860.

87 Sarah Tuthill Howell,[8] (*Sophia Woodruff*, 5,) b. July 18th, 1834.
   M. John Williams, March 16th, 1855.
   Ch. Henrietta S.

88 Sarah North,[8] (*Mehetable Campbell*, 10,) b. July 15th, 1827.
   M. Henry Carter.

89 Sarah Taylor,[9] (*David J. Taylor*, 30,) b. September, 1842.

90 Sarah Ann Wright,[8] (*Elizabeth Corwin*, 16,) b. ——, 1820–30.
   M. Dr. Onderdonk.   (*Brooklyn, N. Y.*)

91 Sarah F. Brown,[8] (*Silas C. Brown*, 16.)

92 Sarah M. Mitchel,[8] (*Maria Corwin*, 3.)

93 Sarah M. King,[8] (*Mary Corwin*, 35.)

94 Sarah Bethia Howell,[8] (*Elizabeth Corwin*, 20.)
   M. James Minor Petty.

95 Sarah Concklin,[9] (*Sarah M. Corwin*, 34,) b. December 9th, 1843.

96 Sarah Abigail Durkey,[9] (*Philenia Corwin*, 2,) b. April 8th, 1846, d.
   May 1st, 1859.   (*Vermont.*)

97 Sarah Allen,⁹ (*Susannah Hobart*, 8,) b. June 4th, 1849.

98 Sarah C. Boyd,⁹ (*Emeline Corwin*, 7,) b. July 24th, 1865. (*Vermont.*)

99 Sarah L. Chatfield,⁹ (*Melissa H. Hobart*, 1,) b. December, 1861.

100 Sarah J. Chamberlin,⁹ (*Jennette Corwin*, 1,) b. December 7th, 1843, d.
April 25th, 1868. (*Vermont.*)

101 Sarah E. Lankton,⁹ (*Phebe A. Padden*, 10,) b. August 23d, 1857, d.
February 14th, 1860.

102 Sarah Horton,⁹ (*Henry Horton*, 29.)

103 Sarah L. Noyes,⁹ (*Louisa Corwin*, 8,) b. November 18th, 1837, d. February 27th, 1839. (*Vermont.*)

104 Sarah Lewis,⁹ (*Emeline Corwin*, 2.)

105 Sarah E. Kyte,⁹ (*Clarinda Corwin*, 2,) b. April 2d, 1854.

106 Sarah Harvey,¹⁰ (*Elizabeth Corwin*, 40,) b. ——, 1858.

107 Sarah Jefford,¹⁰ (*Stella L. Baker*, 1,) b. October 9th, 1870.

108 Sarah E. Cram,⁹ (*Hannah Hackett*, 26,) b. March 29th, 1846.
(*Vermont.*)

(1) SARAH,³ [*Jonathan*, (1),] b. August 12th, 1680, d. December 19th,
1689.

(2) * SARAH,⁶ [*George*, (5),] b. January, 1742, d. February 26th, 1773.
Unm.

(3) † SARAH,⁶ [*Richard*, (2),] b. June 3d, 1797, d. October 29th, 1869, at
Middletown, Butler Co., O.
M. (1) Samuel Stout. He d. 1832 ; (2) Arthur Lefferson, in 1855. He
d. 1869.
Ch. Two, both died young. (*Kentucky, O.*)

(4) † SARAH,⁷ [*Samuel*, (4),] b. July 2d, 1787.
M. John Ryan, (Rien,) December 7th, 1805.

(5) † SARAH ANN,⁸ [*Richard*, (5),] b. in Turnersville, Tenn., December
15th, 1826.
M. Rev. Daniel Stevenson, August 5th, 1849, in Bourbon Co., Ky.
Ch. Lacy C., Daniel C., Richard T., Mary J., Sarah E., Harriet,
Jesse F.
In 1849, Rev. Mr. Stevenson taught school in Centreville, Indiana. In
1850, he removed to Kentucky, and opened a girl's school in Versailles,
his native place. In 1851, he removed to the vicinity of Maysville, Ky.,
and united with the Kentucky Conference of the Methodist Episcopal
Church, South, and itinerated for a year, after which he had the following
settlements successively in Kentucky, namely, Taylorsville, 1852, Danville,
1853, Newport, 1855, Carrollton, 1857, Millersburg, 1858, Shelbyville, 1860,
Frankfort, 1862. Superintendent of Public Instruction, Kentucky, September, 1863, to September, 1867 ; (in the fall of 1865, he withdrew from the

* Curwen. † Corwine.

Methodist Church, South, and united with the Methodist Church, North ;)
Parkersburg, (W. Va.,) 1867, Lexington, Ky., 1868.

(6) * SARAH HOUGHTON,[8] [*Joab*, (1),] b. September 1st, 1821, in Maysville,
Ky.

    M. (1) William Thomas Fountain, in Benton, Miss., December 31st,
1840. He was clerk in the Circuit Court in that place. He d. June
21st, 1844 ; (2) John Godney Miller, October 10th, 1847, of New-
York. He was born June 11th, 1817.

    Ch. (1) Thomas C., Frederick M. ; (2) Samuel C., Fanny, Grace, Belle,
Mae B.

Mrs. Fountain removed to New-Orleans, where she married her second
husband. They soon removed to Cincinnati, O., and thence to New-York
City in 1849 ; to Brooklyn in 1856, and to White Plains, N. Y., in 1862.

(7) Sarah Elizabeth Stevenson,[9] [*Sarah A. Corwine*, (5),] b. at Newport,
Ky., May 15th, 1857, d. at Lexington, Ky., March 14th, 1870.

(8) Sarah Barkalow,[8] [*Rachel Corwine*, (2).]

    M. (1) John Ten Eyck. He d. about 1858 ; (2) William Maxwell.

                                                       (*Franklin, O.*)

(9) * SARAH,[8] [*Samuel*, (6½),] b. March 11th, 1815, d. December 9th, 1843.

    M. John Sharp, November 5th, 1835.

    Ch. Samuel, Mary, John T., Elizabeth. (*Kingston, Pickaway Co., O.*)

(10) * SARAH,[8] [*Amos*, (1½),)] b. ——.   (*Illinois.*)

Sarepta, (Hebrew,) smelting-house.

1 SAREPTA,[6] (*Robert*, 1.)

2 SAREPTA,[8] (*William*, 29.)

1 SARIETTA,[8] (*Gilbert*, 4,) b. May 30th, 1838.

    M. David M. Edwards, January 20th, 1859.

2 Sarietta Smith,[8] (*Sarah Corwin*, 24.)

Selah, same as Salah.

1 SELAH,[6] (*Jonathan*, 1,) bapt. at Mattituck, January 16th, 1757.

    M. Joanna Halleck, who was b. 1757, d. May 11th, 1846.

    Ch. Rachel, Joanna, Lydia, Selah, Elizabeth, Lydia, Lucretia, Maria,
Peter.

In 1776, after the battle of Long Island, he went to Lyme, Ct. In
1780, he was permitted to visit L. I. for supplies. Perhaps he afterward
removed to Orange Co., N. Y.

2 SELAH,[7] (*Selah*, 1,) b. ——, 1790 ?

    M.

    Ch. Peter.

3 SELAH REILEY,[8] (*Joshua*, 6.)

    M. Fanny Sarc. *No Ch.*   (*Middletown, N. Y.*)

* Corwine.

4 SELAH,[9] (*Alsop L.*, 3,) b. December 26th, 1809.

5 Selah E. Grinolds,[10] (*Elizabeth Corwin*, 43½,) b. ——, 1857.

Selina, (Greek,) the moon; or parsley.

1 Selina Smith,[9] (*John E. Smith*, 92,) b. ——, 1845–50.

(*St. Lawrence Co., N. Y.*)

1 SELLA,[6]? (——,) b. ——, d. October 22d, 1805.
Mother's name, Mary. (*Aquebogue Records.*)

1 SEPARATE,[5] (*Edward*, 1, ?) b. ——, 1735–40.
An uncle of Edward, 3. Separate is called by Daniel, 2, a grandson,
his father being dead probably before 1747.

Seth, (Hebrew,) substitution.

1 SETH,[6] (*Abner*, 1,) b. in Southold, L. I., March 5th, 1783, d. May 1st,
1858.
M. (1) Angeline (Schoonmaker?) of Orange Co., N. Y., September
10th, 1808; (2) Sarah Post, of Southampton, L. I., about 1811. She
died July 9th, 1854.
Ch. Jemima; Angeline, Joseph R., Lewis J., Sarah J., Seth R.
About 1808, he removed to Orange Co., N. Y., and in 1810, back to
Southold. He served in the war of 1812.

2 SETH,[7] (*Abel*, 1,) b. ——, 1789.

2½ SETH R.,[7] (*Seth*, 1,) b. January 19th, 1823, d. September 13th, 1851.
He was knocked overboard and drowned by being struck by the main
boom of the ship Cabinet, in East River, opposite New-York City.

3 SETH MONROE,[8] (*Nathaniel*, 7,) b. May 13th, 1827.
M. Silas Silleck, December, 1870. (*New-York City.*)

4 SETH,[8] (*Grover*, 1,) b. ——, 1820–30.

5 Seth Squires,[8] (*Patty Corwin*, 1.)

6 Seth Clark,[9] (*Helen A. Colgrove*, 10,) b. about 1863.

1 SEVELLEN,[8] (*George F.*, 13,) b. ——, 1830–40.

1 SEYMOUR,[7] (*Stephen*, 7½,) b. August 4th, 1825. (*Dunkirk, N. Y.*)

1 SHEPHERD,[8] (*John*, 25.)

Shubal, (Hebrew,) a shoot.

1 SHUBAL,[7] (*Rev. Joseph*, 5,) b. March 25th, 1792, yet living 1871.
M. (1) Abby Horton, who was b. February 25th, 1795, d. March 10th,
1846; (2) Sarah Griffin, widow of David Downs, October 20th, 1850.
Ch. Joseph H., Mary A., Benjamin H., Charlotte M.
(*Upper Aquebogue, L. I.*)

Sibyl, (Greek,) a prophetess.

1 SIBYL,[6] (*Phineas*, 1,) b. May 25th, 1773.
M. Joseph Woodruff.
Ch. Marietta, Silas, Erastus, Sophia, Dr. Elias.

2 SIBYL,[7] (*Samuel,* 9,) b. September 9th, 1818.
   M. Denton Cosman, September 15th, 1841.   (*Middle Hope, N. Y.*)

3 SIBYL CAMPBELL,[7] (*Sarah Corwin,* 6,) b. August 12th, 1812.
   M. John North, March 7th, 1832.
   Ch. Elias W., Fitz-Alan, Crissa P.

4 Sibyl T. Padden,[3] (*Anna Campbell,* 29,) b. ——, 1830 ?
   M. Tobias Van Wagenen, August 8th, 1848.
   Ch. Tobias, William A., Anna B., Douglass.   (*Michigan.*)

       Silas, contracted from Sylvanus, (Latin,) living in the woods.

1 SILAS,[6] (*Daniel,* 2,) b. on Long Island, ——, 1731, d. March 1st, 1806.
   M. Elizabeth Halleck, January 13th, 1753. She was b. September,
   1731, d. February 12th, 1831.
   Ch. Silas, Azubah, Ezra, Mary, Peter, Jabez, Elizabeth, Ebenezer,
   Daniel, Elizabeth.   (*Jamesport, L. I.*)
   It will be noticed that his wife reached very nearly a century of age.
In 1775, he signed engagement to support Congress. In 1776, his name
occurs on census list, 94, having in his family 3 males, 1 female over 16,
and 5 children under 16. Both he and his wife, as well as his father, are
said to be buried at Jamesport, L. I. His wife's mother's name is said to
have been Booth. Will in Lib. B, 365, Suffolk Co., N. Y. He united
with the church of Aquebogue, June 25th, 1769. His wife, March 31st,
1765.

2 SILAS,[6] (*Eli,* 1,) b. July 21st, 1786, d. April 10th, 1865.
   M. Sarah Little, September 4th, 1806. She b. June 20th, 1786.
   Ch. Rosetta, Isaac, George, Alfred E., Parmenas H., William T.
               (*New-Milford or Montrose, Pa.*)
   Perhaps this is the Silas who bought land in Orange Co., N. Y., in
1809 and 1824, or possibly No. 4 or 5.

3 SILAS,[6 or 6] (——,) b. ——, d. October 20th, 1779.
   A private in Second Artillery, Rickers's Company, in Revolutionary war.
He died in service. On October 13th, 1790, according to the bounty law then
passed, he received (or rather his heirs) a tract of land, by letters-patent,
in Cayuga Co., N. Y., being lot 61, in town of Milton, containing 500 acres.
It fell to the possession of his son Silas, say the records. A Silas Corwin,
of Southampton, L. I., (probably this son,) buys additional land in same
county, in 1791, namely, lot 21, in town 18 ; again he buys land in 1794
in the same town, and in town of Homer, in 1803.

4 SILAS,[6 or 7] (*Silas,* 3,) b. ——, d. ——.
   A Silas Corwin sells land in Tompkins Co., N. Y., in 1836. (See
   Silas, 6.)

5 SILAS,[6] (*Thomas,* 1,) b. at Mattituck, L. I., July 5th, 1788, yet living
   1870.

M. Jerusha Reeve, January 10th, 1813, daughter of Hezekiah and Mary Reeve, of Aquebogue, L. I.

Ch. Silas R., Hannah J., Amanda, Deborah, Frances F., Mary E., Catharine.

In 1799, he made his home with John Corwin, No. 5, a distant relative of his father. In 1809, he took up his residence at New-Windsor, Orange Co., N. Y., where he yet resides. A ship-carpenter. On November 12th, 1869, he dug up a stone pot in Newburg, with 650 silver Spanish dollars in it. Probably it was buried there in the Revolution.

6 SILAS,[6] (*Silas*, 1,) bapt. at Mattituck, L. I., August 10th, 1755, d. April 17th, 1837.

M. Elsie Corwin, (No. 1,) April 30th, 1781. (*Aquebogue Recs.*)

Ch. Silas, Jabez.

Both said to be of Ketchobononck, when married. Perhaps this is the Silas who signed engagement to support Congress, in 1775. He removed from L. I. to Orange Co., N. Y., probably between 1810–20, and lived at Minisink. He was also a deacon of the Presbyterian church of West-town. He says in his will that a George Corwin must live in his house after his death. This also may be the Silas Corwin of Southampton, L. I., in 1809, who sold land to Ezra Corwin, in Homer, Cortland Co., N. Y., in that year. We find also a Silas Corwin, Jr., buying land in Onondaga Co., N. Y., in 1791, and he is described as a blacksmith, of Setauket, L. I. Again, he buys land in same county in 1799. In 1803, a Silas Corwin, of Southampton, L. I., buys one soldier's right of land in Onondaga Co., N. Y. (See *Silas, Nos.* 3 *and* 4.)

6½ SILAS,[7] (*Silas*, 6,) bapt. at Aquebogue, October 20th, 1782.

7 SILAS HORTON,[7] (*Jabez*, 1,) b. February 25th, 1800.

M. Fanny Stringham.

Ch. Thomas, John S., William T., Silas H., Julia F., Rosetta E., Edwin H., Mary E.

Lived in New-York City for many years, and fifteen or twenty years ago moved to Michigan.

8 SILAS REEVE,[7] (*Silas*, 5,) b. January 1st, 1815, d. January 9th, 1837.

9 SILAS BENJAMIN,[8] (*Isaac S.*, 5,) b. May 20th, 1832.

10 SILAS R.,[8] (*Francis F.*, 1,) b. ——, d. young.

11 SILAS R.,[8] (*Francis F.*, 1,) b. ——.

12 SILAS GILBERT,[8] (*Joshua*, 6,) b. October 21st, 1821.

M. Charity Corwin, (No. 4,) October 22d, 1846. She b. April 22d, 1826.

Ch. Sarah F., George H., Alice E., Annesta M., Mary E., Charles W., Carrie G., Daniel J., George W.

13 SILAS D.,[8] (*Ebenezer*, 3,) b. February 22d, 1827.

*14 SILAS HORTON,[8] (*Silas*, 7.)

M. Jennie Clark. (*Bedford, Mich.*)

15 Silas Woodruff,[7] (*Sybil Corwin*, 1,) b. ——, 1800 ?
   M. Maria Belknap.
   Ch. Abigail A., Charles, Joseph, Susan K., Elmer, Elias, Theodore,
   Sarah J.

16 Silas C. Brown,[7] (*Mary Corwin*, 8,) b. March 6th, 1799, d. August 25th,
   1861.
   M. Sarah Holbert.
   Ch. Sarah F., Mary C., Amarintha, Emily S., Lewis M., Hannah E.,
   Amanda J., Louisa C., Silas C., John H., Dorastus.

17 Silas C. Brown,[8] (*Silas C. Brown*, 16.)

18 Silas Y. Brown,[8] (*Daniel C. Brown*, 36.)

19 Silas Terry,[8] (*Charlotte Corwin*, 5.)

20 Silas Galloway,[8] (*Amanda Corwin*, 2,) b. ——, d. aged 16.

21 Silas Galloway,[8] (*Amanda Corwin*, 2,) b. ——, d. aged 2.

Simeon, (Hebrew,) acceptable hearing.

1 SIMEON,[4] (*Daniel*, 1,) b. ——, 1710–18.
   M. Mary ——.
   Ch. Jemima, Rachel, Theophilus.   (*Mattituck Records.*)
   A Simeon is mentioned, 1739, 1749, 1753, 1755.

1 SOLON C.,[9] (*Jabin*, 2,) b. June 26th, 1864, at Chelsea, Vt.

Sophia, (Greek,) wisdom.

1 SOPHIA,[7] (*Joseph*, 8,) b. August 26th, 1778, d. March 23d, 1853.
   M. Augustus Reed, March 25th, 1798.   He b. October 7th, 1793, d.
   January 8th, 1824.
   Ch. Mary, John, Elizabeth, Margaret, Joseph, George R., Jaspin S.,
   Sarah, Jacob B., Nathaniel, Mary, Augr͟ ͟ e, Catharine C.
                                        (*Morris Co., N. J.*)

2 SOPHIA,[7] (*Matthias*, 6,) b. ——, 1790–1800.
   M. —— Ketchum.
   Ch. Julia, Gilbert H., Margaret, John, Sophia.

3 SOPHIA SMITH,[7] (*Stephen*, 4.)   (*Pennsylvania.*)

4 SOPHIA,[8] (*Charles*, 3,) b. December 18th, 1861.

4½ SOPHIA ADELL,[8] (*Gilbert*, 4,) b. November 14th, 1849.

5 Sophia Woodruff,[7] (*Sibyl Corwin*, 1,) b. ——, 1800–10.
   M. Charles Howell.
   Ch. Henrietta S., Frances M., Sarah T., Charles J., Philetus W., Har-
   riet W.   (*Newburg, N. Y.*)

6 Sophia Ketchum,[8] (*Sophia Corwin*, 2.)

7 Sophia Osborn,[9] (*Amelia Corwin*, 2,) b. August 31st, 1842.

8 Sophia Elizabeth Embler,[9] (*Frances M. Howell*, 13,) b. May 18th, 1855.

9 Sophia Thatcher,[8] (*Jesse Thatcher*, 10,) b. July 23d, 1818.

Sophronia, (Greek,) of a sound mind.

1 SOPHRONIA,[8] (*Benjamin W.*, 1½,) b. June 13th, 1833.
M. George Beyea, October 19th, 1851.
Ch. Fanny, Winfield S., Emma. (*Near Otisville, N. Y.*)

2 Sophronia Durkee,[9] (*Philenia Corwin*, 2,) b. May 21st, 1850.
(*Vermont.*)

1 SPENCER,[7] (*Hubbard*, 1,) b. May 19th, 1808.

2 SPENCER W.,[8] (*Benjamin W.*, 11,) b. May 7th, 1844, at Mt. Hope, N. Y.
M. Ann Gumaer, 1861.
Ch. Martha. (*Peoria, Ill.*)

Stella, (Latin,) a star.

1 Stella L. Baker,[9] (*Rosalinda Corwin*, 1,) b. August 25th, 1848.
M. Charles, son of Horace Jefford and Clarissa Partridge, of Niagara
Co., N. Y., November 11th, 1866.
Ch. Rosa, Sarah. (*Youngstown, N. Y.*)

Stephen, (Greek,) a crown.

1 STEPHEN,[5] (——,) b. ——, 1730–40.
M. Miss —— Haines.
Ch. Phebe, Hannah, Rachel, Samuel, Stephen.
Was a quartermaster in the Revolutionary war. Removed from South-
old, L. I., to Essex Co., N. J. Afterward moved to Ohio.

1½ STEPHEN,[5] (*Samuel*, 3,) b. ——, 1730–40, d. about 1808.
Unm.

2 STEPHEN,[6] (*Stephen*, 1,) b. ——, 1764, in N. J., d. ——, 1849, in Yates
Co., N. Y.
M. (1) Rachel Wilcox, a widow, who had a son, Cornelius Wilcox. She
had no children by Stephen Corwin. He *M.* (2) Betsy Drew.
Ch. John, Noah, Isaac, Nathan, Anna Polly, Abigail.
He was a soldier of the Revolution, (entering the service at the age of
fifteen,) and also of the war of 1812. He was engaged in the battle of
Valley Forge. He bought land near Springfield, N. J., in 1801, and sold it
in 1810. (*Essex Co. Records.*) In 1814, probably, he removed to Starkey,
Yates Co., N. Y., and in 1826, to Jerusalem, in same county. In 1812,
Rachel Corwin, and Cornelius Wilcox, her son, sold land at New-Provi-
dence, N. J. Probably the same land referred to above.

3 STEPHEN,[6] (*Nathaniel*, 1,) bapt. September 22d, 1782, at Southold.

4 STEPHEN,[6]? (——,) b. ——, 1760, d. ——, 1826, in Washington Co., Pa.
M. ——.
Ch. William, Stephen, John, James, Eunice C., Asenath S., Ruth R.,
Mary B., Sophia S., Elizabeth B., Rachel M., Hannah C.
(Perhaps a descendant of Jesse, 2.)

5 STEPHEN,[7] (*James*, 9.)

6 STEPHEN,[7] (*Samuel*, 12?.) (*Ohio?*)

14

7 STEPHEN,[7] (*Stephen*, 4.)　　　　　　　　　(*Marion Co., O.*)

7½ STEPHEN,[6] (*Phineas*, 1½,) b. in Norwich, Ct., September 10th, 1783, d.
September 10th, 1863, in Dunkirk, N. Y.

M. Priscilla Cobb, January 11th, 1808.　She b. November 25th, 1787.
(Yet living, in Dunkirk.)

Ch. Rev. Ira, Alonzo, Seymour, Ruth M., Hannah, Louisa, Rosetta.

He is buried near Dunkirk, N. Y.　He removed thither from Cazenovia, N. Y.

8 STEPHEN OVERTON,[7] (*Isaac*, 1,) b. September 29th, 1806.

M. Lydia Baker.　She b. February, 1809, d. November 2d, 1857.

Ch. Experience A., George, Hannah, Sarah.　　　　(*Oskaloosa, Iowa.*)

8½ STEPHEN OSBORNE,[8] (*David*, 18,) b. May 24th, 1820.

M. Nancy Ann Hurburt, August 10th, 1844.

Ch. Antoinette, James Buchanan, William Totten, Ellen, Henry Wisner.
　　　　　　　　　　　　　　　　　　　　　　(*Scranton, Pa.*)

9 STEPHEN F.,[8] (*James Y.*, 11,) b. ——, 1820–30.　(*Niagara Co., N. Y.*)

10 STEPHEN,[8] (*Manasseh R.*, 1,) b. ——, 1805–15.　(*Near Rochester, N. Y.*)

11 STEPHEN G.,[8] (*Lemuel*, 1,) b. October 27th, 1849, d. December 22d,
1850.

12 STEPHEN,[8] (*Noah*, 2,) b. July 22d, 1822, d. August 3d, 1864.

M. Lydia Finch, 1843.

He enlisted in 107th Regiment, N. Y. V., July, 1862.　Served at Antietam, and under General Sherman in the West; was killed by a rebel sharpshooter, near Atlanta, while on picket duty.

13 Stephen Halsey,[7] (*Elizabeth Corwin*, 2,) b. ——, 1812 ?

14 Stephen Bowers,[8] (*Elmina Corwin*, 1,) b. November 22d, 1832.

M. Caroline Osborn, of Haverstraw, N. Y.　　　(*Haverstraw, N. Y.*)

15 Stephen D. Sperry,[8] (*Mary Corwin*, 40,) b. ——, 1825.

M. Mary Burst.

Susan or Susannah, (Hebrew,) a lily.

1 SUSAN G.,[7] (*Benjamin*, 4,) b. July 6th, 1786, d. March 12th, 1860.

M. John Honnell, April 12th, 1816.　He b. March 20th, 1791.

Ch. William C., Benjamin B., Adam S., John A., Mary E.
　　　　　　　　　　　　　　　　　　　　　　(*Succasunna, N. J.*)

2 SUSAN C.,[7] (*Jason*, 1,) b. July 12th, 1813.

M. Stephen Harlow, December 18th, 1832.

Ch. Arthur, Addison.

3 SUSAN,[7] (*Jemuel*, 1,) b. September 20th, 1831.

M. —— Fordham.　　　　　　　　　　　　　　　(*Montrose, Pa.*)

4 SUSAN,[7] ? (——,) b. about 1770–75, d. ——, 179-.

M. Silas Horton.

Her tombstone is at Chester, N. J., but the unit, in the date of her

death, is illegible. She died at the age of 27. The inscription says, Susan
Corwin, wife of Silas Horton.

5 Susan,[8] (*Hector*, 1,) b. ——, 1830–40.

6 Susan,[8] (*William*, 31,) b. ——, 1840–50.

7 Susan B.,[8] (*Robert G.*, 2,) b. February 26th, 1847.
   M. David M. Zeller, July 16th, 1867.

8 Susan Brown,[8] (*Parmenas Brown*, 2.)

9 Susan Keziah Woodruff,[8] (*Silas Woodruff*, 15,) b. ——, 1825–35.
   M. Joseph Goldsmith.   (*Newburg, N. Y.*)

10 Susan Losee,[9] (*Ann Kate Corwin*, 21.)

10½ Susan M.,[9] (*Charles L.*, 19,) b. August 28th, 1870.

11 Susan Ellen Cram,[9] (*Hannah Hackett*, 26,) b. November 3d, 1843, d.
   May 12th, 1863.   (*Vermont.*)

12 Susan Blinn,[9] (*Adam Blinn*, 3,) b. about 1855.

(1) * Susan,[8] [*Amos*, (1½),] b. ——.   (*Kinmundy, Ill.*)

1 Susannah,[6] (*Benjamin*, 2,) b. ——.   (*Morris Co., N. J.*)

2 Susannah,[6] (*Samuel*, 6,) b. ——, 1764.
   M. Joseph Terry, March 22d, 1785, at Southold.

3 Susannah Hobart,[8] (*Peter Hobart*, 9,) b. July 13th, 1829.
   M. George Allen, February 22d, 1848.
   Ch. Sarah, Frances, George H., Hobart E.

(1) † Susannah,[2] [*George*, (1),] b. December 10th, 1672, d. ——, 1699.
   M. (1) Edward Lyde, November 29th, 1694; (2) Benjamin Wads-
   worth.   (*Boston, Mass.*)
   Loyd and Lynde are also found for Lyde, but are probably errors.
(*N. E. Ant. and Gen. Reg.* xvii. 239.)

Sylvanus, see Silas.

1 Sylvanus Knapp Miller,[8] (*Sarah J. Corwin*, 32,) b. ——, 1825–35.
   M.
   Ch. Clara.

2 Sylvanus Drake,[8] (*Margaret Corwin*, 1,) b. ——, 1825 ?

Sylvester, (Latin,) rustic.

1 Sylvester B.,[7] (*William*, 4,) b. January 12th, 1825, d. at the age of
   12 years.

2 Sylvester,[8] (*Richard*, 4.)

3 Sylvester Elwood Ayres,[9] (*Experience A. Corwin*, 2,) b. October 4th,
   1860.

4 Sylvester Douns Concklin,[9] (*Lucretia R. J. Corwin*, 3,) b. ——.

5 Sylvester Harvey,[10] (*Elizabeth Corwin*, 40,) b. December, 1855.

* Curwen.   † Corwine.

1 Tabitha Case,[4] (*Henry Case*, 1,) b. about 1690. (*Moore's Index of Southold*, p. 65.)

1 TEMPERANCE,[6] (*Daniel*, 3,) b. ——, 1770–80.

A Temperance Corwin bought land in Chemung Co., N. Y., about 1840–50, of John H. Corwin and Elizabeth, his wife. Temperance was of Southport, Chemung Co., N. Y.

2 TEMPERANCE,[7] (*Henry*, 6,) b. May 7th, 1781.

Thaddeus, (Syriac,) the wise.

1 THADDEUS,[8] (*Harrison*, 1.)

Theodora, Theodore, (Greek,) the gift of God.

1 THEODORA ELIZABETH,[9] (*Theodore*, 4,) b. about 1866.

2 THEODORE,[8] (*Gabriel*, 2.)   (*Marathon, Cortland Co., N. Y.*)

3 THEODORE,[8] (*Archibald*, 1,) b. ——, 1840–50.

4 THEODORE,[8] (*Henry*, 11,) b. November 4th, 1840, d. ——, 1866.
M. Sarah Hayes, November 20th, 1864.
Ch. Theodora E.

4½ THEODORE,[8] (*John Eli*, 19,) b. ——, 1844?

5 THEODORE,[9] (*William O.*, 14,) b. March, 1832, d. ——, 1857.

6 THEODORE WELLINGTON,[9] (*Joseph*, 16,) b. June 1st, 1857.

7 THEODORE FLETCHER,[9] (*George W.*, 23,) b. October 9th, 1850.

7½ THEODORE CHAUNCEY,[9] (*Alva*, 1,) b. December 9th, 1842.
M. Emilie A. ——. She b. December, 1840, d. March 23d, 1870.

8 Theodore W. Little,[7] (*Desire Corwin*, 1,) b. November, 1827.
M. Louisa Baker, November, 1863.   (*Syracuse, N. Y.*)

9 Theodore Brown,[8] (*Dorastus Brown*, 1.)

10 Theodore Woodruff,[8] (*Silas Woodruff*, 15,) b. ——, 1825–35.

11 Theodore C. Campbell,[8] (*Daniel Campbell*, 34,) b. July 23d, 1838, d. November 23d, 1860.
M. Harriet E. Wood, May 13th, 1858.
Ch. William L.   (*Savannah, N. Y.*)

12 Theodore Milnor,[8] (*Harriet A. Millspaugh*, 19,) b. December 19th, 1838.
                (*Covington, O.*)

13 Theodore Mullock,[8] (*Mary Corwin*, 22.)

14 Theodore M. Gray,[10] (*Sarah Corwin*, 56,) b. August 11th, 1860.

Theodosia, (Greek,) the gift of God.

1 Theodosia Case,[4] (*Benjamin Case*, 27.) (*Moore's Index*, p. 64.)

Theophilus, (Greek,) a lover of God.

1 Theophilus,[2] (*Matthias*, 1,) b. in or before 1634, d. before 1692.
   M. Mary ——.
   Ch. Daniel, Theophilus, David, Mary, Mehetable, Bethia, Phebe.

In 1655, he had lands at Southold, L. I., and also lands and a meadow at Aquebogue. Hence the above proposed date of his birth. He must therefore have been born very soon after his father's arrival at Ipswich, Mass., if born in America. That the above date of death, 1692, belongs to this Theophilus, is probable, as he had also a son Theophilus, but only one such name occurs on census list in 1698, whereas if the father were yet living, there would have been two. In 1660, he was engaged in a suit with Henry Case. In 1683, he was rated at £84. (*Doc. Hist. N. Y.* ii. 310.) In 1658, a Theophilus Corwin and Mary are witnesses in a deed of Tustin to Franklin. In 1686, he had 4 males and 3 females in his family, and in same year, sells land to A. Whittier. The above names of children, who evidently belong to the third generation, (as their names are enumerated in the census lists of the *Documentary History of New-York* for 1698,) probably all belong to this Theophilus, as the children of his only brother, John, are mentioned in the will of the latter. (See *John*, No. 1.) These names are also scattered in the said census list, indicating thereby that the family was broken up, the father being deceased. On the other hand, the well-known names of the children of his brother John, who was yet living in 1698, are found together. (*Doc. Hist. N. Y.* i. 450–455.)

2 Theophilus,[3] (*Theophilus*, 1,) b. ——, 1678, d. March 18th, 1762.
   M. Hannah Ramsay, after 1698. She was b. ——, 1684, d. March 11th, 1760.
   Ch. Timothy, John, Samuel, Theophilus?

This Theophilus and wife are buried at Mattituck, on the east side of the cemetery, beneath a small blue-stone slab, still remaining, (1871.) His name stands alone in census list of 1698, (*Doc. Hist. N. Y.* i. 450,) indicating that he was not then married. In 1702, sells land to Peter Simons. In 1703, gives land at Aquebogue to Daniel Corwin, no doubt his brother. In 1704, sells land to his cousin, John Corwin, No. 2. (*Moore's Index of Southold*, p. 74.) His will may be found in Lib. 23, New-York City.

3 Theophilus,[5] (——,) b. ——, d. December 27th, 1734.
   (Probably a grandson of Theophilus, 2.)

4 Theophilus,[4] (*Theophilus*, 2?) b. about 1700.
   M. Anna Jaynes? and perhaps also, before or after, Hannah Youngs, in 1730.
   Ch. Mary? David? George? Jonathan?

A Theophilus Corwin (Curwin, Curran) bought land of William Hudson, at Minisink, Orange Co., N. Y., in 1765, (or possibly as early as 1758.) (See *Records of Deeds at Goshen*.) He was a harness-maker by trade. There was a Theophilus Corwin who fell in battle at the taking of Fort

Montgomery by the British in the Revolutionary war. (*Letter of Silas Corwin, of New-Windsor, N. Y., to the author.*) David Corwin of New-Paltz, says that his grandfather fell at Fort Montgomery, but he thinks the name was Timothy. The family records are lost. At any rate, his grandmother's name, before marriage, was Anna Jaynes. It may be, that a Timothy (with wife Anna Jaynes) should be put down as son of Theophilus, 4, and Mary, David, George, and Jonathan, perhaps, as the children of this Timothy.

4½ THEOPHILUS,[5] (*Simeon*, 1,) bapt. at Mattituck, May 8th, 1757.

5 THEOPHILUS,[7] ? (——.)
    M. Jemima ——.

In 1845 and 1853, the names of Theophilus Corwin and Jemima, his wife, are found in the Cayuga Co., N. Y., records, as selling land. They were of the town of Genoa.

6 Theophilus Case,[2] (*Martha Corwin*, 1,) b. ——, 1661 ? d. October 26th, 1716 ?
    M. Hannah ——, who afterward married ? Jabez Mapes, 1717.
    Ch. Prob. William, Ichabod, John. (See *Moore's Index of Southold*, p. 63.)

Theresa, (Greek,) carrying ears of corn.

1 THERESA,[8] (*Moses H.*, 2,) b. ——, 1850–5.
2 THERESA,[5] (*Polydore B.*, 1,) b. November 2d, 1835.
    M. Miles Howes.
    Ch. Lula F., Charles O.   (*Owego, N. Y.*)
3 THERESA,[9] (*Ezra*, 3,) b. ——, 1845.
    M. Alfred Huntington, 1866.
4 THERESA BELLE,[10] (*James M.*, 39,) b. March 3d, 1859.
4 Theresa Crego,[8] (*Clarissa Taylor*, 4,) b. January, 1864.
5 Theresa M. Concklin,[9] (*Sarah M. Taylor*, 34,) b. December 6th, 1837.

Theron, (Greek,) a hunter.

1 THERON,[*] (*William*, 10,) b. ——, 1835–40.

Thomas, (Hebrew,) a twin.

1 THOMAS,[5] (*Timothy*, 1,) b. at Mattituck, L. I., 1752, bapt. July 28th, 1754, d. February 18th, 1826.
    M. Elizabeth Clark, July 11th, 1776, who was b. 1757, d. September 15th, 1832.
    Ch. Thomas, Silas, Barnabas, James, Joshua, Timothy, Deborah, Ann, Elizabeth, Mary.

In 1776, after the battle of Long Island, he went to Lyme, Ct. In 1780, he was permitted with others to go to Long Island for supplies.

(*Franklinville, L. I.*)

Yrs very truly
Thos. Corwin

2 THOMAS,[6] (*Thomas*, 1,) b. ——, 1785, d. March 7th, 1856.
    M. (1) —— Roberts; (2) Betsey Roberts, her sister, (and mother of
    all his children.) She was born 1794, d. December 2d, 1828; (3)
    Fanny Webb; (4) Widow Abigail Youngs.
    Ch. George W., Charles, Albert P., James C., Elizabeth, Thomas, Ben-
    jamin, Abigail. (*Franklinville, L. I., till* 1830; *then at Greenport.*)
3 THOMAS,[7] (*David*, 6,) b. ——, 1790-1800. (*Ohio.*)
4 HON. THOMAS,[7] (*Matthias*, 5,) b. in Bourbon Co., Ky., July 29th, 1794,·
    d. in Washington, D. C., December 18th, 1865.
    M. Sarah A. Ross, November 13th, 1822.
    Ch. Catharine F. R., William H., Eveline C., Maria L., Carrie R.

(*Lebanon, O.*)

His early education was obtained under such limited advantages as
were common in a newly settled country.[*] He first entered school in the
fall of 1798. The building was a low, rough log-cabin, put up in a few
hours by the neighbors who formed the little settlement, and stood on the
north bank of Turtle Creek, about half a mile west of the place where the
town of Lebanon, O., now stands. This school was taught first by the
late Judge Dunlevy. In 1806, Rev. J. Grigg commenced a school in Lebanon, and taught there for two years. Thomas could only attend during
the winter, and these two winters, when he was between twelve and fourteen years of age, completed most of his *college education*. He was, however, an industrious reader at home.

At the call for troops in 1812, after the surrender of General Hull,
Thomas's father sent his team to aid in furnishing supplies to the suffering
American army of the north, and Thomas, a lad of seventeen, became the
teamster.

While subsequently laboring on the farm, he received an injury on his
knee, which disabled him for some time. During these tedious months he
had recourse to books belonging to his brother Matthias, and made rapid
attainments in knowledge. After assisting his brother for a while as Clerk

---

[*] For the following sketch of Hon. Thomas Corwin, the author is indebted chiefly to the
memoir prefixed to the volume of "Corwin's Speeches," by Isaac Strohm. He has freely
used or abridged the language of this article, as seemed expedient. Mr. A. H. Dunlevy
wrote a short sketch of Mr. C., for Appletons' *Cyclopedia.* Mr. William Green, (formerly
of Cincinnati, and more recently Lieutenant-Governor of Rhode Island,) in *The American
Review*, for September, 1847, p. 310, gives a brief sketch of his life, and a criticism upon his
Mexican war speech. It concludes with a glowing eulogy on his personal character. A
writer in the Boston *Journal*, (March, 1870,) gives political and personal recollections of
him, recounting a number of anecdotes, and querying whether he is not of Massachusetts
descent. This drew forth a letter from Hon. Charles W. Upham, in a subsequent issue of
the *Journal*, who endeavored to show the probability of his descent from the Salem Curwens. His arguments had every appearance of probability, but were nevertheless errone-
ous, as the present volume shows. The author would suggest the propriety of a new edition
of "Corwin's Speeches," with the continuance of his biography to the close of his life. Such
an edition, he believes, would be at once exhausted by the members of the Corwin family,
scattered through the United States. But few of them, outside of Ohio, ever saw the
volume alluded to. An enterprising publisher, with the addresses of the members of the
family as herein given, could easily dispose of a new edition.

of the Court of Common Pleas, in the year 1816, he commenced the study of law under Joshua Collett, Esq., who was afterward successively president of that judicial circuit, and one of the judges of the Supreme Court of the State. Mr. Corwin was admitted to the bar in 1817. He at once took a commanding position as an advocate, and was looked upon as a leading spirit by the legal profession.

His public career began in 1822, when he was elected a member of the lower branch of the General Assembly of Ohio. On retiring from the Legislature, for several years he devoted himself assiduously to his profession, which became extensive and profitable. In 1829, he was returned to the Legislature under circumstances much to his credit, as a man worthy of public confidence. A merely partisan "Jackson" ticket was nominated. But the sturdy yeomanry of that district rebuked the attempt, and Mr. Corwin, (against his inclination,) and ex-Governor Morrow, as the people's candidates, irrespective of party, were elected by decided majorities.

In 1830, he was sent to the Congress of the United States. Before this, he had often been solicited to become a candidate, but had declined. The district (Warren and Butler Counties) had given a decided majority for Jackson, to whose elevation to the presidency Mr. Corwin had always been opposed. But by his personal popularity he overcame this partisan opposition, and was elected by a large vote. He was reëlected four times successively, after this, to the same high office, (1830–40.) He resigned on February 22d, 1840, to take effect in May following, because of his nomination to the governorship of Ohio.

During the campaign of 1840, he became extensively known as an effective public speaker, and was a principal instrument in swelling the unexampled majority of General Harrison. His own majority for governor was greater than was ever attained before in Ohio. He served but one term, however, as his former competitor, through a diversion in favor of a (then) ultra anti-slavery candidate, led him in the contest of 1842 by a plurality of some five thousand votes.

In 1844, he was placed at the head of the Clay electoral ticket in Ohio. He was soon after elected United States Senator, by the State Legislature, both branches of which were then, for the first time in several years, anti-Democratic. He took his seat with the accession of Mr. Polk to the Presidency in 1845, and served until July 22d, 1850, when, at the invitation of President Fillmore, he entered upon the duties of Secretary of the Treasury. At the end of that administration, in 1853, he resumed his professional duties in Cincinnati, in which he continued till the fall of 1858, having his residence at Lebanon. Assenting to pressing solicitations, and under a deep sense of duty to his country, he consented to be a candidate again for Congress, and was triumphantly elected. He was in the heat of the great discussion, pending the breaking out of the Rebellion. In 1861, he was appointed Minister to Mexico, by President Lincoln, but declining health compelled him to resign in 1864.

Thus, without the advantages of a liberal education, he attained a high degree of eminence. The fire of true genius and the possession of real merit were unmistakably evident. His talent for effective oratory early displayed itself. This, with his keenness of discrimination, was the cause of his rapid strides to distinction as an advocate at the bar. His eloquence was unequaled, in his day, in Ohio.

"When addressing a jury, he exhibited all that opulence of legal and miscellaneous information, and philosophic reflection, which are the result of a life of study of books and of human nature. No one could more readily or more delicately sweep every chord by which the sympathies of the heart could be touched, or find a more direct avenue to whatever feeling, passion, sentiment, or prejudice it might be necessary to reach to obtain consent to his reasoning. The infinite humor playing in his countenance, and ever anticipating the utterances of his eloquent tongue, but added to his power to overwhelm with ridicule any unfortunate position taken by an opponent; and though this sometimes was calculated to divert the auditor from the more ponderous logic which he used, it carried with it, nevertheless, an undercurrent of sound argument which did not fail to convince. All his varied acquirements he used, wherever it might be desirable, to give strong probability to a mere possible fact; and he brought up the known to prove the unknown by a masterly comparative course of illustration."

While in the Ohio Legislature he distinguished himself by a speech against public whippings for petty larcenies. It was proposed to reënact such a law, which had recently been repealed. It was maintained by its advocates that nothing was more appropriate as a punishment for theft than whipping, and that it would save the tax, which was becoming onerous, of supporting such prisoners. His speech presented the repugnance felt by every person of refined sensibility to such a penalty. The bill was defeated.

As a representative in Congress he was industrious and faithful to the trust reposed in him. His course was that of a careful, thoughtful, conscientious man. His appearance in debate was rare, but always effective. The announcement of his name produced profound stillness in the House, except as it was broken by demonstrations such as wit, argument, and eloquence like his must occasionally produce. His impromptu speech on the Surplus Revenue is a masterly effort. His well-remembered reply to General Crary is unapproachable for overwhelming sarcasm and ridicule, and its abounding good humor. His speech on the constitutionality of Internal Improvements, by the general government the reporters failed to write out, as they could not do him justice, requesting him to write it out himself from their imperfect notes. This, however, he did not find it convenient to do, and hence one of his best speeches has never been published.

When canvassing the State in behalf of himself and General Harrison, in 1840, his power to sway the masses became widely known. From the circumstance of his driving his father's team with provisions for the army in 1812, he acquired the *sobriquet* of "The Wagoner Boy," and this now became a rallying-word of the campaign. There was a happy coincidence in the object accomplished by the wagoner candidate for Governor and the achievements of the military candidate for President. It was to succor the army under command of Harrison that the lad, Tom Corwin, performed his humble mission. There was much in it to touch the sympathies of a grateful people, and their feelings burst forth in song during that exciting campaign.

> "Success to you, Tom Corwin!
> Tom Corwin, our true hearts love you!
> Ohio has no nobler son,
> In worth there's none above you;
> And she will soon bestow
> On you her highest honor,
> And then our State will proudly show
> Without a stain upon her.
>
> "Success to you, Tom Corwin!
> We've seen, with warm emotion,
> Your faithfulness to freedom's cause,
> Your boldness, your devotion;
> And we will ne'er forget
> That you our rights have guarded;
> Our grateful hearts shall pay the debt,
> And worth shall be rewarded."
> etc.,   etc.,   etc.

But the office of Governor of Ohio, did not afford much scope to exhibit the talents for which Mr. Corwin was most remarkable. His annual messages to the General Assembly, however, are model documents of the kind, brief in form, clear and perspicuous in style. The following extract is from his first message:

"It is in times of profound tranquillity, when the people are undisturbed by the tumults of war, that the duties of enlightened patriotism invite us to the grateful task of giving depth and permanency to our free institutions. It is only at such periods that a commonwealth can hope to deliberate calmly and successfully upon systems of policy calculated to stimulate industry, by giving it legal assurance that it shall be protected in the enjoyment of its acquisitions; to strengthen general morality by laws which shall tend to suppress vice and crime in all their forms; to give energy and independence of character to all classes by measures which will promote, as far as practicable, equality of condition, and thus establish rational liberty for ourselves, and give hope of its continuance for ages to come.

"Of measures which contribute to this end, education, comprehending moral as well as intellectual instruction, is of the first importance. Under a constitution like ours, which imparts to every citizen the same civil rights, education must remain a subject of vital interest in reference to the general welfare of the state. If we are to trust the lessons of history, we are brought to the conclusion that government is, and always has been, the most efficient of all the causes which operate in forming the character and shaping the destinies of nations. Where the right of suffrage is so unrestricted as with us, government is necessarily the offspring of all the people, and will reflect the moral and intellectual features of its parent with unvarying fidelity.

"If the operations of the most profound thinkers had left us in doubt upon this interesting subject, the familiar history of the last century alone has furnished numerous and melancholy proofs that no people to whom moral and intellectual culture have been denied are capable of achieving or enjoying the blessings of rational liberty, founded upon any system which tolerates popular agency in the conduct of public affairs. So profoundly impressed with this great truth were the framers of our constitution that they did not leave it to the judgment of the future to decide. They did not allow the subject of education to remain in that class which might be, in after times, adopted or rejected upon the doubtful test of expediency. They incorporated it into the constitution. In the 3d section of the 8th article of the Constitution it is expressly declared that 'religion, morality, and knowledge being essentially necessary to good government and the happiness of mankind, *schools and the means of instruction shall forever be encouraged by legislative provision*, not inconsistent, with the rights of conscience.' In the schools, the encouragement of which is thus enjoined as a proper subject of legislative provision, it is apparent that the makers of the Constitution intended to combine moral with intellectual instruction. All experience and observation of man's nature have shown that merely intellectual improvement is but a small advance in the accomplishment of a proper civilization. Without morals, civilization only displays energy, and that the more fearful in its powers and purposes as it wants the restraining and softening influences which alone give it a direction to objects of utility or benevolence."

At the Whig State Convention, held at Columbus in January, 1844, over which Mr. Corwin presided, a unanimous nomination was tendered him to be the candidate for governor a third time. This he declined in a masterly speech, never to be forgotten by those who heard it. Ex-Governor Morrow then submitted the following resolution, which was adopted with loud acclaim :

" *Resolved*, That in THOMAS CORWIN we recognize a patriot, a statesman, an orator, a man of the people, and a champion of their rights—a man whom Ohio is proud to call her own. We esteem him and we love him."

This resolution and the enthusiastic call of the vast multitude brought forth a response from Governor Corwin, who for an hour and a half enchained his audience in breathless attention, or called from them deafening shouts of applause, as he overwhelmed the pretended friends of the people, who, in the name of democracy, were attempting to subvert our free institutions.

When in the Senate, during President Polk's administration, he did not take much part in debate, until the Mexican war arose. He favored the proposition to give our soldiers bounty-land, and of always dealing most liberally with them. But his great Mexican war speech, on the Three Million bill, is one of the finest specimens of American eloquence. He boldly asserted what he esteemed to be the truth, regardless of the unpopularity of his sentiments. He dared to think as his conscience bade him, and opposed any further appropriations for the prosecution of the war. His friends thought his course impolitic, though correct, and his enemies did not fail to use his utterances to his political injury, distorting his sentiments to suit their own purposes. But all upright men commended the honest expression of his conscientious convictions upon a subject so fraught with momentous consequences.

Upon the conclusion of the Mexican war, the question of the extension of slavery over the new territory was the one exciting theme. On July 24th, 1848, Mr. Corwin discussed the powers of Congress in reference to this subject, with all the ability of a profound constitutional lawyer and well-read statist. Upon this speech the Louisville *Journal* then remarked :

"In the speech on the Compromise bill, Mr. Corwin has displayed in an eminent degree all the great features of his massive mind. The oft-repeated sophistries of slavery, the formulas that have passed from mouth to mouth among those who love to be deceived, those paltry, rickety things that seem to be heir-looms of perpetualism, are trampled into dust by Mr. Corwin with as much disdain as Mirabeau spurned and trampled on the formulas of royalty. When did falsehood ever receive a quietus more effectually than this mendicant plea of the ultras for more slave territory, on account of their worn-out lands ?" . . .

"A leading feature of Mr. Corwin's character is devotion to truth ; and he follows her paths with the enthusiasm, the spirit, and fortitude of a martyr. In the worship of that heavenly essence, he discards all personal considerations ; he never seems to pause to ask what may be the personal consequence of any thing he may feel required to do—it is enough for him to learn where truth marks the line of duty, to secure his obedience.

"This trait of Mr. Corwin was eminently displayed on the Mexican war question. His lucid mind as clearly saw the consequences of his course upon that question as his experience now feels them. He deliberately surveyed his ground, and duty made him brave the fires of persecution and the anathemas of party. It would be as easy to make a slave of Mr. Corwin as it would be to make him a demagogue. He stood up before the country as a man who dared to do what he considered right ; he made no appeal to the *aura popularis ;* he felt the spirit of the ancient sage expressed in the sentiment, ' I love Socrates, I love Plato ; but I love truth more than either.' " . .

"He towered in the Senate like a giant among pigmies. Every one that couched a lance with him seemed to be ridiculously small. . . . There is not a man in the country who can strip a subject of its verbiage more effectually than Mr. Corwin. With a keen and bold eye, and a steady hand, he dissects the sophistries and false logic of the advocates of error, and he follows falsehood through all its doubling and involutions with a practical skill which no ingenuity can elude.

"Nothing seems to escape his eye in analyzing an untenable position. His powers of sarcasm are great, and they sparkle amidst the strength of logic like gems that deck the brow of purity and beauty."

In the Treasury Department, he pursued in the main the line of policy of his predecessor, Mr. Meredith. But during Fillmore's administration, Congress and the Executive did not agree upon the financial policy of the nation. Hence Mr. Corwin's suggestions that there be either a change in the *ad valorem* system, imposing specific duties on all articles to which they might be safely applied ; or that the home, instead of the foreign valuation, be introduced ; or that the rates upon a variety of proper articles should be increased—were not adopted. The influx of gold from California stimulated the importation of foreign goods, which filled the treasury for the time, though the prosperity was illusive. A large surplus over the annually accruing liabilities of the government remained in the sub-treasury vaults when the term of the administration expired. He then (1853) resumed his profession.

One characteristic of all his speeches was his earnest exhortations to the proper and unfailing exercise of the elective franchise. He never failed to admonish the people that they had the power to make and unmake rulers, and that the evils in government are the results either of their own remissness in refusing to vote, or in their not taking pains to vote intelligently. He appealed to the people to exercise the reason and the conscience which God had given them, to decide how they should vote.

5 THOMAS,⁷ (*John*, 14,) b. August 16th, 1812.

    M. Dorcas Dalsen.

Ch. Langford, Eliza J., Ann E., Lydia, Isaiah. (*Illinois.*)

6 THOMAS C.,[7] (*Thomas*, 2,) b. ——, 1825. 340 [illegible]
M. Cora A. Losee. 83 Irving Place
No ch. She died [illegible] (*Brooklyn, N. Y.*)

7 THOMAS,[7] (*David*, 8.)
(There was a Thomas Corwin of Scipio, Cayuga Co., N. Y., in 1803, who purchased land at said date.)

8 THOMAS,[8] (*Jesse*, 6.) (*Ohio.*)

9 THOMAS,[8] (*Benjamin*, 15.) (*Kentucky.*)

10 THOMAS,[8] (*David*, 19,) b. March 12th, 1841.

11 THOMAS,[8] (*Robert G.*, 2,) b. October 21st, 1854. (*Ohio.*)

12 THOMAS T.,[8] (*Silas*, 7.)

13 THOMAS MAYO,[8] (*Ira C.*, 1,) b. ——, 1866.

13½ THOMAS CLAY,[8] (*Ira*, 4½,) b. ——, d. young.

14 THOMAS J.,[7] (*David*, 10,) b. February 27th, 1852, d. March 15th, 1852.

14½ THOMAS DUKE,[8] (*Rev. Jason*, 4,) b. at Clinton, N. Y., March 30th, 1845.
Graduate of Commercial College, Poughkeepsie, N. Y. (*Chicago, Ill.*)

15 THOMAS SCOTT,[9] (*Barnabas*, 5,) b. December 18th, 1860.

16 THOMAS Horton,[9] (——.)

17 Thomas Hutchinson,[3] (*Martha Corwin*, 1,) b. ——, 1667, d. January, 1749. (See *Moore's Index of Southold*, p. 94.)

18 Thomas Drake,[8] (*Mary Corwin*, 20.) (*Ohio.*)

19 Thomas Morris,[8] (*Rhoda Corwin*, 1.) (*Wilmington, O.*)

20 Thomas De Boyce,[8] (*Sarah Corwin*, 20,) b. ——, 1820–30.

21 Thomas Terry,[8] (*Charlotte Corwin*, 5,) b. ——, 1820–30, d. about 1850.

22 Thomas Horton Easton,[9] (*Hannah E. Corwin*, 16,) b. January 25th, 1853. (*Howells, N. Y.*)

23 Thomas Chalmers Penney,[9] (*Helen M. Corwin*, 3,) b. March 6th, 1857.

24 Thomas Paris,[9] (*Julia A. Blinn*, 18,) b. ——, 1864?

25 Thomas J. Padden,[9] (*William A. Padden*, 94.)

26 Thomas Corwin Cropper,[9] (*Carrie R. Corwin*, 8,) b. November 21st, 1866, d. October 18th, 1870.

27 Thomas D. Clark,[9] (*Helen A. Colgrove*, 10,) b. about 1869.

(1) * THOMAS R.,[9] [*Rev. Jesse*, (1),] b. in Ohio, September 22d, 1869.

(2) Thomas Corwine Fountain,[9] [*Sarah H. Corwin*, (6),] b. May 6th, 1842, d. in New-York City, July 12th, 1849.

(I) Rev. Thomas Smith,[3] [*Mary Corwin*, (I),] b. March 10th, 1702.
M. (1) Sarah Tyng, September 12th, 1728; (2) Widow Jordan, of Saco, Maine, March 1st, 1744; (3) Widow Elizabeth Wendall, August 10th,

* Corwine.

1766.   The first wife had eight children; names, etc., given in *N. E. Gen. and Ant. Reg.*, Vol. ii.

He graduated from Harvard, 1720, and was ordained to the ministry, in Falmouth, (near Portland,) Maine, 1727.   (*Glover Memorials*, 129.)

(*A.*) Thomas Curwin, with Alice his wife, Quakers, were whipped for their faith in 1676, having recently arrived in Boston.   No descendants of theirs are known of by the writer.   He may have returned to England. (See *Introduction*.)

1 THORNTON J.,[8] (*William*, 12,) b. September 6th, 1850.   (*Ansonia, Ct.*)
   C.· ·——

Timothy, (Hebrew,) fearing God.

1 TIMOTHY,[4] (*Theophilus*, 2,) b. ——, 1720 ? d. August 22d, 1792.
   M. Mary Webb, daughter of Ebenezer Webb, January 24th, 1750.
   Ch. Amaziah, Timothy, Sarah, Thomas.

In 1775, he signed engagement to support Congress.   In 1776, on census list, 121, having in his family 1 male over 50, and 2 males and 2 females over 16, and 2 children.   One correspondent thinks his father's name was John.   His will may be seen in Liber A, p. 265. (*Suffolk Co., N. Y.*)

2 TIMOTHY,[5] (*Timothy*, 1,) b. about 1750, d. of small-pox, 1792.
   M. Jemima Brown, April 13th, 1775. (*Aquebogue Records*.)
   Ch. Martin, Samuel, Richard, John, Orange, Elizabeth.
                        (*Balmville, Orange Co., N. Y.*)

In 1775, signed engagement to support Congress.   In 1776, he was probably living with his father.

3 TIMOTHY,[6] (*Amaziah*, 1,) b. about 1790.

4 TIMOTHY,[6] (*Thomas*, 1,) b. ——, 1790–1800.

5 TIMOTHY AUGUSTUS,[7] (*Barnabas*, 2,) b. April 22d, 1825.

6 TIMOTHY,[7] (*Richard*, 2,) b. ——, 1810–20.
   M.
   Ch. James, Mary E., Rachel.

(A Timothy Corwin died at Mobile, Ala., August 20th, 1825; a member of this church, say the Aquebogue Records.)

7 TIMOTHY B.,[8] (*Rev. Ira*, 4,) b. at Erie, Pa., May 3d, 1842, d. at Marietta, Ohio, February 21st, 1845.

Tobias, (Hebrew,) distinguished of the Lord.

1 Tobias Van Wagenen,[9] (*Sibyl T. Pudden*, 4,) b. February 22d, 1851.

1 TUTHILL,[7] (*Jabez*, 1,) b. April 17th, 1795.
   M. Sarah Ann Smith.
   Ch. Jabez, Elizabeth, Sarah A.
                        (*New-York City*, 1820–32; *again*, 1841–56

Ulysses, (Greek,) a hater.

1 ULYSSES GRANT,[9] (*William H.*, 38,) b. March 20th, 1864.

Uriah, (Hebrew,) light of the Lord.

1 URIAH,[7] (*David*, 9,) b. ——, 1790–1800.
M. Mary ——.
Ch. Aletta, Lydia ?

2 Uriah Mitchel,[8] (*Maria Corwin*, 8.)

3 Uriah Hulse Goble,[8] (*Elizabeth H. Corwin*, 19.)
M. Jennette A. Palmer.

1 VALERIA CHARLOTTE,[8] (*Isaac S.*, 5,) b. March 16th, 1842.

1 Vangelia Maud Wetherby,[9] (*James H. Wetherby,134*) b. February 4th, 1870.

1 VENTON ALFRED,[9] (*John A. R.*, 60,) b. ——, 1855–60.

1 VINCENT,[6] (*Nathaniel*, 1,) bapt. July 15th, 1770, at Southold.

Viola, (Latin,) a violet.

1 Viola King Baker,[10] (*Herbert Baker*, 2,) b. December 1st, 1868.

Virginia, (Latin,) virgin, pure.

1 Virginia Clark,[8] (*Sarah J. Millspaugh*, 66,) b. January 17th, 1846.

2 Virginia E. Aldrich,[8] (*Deborah Corwin*, 6.)

1 WALLACE,[8] (*Nathan*, 5,) b. ——, 1825–30.

Walter, (Old High-German,) ruling the host.

1 WALTER EUGENE,[8] (*Nathaniel*, 7,) b. September 27th, 1832.
M. Mary Brown, December 24th, 1867.   (*New-York City.*)

2 WALTER S.,[8] (*Richard W.*, 5,) b. ——.
M. Rhoda Little.
Ch. Florence, Richard.   (*Narrowsburg, N. Y.*)

3 WALTER BARTLETT,[8] (*Rev. Eli*, 4,) b. March 24th, 1867, at Honolulu, Sandwich Islands, d. May 24th, 1870, at Oakland, Cal.

4 WALTER FREDERIC,[9] (*Joseph*, 20,) b. February 18th, 1857.

5 WALTER FRANCIS,[9] (*Charles D.*, 22,) b. April 4th, 1869. (*Massachusetts.*)

6 Walter Collins Everett,[7] (*Abigail Corwin*, 2,) b. September 2d, 1814, d. January 19th, 1842.

7 Walter Murray,[9] (*Lucinda Corwin*, 1.)

8 Walter Paris,[9] (*Julia A. Blinn*, 18,) b. ——, 1865 ?

9 Walter S. Cushman,[9] (*Lydia A. Smith*, 11,) b. November 4th, 1859.
          (*Vermont.*)

1 WARREN,[8] (*Henry*, 11,) b. October 16th, 1846.
M. Elizabeth Philips, April, 1867.
Ch. Eddy.   (*Lake City, Minn.*)

2 WARREN,[8] (*Rev. David*, 15.)

3 WARREN,[8] (*Jesse*, 6.)

4 Rev. Warren L. Noyes,[9] (*Louisa Corwin*, ?) b. December 25th, 1842.
(*Vermont.*)
He graduated at the New-Hampton Theological Seminary in New-Hampshire, July 9th, 1868.

1 WATTS,[6] (*Nathaniel*, 1,) bapt. August 13th, 1775, at Southold.
A Watts Corwin bought land in Suffolk Co., N. Y., 1815, and lived in
New-York City, 1819–20.

1 Wellington Connor,[8] (*Caroline Corwin*, 2,) b. December, 1846.

2 Wellington H. North,[9] (*Fitz-Alan North*, 2,) b. November 11th, 1856.

1 Wessel Girard,[8] (*Jerusha Corwin*, 2.)

1 WICKHAM,[7] (*Eli*, 3,) b. December 22d, 1806, d. May, 1807.

2 Wickham Moffatt,[8] (*Sarah Corwin*, 11,) b. ——, 1810–20.
M. ——.
Ch. William.

3 Wickham Moffatt,[8] (*Sarah Corwin*, 31.)

1 Wilberforce Dunlevy,[8] (*Lucinda Corwin*, 4,) b. March 20th, 1834.
(*Chicago, Ill.*)

1 WILBUR M.,[8] (*Lemuel*, 1,) b. July 17th, 1868, d. November 10th, 1868.

1 Willard Blinn,[9] (*Amos T. Blinn*, 3,) b. August, 1855.

2 Willard A. Padden,[9] (*Phineas Padden*, 7.)

3 Willard Lewis,[8] (*Emeline Corwin*, 2.)

William, (Old High-German,) resolute helmet; protector

1 WILLIAM,[4] (*Matthias*, 2,) b. ——, 1700–10.
Hon. Thomas Corwin, of Ohio, informed the writer in 1859, that his
great-grandfather was William. But his uncle, David, (8,) is quite sure
that his grandfather's name was Jesse, and upon the whole this seems the
most probable. Possibly the two names should be combined, as Jesse-
William. (See *Jesse, No.* 1.)

2 WILLIAM,[5] (*John*, 3,) b. February 21st, 1744, d. December 1st, 1818.
M. Hannah Reeves, of Mattituck, January 14th, 1768. She was born
May 23d, 1747, and died about 1840.
Ch. John C., Sarah, Hannah, William, James, Joseph, Nathaniel,
Elizabeth, Daniel, Ebenezer, Joshua G.
He moved to Chester, N. J., from Long Island, about 1774. He was a
soldier in the French and Indian war, a lieutenant in the Revolution, and a
Representative in the New-Jersey Legislature. His original homestead,
(1¼ miles north of Chester, N. J.,) is now possessed by a Mr. Kelsey. As
early as 1776, he takes a mortgage of John Dickerson, of Roxbury, Morris Co., N. J., for lands on Black River lying next to Joseph Corwin's
land. In 1783, he takes a mortgage of Silas Reeve, for lands in Roxbury,

and in 1787, another of Samuel Morris. In 1800, he buys 111 acres of land of Aaron Stark, in Roxbury, and after this his name occurs very frequently on the records. His will is recorded at Morristown.

3 WILLIAM,[5] or [6]? (*George,* 1,) b. ———, 1770–80.

(A William Corwin, (or Curwin,) and Rebecca, his wife, were of New-Windsor, Orange Co., N. Y., in 1795; in 1801–7 and 1810, they were of Milton, (Genoa,) Cayuga Co., N. Y. Perhaps identical with William, 3.)

4 WILLIAM,[6] (*Joseph,* 1,) b. December 12th, 1779, d. April 17th, 1852.
M. Polly Brown, 1798. She b. 1781.
Ch. Anson, Epenetus H., Maria, Joseph W., Elma, Polly A., Abner,
Aletta, Gilbert, Sylvester. (*Aquebogue, L. I.*)

5 WILLIAM,[6] (*James,* 1,) b. at Southold, L. I., March 19th, 1764, d. May 17th, 1800.
M. Leah Johnson.
Ch. Elizabeth, Sarah.

He was a seafaring man. He lived in New-York City from 1790 till his death, after which his widow kept store at 8 Front street, till 1830. The writer remembers a vault in front of the Presbyterian church in Wall street, (which church was removed to Jersey City, in 1844,) which was inscribed as "Leah Corwin's Vault," indicating their place of worship and burial.

6 WILLIAM,[6] (*William,* 2,) b. near Chester, N. J., October 9th, 1776, d. September 30th, 1821.
M. (1) Martha Vance, December 12th, 1801.
Ch. Joseph, William V., Eliza A. (*Sparta, N. J.*)
In 1817, he resided in New-York City, in partnership with James Corwin, 6.

7 WILLIAM,[6] (*David,* 5.) b. September 14th, 1813, d. December 23d, 1845.
M. (1) Althea, daughter of Bartholomew and Sarah Besley. She d. December 23d, 1846; (2) Catharine Cantine, September 4th, 1849.
Ch. Sarah, John D.; Ann Maria.

8 WILLIAM,[6] (*David,* 4.)
M. Martha Overton.
No ch.

9 WILLIAM S.,[7] (*Elisha,* 1,) b. ———, 1820?
M. Cornelia Hawkins.
Ch. Helen, Clara. (*New-York City.*)

10 WILLIAM,[7] (*John,* 7,) b. ———.
M. (1) Emeline Weed; (2) ——— Thorn.
Ch. Therin, etc.

11 WILLIAM,[7] (*Benjamin,* 5,) b. February 19th, 1811.
M. Lydia A. Smith, December 22d, 1831.
Ch. Ira M., Mary E., Charles E., Sarah A., Benjamin, Clarissa.
(*Middletown, N. Y.*)

15

12 WILLIAM HORTON,[7] (*Jason*, 1,) b. March 16th, 1818.
    M. (1) Harriet, daughter of Charles P. Baldwin, January 13th, 1841.
    She was granddaughter of Methuselah Baldwin, who was for nearly
    half a century pastor of the Presbyterian church at Scotchtown,
    Orange Co., N. Y.   She d. July 31st, 1867.   He *M.* (2) Conissa A.
    Smith, December 31st, 1868.
    Ch. Sarah A., Jennette M., Thornton J., Harriet.

    He was first a farmer, at Scotchtown, but moved to Ansonia, Ct., in
1854, and is now engaged in that place in the manufacture of sheet-brass.
In 1862, he enlisted in Company B, Regiment 20, Connecticut Volunteers,
and was a sergeant throughout the war.   Was in the army of the Potomac
from 1862 to the fall of 1863 ; then in the army of the West, and marched
with General Sherman to the sea.   This regiment has the names of fourteen
battles on its flag, among which are Chancellorsville, Gettysburg, and Atlanta.   He never drank a drop of liquor nor saw a single sick day during
the war.

13 WILLIAM THOMPSON,[7] (*Silas*, 2,) b. March 11th, 1822.
    M. Caroline Sutliff, February 23d, 1843.   (*New-Milford, Pa.*)

14 WILLIAM OWEN,[7] (*James*, 5,) b. August 10th, 1800, d. May 20th, 1835.
    M. Amanda Chappel.
    Ch. Theodore.   (*New-York City.*)

15 WILLIAM HENRY,[7] (*Moses*, 1,) b. September 22d, 1833.
    M. Rebecca B. Williams.   (*Brooklyn and New-York City.*)
    No ch.

16 WILLIAM,[7] (*John Calvin*, 11,) b. ——, 1803.
    M. (1) —— ; (2) Maria Outwater.
    Ch. Edward, Peter, Peter F.   (*Ithaca, N. Y.*)

17 WILLIAM,[7] (*James*, 6,) b. February 2d, 1808.
    M. Mary Hyatt.
    Ch. William, Henry, Frederick.   (*New-York City.*)

18 WILLIAM HENRY HARRISON,[7] (*Nathaniel*, 2,) b. December 7th, 1813.
    M. Elizabeth Jenkinson, February 13th, 1838.
    Ch. Joseph, Moses, Monroe, Edgar.   (*Succasunna, N. J.*)

19 WILLIAM,[7] (*Joshua G.*, 3,) b. October 21st, 1824.
    M. Harriet Doremus.
    Ch. Marcus W., Mary E., Emeline C., Philip F., Edmund D.
                                               (*Succasunna, N. J.*)

20 WILLIAM,[7] (*Richard*, 1,) b. ——, 1790?
    M. Philothea Homan.
    Ch. William, Barnabas.

21 WILLIAM,[7] (*Martin*, 2,) b. ——, 1800–10, d. at New-Orleans.
    M. ——.
    No ch.

22 WILLIAM BLOOMER,[7] (*Richard*, 2,) b. ——, 1810–20.

23 WILLIAM B.,[7] (*John*, 14,) b. ——, 1809.
    M. Phebe E., daughter of William and Rachel Rose, of Haverstraw, N. Y.
    Ch. Emma L., John W., Abigail A., Elmina C., John W.
                                                     (*Peekskill, N. Y.*)

24 WILLIAM,[7] (*Asa*, 2,) bapt. at Aquebogue, October 16th, 1791.

25 WILLIAM,[7] (*Hubbard*, 1,) b. October 13th, 1787.

26 WILLIAM G.,[7] (*Ichabod*, 1,) b. January, 1792, d. August, 1850.
    M. Narcissa Beach, 1829.
    Ch. William H.   (*Lebanon, O.*)

27 WILLIAM,[7] (*Joseph*, 10.)

28 WILLIAM,[7] (*Benjamin*, 4,) b. ——, 1795 ? d. young.

29 WILLIAM O.,[7] (*Eli*, 3,) b. July 16th, 1799, d. ——, 1837.
    M. Matilda Benjamin.
    Ch. Mary, Oliver B., Sarepta, Adaline.

30 WILLIAM,[7] (*Stephen*, 4.)   (*Marion Co., O.*)

31 WILLIAM,[7] (*Alexander*, 1,) b. ——, 1815 ?
    M. ——.
    Ch. Susan, Hector.   (*Chester, N. Y.*)
    He was in the Federal army during the rebellion.
    (There was a William Corwin, and Rhoda, his wife, of Lansing, Tompkins Co., N. Y., from 1819–24 and 1833.)

31½ WILLIAM P.,[7] (*Joseph*, 10¼,) b. ——, d. August 15th, 1870.
    M. Eliza Hill, (widow of —— Clark,) of Pennsylvania, May, 1851.
    No ch.
    She had four children by a former marriage. William P. was in the Mexican war.

32 WILLIAM FRANKLIN,[8] (*Horton*, 1,) b. April 28th, 1843, d. April 13th, 1861, by lightning.

33 WILLIAM ALBERT,[8] (*Joseph*, 16,) b. March 12th, 1843.

34 WILLIAM,[8] (*William*, 17,) b. December 16th, 1831.
    M. Sarah Ladd.
    Ch. Alfred, George, William.

35 WILLIAM,[8] (*Nathaniel*, 3.)

36 WILLIAM HENRY,[8] (*Hon. Thomas*, 4,) b. January 3d, 1829.
    He was Secretary of Legation to his father, when the latter was Minister to Mexico, 1861–4, and remained minister resident after his father's return for about a year. He is now a homeopathic physician in Lebanon, Pa.

37 WILLIAM,[8] (*Richard*, 4.)

38 WILLIAM H.,[6] (*John*, 32,) b. January 24th, 1813, at Tunbridge, Vt.
    M. (1) Eliza Goodwin, March, 1837. She d. February, 1852; (2) Melinda Grant, March 16th, 1853.

Ch. Mary E., George F., Clarissa T., Martha M., Eliza ; Alice M., Marcia E., Arabella, Nellie, Lester, Hattie, Ulysses G.

First lived at Tunbridge, Vt., on his father's farm, 2 years ; then in the village, 12 years ; at Chelsea, Vt., 14 years ; now again at Tunbridge, Vt. While at Chelsea, he lost five of his daughters with diphtheria.

39 WILLIAM,⁸ (*John*, 30,) b. ——, 1826.
> *M.* Sylvia Beman, 1852.
> *Ch.* Jesse, William, Isaac, Emory, John, Alice G., Lillie B., Isabel.
> > (*Genesee Co., Mich.*)

40 WILLIAM HARVEY,⁸ (*William G.*, 26.)

41 WILLIAM H.,⁸ (*James H.*, 22.)

41½ WILLIAM A.,⁸ (*Alonzo*, 1,) b. November 16th, 1833.
> *M.* Lizzie J. Rogers, December 24th, 1863.
> *Ch.* Charles, George, Hester, William.

42 WILLIAM LOUIS,⁸ (*Isaac*, 4,) b. October 7th, 1835, d. September 26th, 1861.

43 WILLIAM HENRY,⁸ (*De Witt C.*, 2,) b. February 16th, 1854.

44 WILLIAM H.,⁸ (*Henry*, 11,) b. November 16th, 1831, at Peconic, L. I.
> *M.* Lucy E. Richmond, May 17th, 1866.
> *Ch.* Burton D.
> (Went to California, 1858 ; to Minnesota, 1860 ; returned to Peconic, 1866.)

45 WILLIAM O.,⁸ (*Francis F.*, 1.)

46 WILLIAM,⁸ (*Alsop H.*, 2.)

47 WILLIAM TUTHILL,⁸ (*Silas*, 7.)
> *M.* Elizabeth ——.
> *Ch.* Two. (*Michigan.*)

48 WILLIAM,⁸ (*Philip*, 1.)

48½ WILLIAM,⁸ (*John*, 32,) b. ——, 1810–20.

49 WILLIAM JACKSON,⁸ (*Benjamin V. R.*, 9,) b. ——, 1845–50.

50 WILLIAM,⁸ (*Alsop*, 1,) b. ——, 1810–20. (*Luzerne Co., Pa.*)

51 WILLIAM HENRY,⁸ (*Epenetus H.*, 1,) b. ——, 1830–40.

51½ WILLIAM HENRY,⁸ (*Rev. Ira*, 4,) b. at Marietta, Ohio, February 16th, 1846, d. at South Bend, Ind., July 31st, 1856.

52 WILLIAM,⁸ (*William*, 20.)

53 WILLIAM MELVILLE,⁸ (*Joseph W.*, 12,) b. ——, 1850–60.

54 WILLIAM,⁸ (*Richard S.*, 7,) b. ——, 1820–30.

55 WILLIAM,⁸ (*Eleazer*, 1,) b. ——, 1835–45.

56 WILLIAM,⁸ (*John*, 22,) b. July 19th, 1819.
> *M.* —— Brown. (*Orient, L. I.*)

56¼ WILLIAM BESON,⁸ (*Henry W.*, 23½,) b. August 21st, 1863.

56½ WILLIAM TOTTEN,⁸ (*Stephen O.*, 8½,) b. ——.

57 WILLIAM,[9] (*Samuel W.*, 25,) b. January 24th, 1864.

57½ WILLIAM E.,[9] (*William*, 41½,) b. October 3d, 1870. —

58 WILLIAM,[9] (*William*, 39,) b. ——, 1853–60.

58½ WILLIAM,[9] (*Daniel A.*, 23½,) b. ——, 1857–65.

59 WILLIAM HANFORD,[9] (*Nicholas*, 1,) b. about 1842.
   M. Cornelia Todd.
   Ch. Ella, Lena.   (*Canton, Ill.*)

59¼ WILLIAM HERBERT,[10] (*John*, 67,) b. August 30th, 1858.

59½ WILLIAM D.,[9] (*Daniel*, 28½,) b. ——, d. young.

59¾ WILLIAM LIVINGSTON,[9] (*John*, 53,) b. ——.

60 WILLIAM WICKHAM,[9] (*Henry W.*, 19,) b. June 17th, 1847.

61 WILLIAM HOLLOCK,[9] (*Barnabas*, 5,) b. December 20th, 1861, d. August 8th, 1862.

62 WILLIAM,[9] (*William*, 34.)

62½ WILLIAM,[9] (*Daniel*, 23.)

63 WILLIAM,[9] (*Ira*, 3,) b. September 1st, 1862.

63½ WILLIAM,[9] (*George*, 42,) b. May 11th, 1859.

64 William Case,[4] (*Theophilus Case*, 6,) b. ——, d. ——, 1769.
   M? Anne ——.
   Ch. James Azubah. (See Moore's *Index of Southold*, p. 63.)

64½ William Case,[6] (*Samuel Case*, 29,) b. ——, 1713, d. ——, 1796.
   (*Moore's Index*, 64.)

65 William Campbell,[7] (*Sarah Corwin*, 6,) b. November 5th, 1816.
   M. Lucy Burch, December 31st, 1843.

65½ William Crawford Little,[7] (*Desire Corwin*, 1,) b. May, 1813.
   M. Sylvia Bowen, October, 1835.
   Ch. James M., Egbert.

66 William Henry Woodhull,[7]? (*Hannah Corwin*, 2,) b. September 17th, 1801, d. March 27th, 1813.

67 William Henry Brown,[8] (*Elizabeth S. Woodhull*, 45,) b. July 13th, 1818.
   M. (1) Aletta Tiger ; (2) Ellen Cartwell.   (*New-York City.*)

68 William Upright,[7] (*Anna Corwin*, 4,) b. ——, 1832–5.
                                  (*New-York City.*)

69 William Philips,[7] (*Hannah Corwin*, 2.)

70 William Thatcher,[7] (*Jemima Corwin*, 1,) b. ——, 1797?
                     (*Valparaiso, Porter Co., Ind.*)

71 William H. Millspaugh,[7] (*Cynthia Corwin*, 1,) b. February 16th, 1823.
   M. Sarah Jane Riggler.
   Ch. Benjamin F., Mary Ann.   (*Columbia, O.*)

72 William E. Millspaugh,[7] (*Dorothy Corwin*, 2,) b. July 10th, 1823?

M. Ann Melinda Corwin, 4, November 21st, 1844 ?
Ch. Cynthia E., Clara, Florence, ——.

73 William Halsey,[7] (*Elizabeth Corwin*, 2,) b. ——, 1812.   (*Wisconsin.*)

74 William Terry,[7] (*Keturah Corwin*, 1,) b. ——, 1780–90.

75 William S. Hurtin,[7] (*Mary Ann Corwin*, 27,) b. ——, 1800–10.

76 William Moffat,[8] (*Sarah Corwin*, 11,) b. ——, 1810–20.

77 William Thompson,[8] (*Mary Corwin*, 25,) b. September 11th, 1836.

78 William Corwin Borgy,[8] (*Harriet N. Corwin*, 1,) b. ——, 1825–35.
                                                  (*Washington, D. C.*)

79 William Louis Coleman,[8] (*Jemima Corwin*, 8,) b. ——, 1810–20.

80 William Henry Peters,[8] (*Mary Corwin*, 30,) b. ——, 1830–40.

81 William Girard,[8] (*Jerusha Corwin*, 2.)

82 William Lippincott,[8] (*Elizabeth Corwin*, 18,) b. ——, 1830–40.

83 William Corwin Honnell,[8] (*Susan Corwin*, 1,) b. ——, 1816.

84 William H. King,[8] (*Mary Corwin*, 35.)

85 William H. Odell,[8] (*Jemima Corwin*, 7.)

86 William Carr,[8] (*Azubah Corwin*, 3.)
   M. Ann Elizabeth Ransom.   (*McGrawville, N. Y.*)

87 William Evans,[8] (*Rosetta Corwin*, 2,) b. January 28th, 1826.

88 William Wood,[8] (*Mary Corwin*, 16.)

89 William Osborn,[8] (*Amelia Corwin*, 3,) b. May 26th, 1845.

90 William Dunlevy,[8] (*Lucinda Corwin*, 4,) b. December, 1832, d. August
   12th, 1833.

90½ William F. Call,[8] (*Bethia Corwin*, 8,) b. November 24th, 1821.
   M. Samantha Concklin, April, 1845.   .
   Nine ch. Charles, J. Anna M., Emma ?   A. (*Smithtown Branch, L. I.*)
90¾ William Totten Bennett,[8] (*Caroline Corwin*, 6½,) b. ——, 1845–53.

91 William H. Wells,[8] (*Nathaniel C. Wells*, 10.)   (*Cold Spring, N. Y.*)

92 William A. Padden,[8] (*Anna Campbell*, 29.) ·

93 William Swina,[8] (*Bethia Hobart*, 11.)

93½ William Skult,[8] (*Fanny Hobart*, 9,) b. ——, 1820–30.

93¾ William Swina,[8] (*Bethia Hobart*, 9,) b. May, 1829.
   M. Maria Chappel, February, 1854.

94 William A. Padden,[8] (*Anna Campbell*, 25,) b. ——, 1829 ?
   M. ——.
   Ch. Mary J., Thomas J.

95 William Millspaugh,[8] (*Gilbert C. Millspaugh*, 6,) b. March 16th, 1840.

96 William Millspaugh,[8] (*Gideon H. Millspaugh*, 1,) b. October 22d, 1844.
                                                  (*Covington, O.*)

97 William Terry,[8] (*Charlotte Corwin*, 5,) b. ——, 1820–30.

William Thomson 77 (Mary Cousin 25)
   born Sept 11. 1836
   married May 13. 1863
      Julia L. Clearwater,
            (Rhinebeck. N. Y.)

   issue
2 Nettie Thomson born April 30. 1864
married May _____ _____
   _____ _____ Cookingham
issue
   Wil_____ Tho_____ Cookingham,
   born March 29. 1893.
      Joseph Timson Cookingham
      born March 71. 1896

M. Jane? Finch.

Ch. William.

98 William S. Milnor,[8] (*Harriet A. Millspaugh*, 19,) b. January 25th, 1847.

99 William Young Howell,[8] (*Harriet M. Corwin*, 3,) b. January 4th, 1843.

100 William Merville Sanders,[8] (*Elizabeth Wells Corwin*, 13,) b. January 28th, 1848, d. December 17th, 1856.

101 William Corwin Beach,[8] (*Eliza Ann Corwin*, 5,) b. August 7th, 1832.

102 William Moffatt,[8] (*Sarah Corwin*, 31.)

102 William Henry Concklin,[9] (*Sarah M. Corwin*, 34,) b. October 27th, 1839.

103 William H. Noyes,[9] (*Louisa Corwin*, 3,) b. June 3d, 1855. (*Vermont.*)

104 William Mattison Fairchild,[9] (*Eliza E. Corwin*, 10.)

105 William Thornton Henderson,[9] (*Jennette M. Corwin*, 2.)

106 William Terry,[9] (*William Terry*, 97.)

107 William M. Chamberlin,[9] (*Jennette Corwin*, 1,) b. January 26th, 1846. (*Vermont.*)

108 William L. Campbell,[9] (*Theodore C. Campbell*, 11,) b. December 20th, 1859.

109 William A. Van Wagenen,[9] (*Sibyl T. Padden*, 4,) b. October 5th, 1852.

110 William Moffat,[9] (*Wickham Moffat*, 2.)

111 William Wilcox,[9] (*Mary Corwin*, 47,) b. May 9th, 1853.

112 William Derrick,[9] (*Mary M. Blinn*, 137,) b. ——, 1864.

113 Willie Atwood,[9] (*Caroline Corwin*, 9,) b. November 22d, 1860.

113½ William Deel,[10] (*Elizabeth J. Concklin*, 61,) b. ——, 1865?

114 William Lockwood,[10] (*Lucretia Corwin*, 4,) b. ——, 1855–60.

115 William Lewis,[10] (*Imogene Davis*, 1,) b. ——, 1865–70.

116 William Elmer Crater,[10] (*Mary Corwin*, 83.)

117 William Archibald Thomson Clark (Anna M. Thomson 33) b. Fely 26. 1876

(1) * WILLIAM,[5] [*Bartholomew*, (1),] b. January 26th, 1726, d. ——, 1726.

(2) * WILLIAM,[6] [*John*, (4),] b. ——.

M. ——.

Ch. ——.   (*Baltimore, Md.*)

(3) * WILLIAM,[7] [*Samuel*, (4),] b. February 27th, 1791.

M. Sarah Cool, (or Kahl,) 1810.

Ch. Leonard.   (*Mill Township, Tuscarora Co., O.*)

(4) * WILLIAM,[7] [*Amos*, (1),] b. ——, 1790–1800.

(5) * WILLIAM ALEXANDER,[8] [*Joab*, (1),] b. ——, 1825–35.

(6) * WILLIAM,[8] [*Amos*, (1½),] b. ——.   (*Illinois.*)

(7) * WILLIAM AUSTIN,[9] [*Rev. Jesse*, (1),] b. in Kentucky, August 21st, 1859.

* Corwine.

(8) * WILLIAM ROSSELL,[9] [*Amos B.*, (2),] b. January 15th, 1855.

(9) William Barcalow,[8] [*Lydia Corwine*, (2),] b. ——.   (*Franklin, O.*)

1 Wilmer Taylor,[8] (*David J. Taylor*, 30,) b. ——, 1855.

1 Winfield Scott Beyea,[9] (*Sophronia Corwin*, 1,) b. September 13th, 1853.

1 WOLVERTON MEREDITH,[8] (*James*, 20,) b. January 6th, 1850.

1 Zada L. Lankton,[9] (*Phebe A. Padden*, 10,) b. September 20th, 1850.

Zechariah, (Hebrew,) remembered of the Lord.

1 ZECHARIAH,[7] (*Isaac*, 1,) b. October 4th, 1799, d ——, 1814.

Zeruiah, (Hebrew,) cleft; wounded.

1 Zeruiah Case,[6] (*Samuel Case*, 29.)
   M. Joseph Corwin, (No. 2.)

APPENDIX.

— ••• —

APPENDIX A.

THE NAME CORVINUS, AND THE ROMAN CORVINI.

It may not be without interest, at least to some, to give a fuller history of this *name*. For although, of course, no pretense of connection with it, as it existed in Roman times, is thought of, yet, since it has been borne by several persons eminent in influence and in virtue, it may be not only interesting, but perhaps useful, to refer to them—the reader ever remembering that the relation is only *nominal*.

Corvinus was a cognomen in the Valeria Gens, and merely a lengthened adjective form of Corvus. It originated in the Valerian family in the following manner:

(From Anthon's Classical Dictionary, p. 1370.)

"Valerius Corvus (or Corvinus) was a tribune of the soldiers under Camillus. When the Roman army was challenged by one of the Senones, remarkable for his strength and stature, Valerius undertook to engage him, and obtained an easy victory by means of a raven (*corvus*) that assisted him, by attacking the face of the Gaul, whence his surname of Corvus, or Corvinus, B.C. 349. Valerius triumphed over the Etrurians and the neighboring states that made war against Rome, and was six times honored with the consulship. He died in the one hundredth year of his age, B.C. 271, admired and regretted for many private and public virtues." (*Val. Max.* 8, 13. *Liv.* 7, 27.)

He was also appointed dictator twice, and had the honors of the curule chair twenty-one times.[*]

But Marcus Valerius Messala Corvinus was a still more eminent man. Gibbon[†] says concerning him, "The fame of Messala has been scarcely equal to his merit. In the earliest youth he was recommended by Cicero to the friendship of Brutus. He followed the standard of the republic till it was broken in the fields of Philippi; he then accepted and deserved the favor of the most moderate of the conquerors; and uniformly asserted his freedom and dignity in the court of Augustus. The triumph of Messala was justified by the conquest of Aquitain. As an orator, he disputed the palm of eloquence with Cicero himself. Messala cultivated every muse, and was the patron of every man of genius. He spent his evenings in philosophic conversation with Horace; assumed his place at table between Delia and Tibullus; and amused his leisure by encouraging the poetical talents of young Ovid."[‡] He was born about B.C. 64, and died B.C. 8.

---

[*] Smith's *Dictionary of Greek and Roman Biography* gives an interesting and much fuller sketch, of this man, with many references to the Roman authors. See also Niebuhr's Rome, iii. 124.

[†] Gibbon's *Decline and Fall of the Roman Empire*, ii. 117, note.

[‡] The above note in Gibbon induced Rev. Edward Berwick to search out and write as full a biography of this man as the material allowed. He has made a very interesting little volume, which was published at Edinburgh, 1813. Anthon condenses it in his *Classical Dictionary*, art. Messala. The Roman writers contain many references to him, all, without exception, of a very complimentary nature.

Cicero's fifteenth epistle to Brutus, who was then in Macedonia, thus describes Messala Corvinus, who was also the bearer of the letter. Cicero is lamenting over the prospective downfall of the republic:

"You have Messala now with you. It is not possible, therefore, for me to explain by letter, though ever so accurately drawn, the present state of our affairs, as exactly as he can, who not only knows them all more perfectly, but can describe them more elegantly than any man. For I would not have you imagine, Brutus, that for probity, constancy, and zeal for the republic, there is any one equal to him; so that eloquence, in which he wonderfully excels, scarce finds a place among his other praises: since even in that his wisdom shines most eminent, by his having formed himself with so much judgment and skill to the truest manner of speaking. Yet his industry all the while is so remarkable, and he spends so much time in study, that he seems to owe but little to his parts, which are still the greatest. But, I am carried too far by my love for him; for it is not the purpose of this epistle to praise Messala, especially to Brutus, to whom his virtue is not less known than to myself; and these very studies, which I am praising, still more: whom, when I could not part with without regret, I comforted myself with reflecting, that, by his going away to you, as it were to my second self, he both discharged his duty, and pursued the surest path to glory."

Corvinus was among the many proscribed to death at that time, for his strong republican principles. But he was saved by his absence from Italy. The traitors afterward sought to induce him to join them by a special proclamation of his pardon, but he spurned their offer. At the battle of Philippi, when the republic was overthrown, he commanded a legion. That fatal day, after supping with some friends, Cassius took Messala by the hand, and pressing it close, said courteously in Greek, "Bear witness, Messala, that I am reduced to the same necessity as was Pompey the Great, of hazarding the liberty of my country upon one battle."[*] But Cassius fell, and with him the liberty of Rome. Messala and Bibulus had the command of the broken republican forces offered them; but they declined, saying, "It has cost Rome too much blood already, and we must yield to the storm, and fall under the power of the strongest."[†] Velleius Paterculus says,[‡] "Messala Corvinus, a young man of shining character, who, next to Brutus and Cassius, possessed the greatest influence of any in the camp, and whom some solicited to take the command, chose to be indebted to safety to Cæsar's kindness, rather than to try any further the chance of arms. Nor did any circumstance attending his victories afford greater joy to Cæsar than the saving of Corvinus; nor was there ever an instance of greater gratitude or more affectionate attachment than Corvinus showed to Cæsar in return." Berwick says, "He retained his freedom and dignity in a despotic court, which his understanding condemned and his heart reprobated." He at first joined the army of Antony, and followed him through his Syrian and Egyptian campaigns, but, disgusted with Antony's life, he afterward left him, and joined the party of Augustus, whose life he saved, after a certain defeat, though Augustus had once proscribed him to death. After a certain victory of Messala, he replied to the praise which Augustus bestowed on him, "I have always taken the best and justest side." Tibullus wrote an elegy§ in praise of Messala, consisting of 211 lines, B.C. 32.

His name also occurs on the Capitoline marbles for his victory in Aquitain, and an ancient medal is described by a French writer, which was struck to his honor.

"He had the singular merit of supporting an unblemished character in a most despotic court, without making a sacrifice of those principles which he fought for in the fields of Philippi; and the genuine integrity of his character was so deeply impressed on all parties that it attracted a general admiration in a most corrupt

* Applan, book iv.   † Plutarch's Life of Brutus.   ‡ Book ii. c. 71.   § Book iv.

age. He was brave, eloquent, and virtuous. He was liberal, attached to letters, and his patronage was considered as "the surest passport to the gates of fame." No writer, either ancient or modern, has ever named Messala without some tribute of praise. Cicero soon perceived he possessed an assemblage of excellent qualities, which he would the more have admired, had he lived to see them expanded and matured to perfection. Messala was his disciple. Quintilian says his style was neat and elegant; and in all his speeches he displayed a superior nobility. In the Dialogue of Orators, he is said to have excelled Cicero in the sweetness and correctness of his style. His taste for poetry and polite literature will admit of little doubt, when we call to mind that he was protected by Cæsar, favored by Mæcenas, esteemed by Horace, and loved by Tibullus. Horace, in one of his beautiful odes,* praises Corvinus in the happiest strains of poetry, calls the day he intended to pass with him propitious, and promises to treat him with some of his most excellent wine. "For," says the poet, "though Messala is conversant with all the philosophy of Socrates and the academy, he will not decline such entertainment as my humble board can afford." The modest Tibullus flattered himself with the pleasing hopes of Messala's paying him a visit in the country, where "My beloved Delia shall assist in doing the honors for so noble a guest,"† says he. The rising genius of Ovid was admired and encouraged by Messala; and this condescension the exiled bard has acknowledged in an epistle to his son Messalinus, dated from the cold shores of the Euxine. In this letter Ovid calls Messala his friend, the light and director of all his literary pursuits. . . . Tacitus seems fond of praising Messala, . . . and considers him among the few great characters who have risen to the highest honors by their integrity and eloquence. Even Tiberius himself, when a youth, took him for his master and pattern in speaking,‡ though not in virtue. Messala made eloquence his chief study, though no specimen remains by which to judge of his excellence in this art.

Messala Corvinus married Terentia, the widow first of Cicero, and then of Sallust. She married again a fourth time a Roman senator, after Messala's death. Messala left two sons, Marcus and Lucius. The elder became consul in B.C. 3, and was called Messalinus. He distinguished himself in the Pannonian war. He also inherited his father's eloquence. Lucius assumed the name of Cotta, from his maternal line. Both brothers were also friends of Ovid.

------- ••• -------

APPENDIX B.

JOHN HUNYADY CORVINUS.

AFTER the death of Albert of Hungary, the States offered the crown to Wladislaus, of Poland; but shortly after, the widow of Albert had a son, called Ladislaus Posthumus. This was the cause of dissension, and Amurath of Turkey prepared at the same time to invade the country. Wladislaus conquered in the struggle, and it was just at this time that Hunyadi began to display his military capacities. He was surnamed CORVINUS. The origin of Hunyadi is shrouded in mystery. Pro-

* "Descende, *Corvino* jubente
Promere languidiora vina."

Horace, book 3, ode 21.

† "Huc veniet Messala, cui dulcia poma
Delia selectis detrahet arboribus."

Tibullus, book i, elegy 5.

‡ "In oratione Latina secutus est Tiberius Corvinum Messalam, quem senem, adolescens, observaverat."
Suetonius, Vita Tiberii, cap. 70.

bably he was a Wallachian, and son of George Hunyadi, vaywod of Wallachia, during the reign of Sigismund. As to his surname Corvinus, some derive it from his estate, Piatre de Corvo; others from his ancestors.

Having been made vaywod of Transylvania by Wladislaus, Hunyadi met Amurath on the plains of Wallachia, and defeated him. The Turks then entered Servia, and were again defeated. The Hungarians then proceeded to Bulgaria, conquered Nissa, and gained a signal victory before Sophia, 1443, and hoped to be able to drive the Turks out of Europe. Treaties were made between the Hungarians and Castriot, Prince of Epirus, and Paleologus, Emperor of the East. The Sultan was afraid, and sought peace, which was concluded for ten years. He gave up his claim to Wallachia, promised to evacuate Bulgaria, and release Christian prisoners. The treaty was sworn to on the Bible and the Koran. But the Pope's legate said it was void, for it concerned all Christendom, and the Pope had not indorsed it. Hunyadi tried to avert this perfidy, but in vain. Wladislaus entered Bulgaria, but the Sultan compelled him to retreat to Varna. Hunyadi thought it unsafe to stand a siege at Varna, from a lack of food; hence the battle was without, on a plain. It raged for three days and three nights, without result. On the fourth day, Hunyadi, charging with his horse, put to flight the Janisaries. The Sultan was flying. But two bishops now, jealous of Hunyadi, left their positions, contrary to his orders, and pursued the flying enemy. This turned the day against him. Hunyadi, with a few followers, escaped. The Sultan gained Wallachia and Servia. Great confusion followed. Hunyadi was made governor. In 1452, the captive prince, Ladislaus Posthumus, was released, and Hunyadi, resigning his position, was made chief general. All Europe wanted to help Hungary, but the succor, when needed, failed. The Sultan determined to conquer Hungary, and in 1456 appeared before Belgrade with 150,000 men. The Hungarian king, by advice of Hunyadi, fled to Vienna. Hunyadi quickly raised 10,000 troops, and hastened to the relief of the city. The fate of Europe hung on this battle. After a siege of several weeks, the haughty Ottoman conqueror left 24,000 men dead on the field, and his army fled to Adrianople. The great fatigue brought on Hunyadi a sickness, of which he died a month later, September 10th, 1456. Of his two sons, Ladislaus and Matthias, the former was cruelly murdered by the king. The king himself soon after died, at Prague. Three foreign candidates appeared for the throne; but the nobles desired a native ruler, and chose Hunyadi's son, Matthias Corvinus. (*British Cyclopædia*, art. Hungary.)

Hunyadi was, at his death, eighty years old. He said to his spiritual comforter, "I am prepared for my journey; my whole life has only been a preparation to receive that Friend who is to lead me to the throne of my Omnipotent King. I have served him faithfully, and, amidst all the storms and perils of life, maintained the post of duty on which He placed me. He will graciously receive his soldier, worn out in his service, and grant him rest in the dwellings of his saints." Turning to his sons, he said, "Let the actions and example of your father bear fruit in your hearts; continue in the path of uprightness and virtue, which I have pointed out to you as the safest; I leave you the fear of God and the love of your country as an abiding inheritance; all else that falls to you from me belongs to fortune." (Kossuth's *Hungary*, p. 86.)

Hallam, in his *Middle Ages*, p. 245, says, that while "the crown was permitted to rest on the head of young Ladislaus, the regency was allotted by the States of Hungary to a native warrior, John Hunniades. This hero stood in the breach twelve years against the Turkish power, and, though frequently defeated, remained unconquered in defeat. If the renown of Hunniades may seem exaggerated by the partiality of writers who lived under the reign of his son, it is confirmed by more unequivocal evidence, by the dread and hatred of the Turks, and by the deference of a jealous aristocracy to a man of no distinguished birth. He surrendered to young Ladislaus a trust that he had exercised with perfect fidelity; but his merit was too great to be forgiven, and the court never treated him with cordiality. The last and most splendid service of Hunniades was the relief of Belgrade, 1456. That strong city was besieged by Mohammed II. three years after Constantinople fell into their hands, and its capture would have laid open all Hungary. A tumultuary army, chiefly collected by the preaching of a friar, was intrusted to Hunniades; he penetrated into the city, and, having repulsed the Turks in a fortunate sally, wherein Mohammed was wounded, had the honor of compelling him to raise the siege in confusion. The relief of Belgrade was more important in its effect than in its immediate circumstances. It revived the spirit of Europe, which had been appalled by the unceasing victories of the infidels."

Janos Hunyadi is the Hungarian form of his name, which is Anglicized by Gibbon to John Huniades. Appletons' *Cyclopædia* (Article Hunyadi) gives his surname "Corvinus, (Hollósi.)" See also *Universal History* and Moreri. Hunyadi was born 1376; died 1456.

Says Gibbon, in his *Decline and Fall of the Roman Empire*, vol. vi., 358: "From an humble or at least a doubtful origin, the merit of John Huniades promoted him to the command of the Hungarian armies. His father was a Wallachian, his mother a Greek; and the claims of the Wallachians, with the surname of Corvinus, from the place of his nativity, might suggest a thin pretense for mingling his blood with the patricians of ancient Rome. In his youth he served in the wars of Italy, and was retained, with twelve horsemen, by the Bishop of Zagrab. The valor of the *White Knight of Wallachia*" (as he was styled) "was soon conspicuous; he increased his fortunes by a noble and wealthy marriage, and in the defense of the Hungarian borders he won, in the same year, three battles against the Turks. By his influence, Ladislaus of Poland obtained the crown of Hungary, and the important service was rewarded by the title and office of Waivod of Transylvania. The first of Julian's crusades" (Julian was legate of the Pope) "added two Turkish laurels on his brow, and in the public distress the fatal errors of Varna were forgotten. During the minority and absence of Ladislaus of Austria, (the titular king,) Huniades was elected supreme captain and governor of Hungary; and if envy at first was silenced by terror, a reign of twelve years supposes the arts of policy as well as of war. Yet the idea of a consummate general is not delineated in his campaigns; the *White Knight* fought with the hand rather than the head, as the chief of desultory barbarians, who attack without fear and fly without shame; and his military life is composed of a romantic alternative of victories and escapes. By the Turks, who employed his name to frighten their perverse children, he was corruptly denominated *Jancus Lain*, or the Wicked. Their hatred is the proof of their esteem; the kingdom which he guarded was inaccessible to their arms; and they felt him most daring and formidable when they fondly believed the captain and his country irrecoverably lost. Instead of confining himself to a defensive war, four years after the defeat of Varna he again penetrated into the heart of Bulgaria, and in the plain of Cossova sustained, till the third day, the shock of the Ottoman army, four times more numerous than his own. As he fled alone through the woods of Wallachia, the hero was surprised by two robbers; but while they disputed a gold chain that hung at his neck, he recovered his sword, slew the one, terrified the other, and after new perils of captivity or death, consoled by his presence an afflicted kingdom. But the last and most glorious action of his life was the defense of Belgrade against the powers of Mohammed the Second in person. After a siege of forty days, the Turks, who had already entered the town, were compelled to retreat; and the joyful nations celebrated Huniades and Belgrade as the bulwarks of Christendom. About a month after this great deliverance, the champion expired; and his most splendid epitaph is the regret of the Ottoman prince, who sighed that he could no longer hope for revenge against the single antagonist who had triumphed over his arms. On the first vacancy of the throne, Matthias Corvinus," (his son,) "a youth of eighteen years of age, was elected and crowned by the grateful Hungarians. His reign was prosperous and long. Matthias aspired to the glory of a conqueror and a saint, but his purest merit is the encouragement of learning; and the Latin orators and historians who were invited from Italy by the son, have shed the lustre of their eloquence on the father's character."*

APPENDIX C.

MATTHIAS CORVINUS.

MATTHIAS CORVINUS was elected and crowned by the grateful Hungarians in reward for his father's services. "His reign was prosperous and long. He aspired to the glory of a conqueror and a saint, but his purest merit is the encouragement of learning," says Gibbon. He had been imprisoned by the preceding king, Ladislaus, but escaped by the help of George Podiebrad, who aimed, and succeeded, in making him his son-in-law. The German Emperor, Frederic III., tried to obtain also the title of King of Hungary; but the people were averse to his character, and to connection with Austria, and chose Matthias instead, 1458. "He reigned above thirty years, with considerable reputation, to which his patronage of learned men, who repaid his munificence with very profuse eulogies, did not a little contribute. Hungary, at least in his time, was undoubtedly formidable to her neighbors, and held a respectable rank, as an independent power, in the republic of Europe." (Hallam's *Middle Ages*, p. 245.)

There was great joy at the arrival of the young king. He spoke, besides Hungarian, Slavonian, Bulgarian, German, and Latin. He had already been interpreter and secretary to his father. He became an impressive orator and an able general. He reorganized his army, drilling his troops in person. He then turned his attention to the state of education. He founded the University of Buda, which could accommodate more than 40,000 students, and provided a library of more than 50,000 volumes. This library was known by the name of CORVINA.

In 1466, he again defeated the Turks. The Papal Legate then induced him to fight Podiebrad and the Hussites, which he did, but reluctantly. He conquered Bohemia and Moravia. In 1480, he married Beatrice of Naples. Civil contentions, through the machinations of the Pope, now troubled him. He imprisoned Varda, a bishop, and kept him in confinement. The last five years of his life he made Vienna his capital, and died there. (*British Cyclopædia*, art. Hungary.)

Rees's *Cyclopædia* has the following remarks on *Corvini, Matthew*, taken from Warton's *History of English Poetry*, vol. ii., p. 417:

"He was a lover and guardian of literature. He purchased innumerable volumes of Greek and Hebrew writers, at Constantinople and other Grecian cities, when they were sacked by the Turks; and, as the operations of typography were then imperfect, he employed at Florence many learned librarians to multiply copies of the classics, both Greek and Latin, which he could not procure in Greece. These, to the number of 50,000, he placed in a tower which he had erected in the metropolis of Buda; and in this library he established thirty amanuenses, skilled in painting, illuminating, and writing, under the conduct of Felix Ragusinas, a Dalmatian, consummately learned in the Greek, Chaldaic, and Arabic languages, and an elegant designer and painter of ornaments on vellum, attending incessantly to the business of transcription and decoration. The librarian was Bartholomew Fontius, a learned Florentine, the writer of many philosophical works, and professor of Greek and

---

* See also John Thrwocz' account of Huniades, and that of Bonfinius,

oratory at Florence. When Buda was taken by the Turks, in 1526, Cardinal Boz-wanni offered for the redemption of this inestimable collection 200,000 pieces of the imperial money, but without effect; for the barbarous besiegers destroyed or defaced most of the books in the violence of seizing the splendid covers and the silver bosses and clasps with which they were enriched."

After Matthias's death, the whole of Hungary rapidly degenerated, and was, for a while, merged in Austria.

In the century succeeding the death of Matthias, the Reformation made rapid progress, especially among the higher classes in Hungary. But the Jesuits having been invited in 1604, persecutions began in earnest. But with Matthias II., (1608,) brother of Rudolph, better days came. At his death, in 1619, persecutions were renewed, but Protestants greatly increased in Bohemia about this time. Gabriel Bethlen, of Transylvania, helped the Protestants. He entered into an alliance with the Sultan, with the Protestants of Austria, and with the Bohemians. The Bohemian prince was son-in-law of James I., of England. (*British Cyclopædia*, art. Hungary.)

Matthias II., of Hungary, 1608–18, seems to have been of no relation to Matthias Corvinus. The only surviving son of the latter, John, after his father's death, retired to private life. Matthias II. left no children.

Several lives of Matthias Corvinus have been written, but none in the English language. The Latin, Hungarian, or German has been employed. The volume of Bonfinius is called *Decades IV. Ungaricarum Rerum*, or *Four Decades of Hungarian Affairs*. This is full of anecdotes, and is thought to be too flattering. Gibbon says, "The observations of Spondanus on the life and character of Matthias Corvinus are curious and critical, (A.D. 1464, No. 1; 1475, No. 6; 1476, No. 14–16; 1490, No. 4, 5.) Italian fame was the object of his vanity. His actions are celebrated in the *Epitome Rerum Hungaricarum* (pp. 322–412) of Peter Ranzanus, a Sicilian. His wise and facetious sayings (*De dictis et factis Mathiæ*) are registered by Galestus (or Galeotus) Martius of Narni, (528–568,) and we have a particular narrative of his wedding and coronation. These three tracts are all contained in the first volume of Bel's *Scriptores Rerum Hungaricarum*."

Other works referring to him are, Turotz's *Chronica Hungarica*, Fessler's *Mathias, König von Hungarn*, 1793; Korn's ditto, Breslau, 1796, 2 vols.; Wenzel's ditto, 1810. The writer possesses only Korn's memoir.

Nicholas Isvanfi, vice-palatine in the reign of Rudolph, wrote in ███████tinity, *The History of Hungary, from the Death of Matthias Corvinus* ████ *of Austria.*

In 1850, a volume was published in London called, *Hungary, its Co███████n and Catastrophe, by Corvinus.* (A copy in Philadelphia Library, 9591, D. I.)

There appeared in London, 1871, a work in three volumes 16mo, entitled, *Corvin, Wiersbitzky, O. J. B., von. A Life Adventure: an Autobiography.*

------------ • • • ------------

APPENDIX D.

THE REFORMATION IN HUNGARY.*

A CORRUPT form of Christianity was firmly established in Hungary about A.D. 1100. Paulicians soon appeared to uphold the purity of the Gospel. Thousands of Hungarians suffered death for the purity of their faith; yet their numbers continually

---

* As Matthias Corwin, the American immigrant, came over with the Puritans, religious persecution, no doubt, was one reason of his emigration. (*See Introduction.*) A short account of the Reformation in Hungary, condensed from a recent volume called *The Faith in Hungary*, Presbyterian Board Publication, is given.

increased. At length the "Golden Bull," or a charter of rights and liberties, was extorted from the king, and which secured, to a good extent, freedom of religion. Refugees from the surrounding countries therefore swarmed to Hungary. Especially the Hussites, from Bohemia, flocked thither, spreading and strengthening evangelical truth.

Rome succeeded in stirring even the famous and learned king, Matthias Corvinus, against them; yet he refused to give the Pope any additional rights within his realm. Matthias appointed his own bishops and abbots, and rather delighted in the satires hurled against the Papal Church by some of his bishops.

Corvinus founded a library at Ofen, or Buda, and established a printing-press very soon after the art was discovered. An Hungarian version of the Bible (made from the Latin, by Ladislaus Bathory, an obscure monk) lay in type, but was not printed; for at this juncture, as before, Hungary became the battle-field between the Turks and Christians.

In no other land did so many forsake, and so rapidly, the Roman Church. The Golden Bull made persecution difficult. Luther's writings were introduced and read everywhere, (A.D. 1530.) Whole villages and towns declared for the Reformation. The converts embraced half the population. But Louis, the king, was only a child of sixteen. The infamous Cardinal Cajetan was sent to be his instructor, and he procured the issue of an edict that heretics should be put to death, and their property confiscated. But another Moslem invasion interfered with the contemplated persecutions, (A.D. 1526.)

For Solyman the Magnificent entered Hungary, and the twenty-six thousand Hungarians who opposed him were as nothing before the Turkish hordes. He marched to Ofen and burned the splendid library which Matthias Corvinus had gathered, containing forty thousand manuscript volumes; for Corvinus had kept thirty secretaries constantly copying books which he borrowed from the universities and princes of other countries. The glory of the kingdom had departed.

The widow of King Louis was a friend of the Reformation. She always carried about with her a Latin Testament, whose margins she herself filled with annotations. Luther wrote to her. She was, however, driven away through the machinations of the priesthood, and she went to live with her brother Charles V. John Henkel, her chaplain, attended her, and expounded the word of God to her. She afterward became regent in Holland, where she assisted the Protestants whenever possible.

The crown of Hungary had been elective since 1301. The royal elections were often scenes of wild confusion. In 1458, when Matthias Corvinus was elected to the throne, he was proclaimed by forty thousand soldiers standing on the frozen Danube, beneath the terraced heights of Buda. In 1505, they resolved that no foreigner should be chosen king of Hungary. After this last Turkish invasion, John Zapolya was elected to the sovereignty by one party; but Ferdinand of Austria was also chosen by the Diet of Presburg. The greatest persecutor should be king. Ferdinand excelled in this crime. The denunciations of heretics were most severe, yet the truth spread and prospered. Many nobles sided with the Reformers. The municipal authorities of Hermannstadt, in Transylvania, banished all monks and nuns from the city. The ever-threatening Turks mitigated the spite of the hierarchy.

Matthew Devay was now the Luther of Hungary, being also Luther's disciple. He was bold and fearless in exposing abuses. Imprisoned by Zapolya, and released by a whim of good nature in the king, he went to the domains of Ferdinand and received the same reward for his fidelity. After two years of prison life at Vienna, he was unaccountably released by the king. The latter had no principles, and seemed, at times, to favor both parties. Devay now went back into Zapolya's dominions. He was joyfully received by nobles and princes. He was a traveling apostle. He printed his own translation of Paul's Epistles in the Hungarian tongue in 1533. Stephen Szantai was another kindred spirit to Devay. Ferdinand befriended him privately, while publicly he befriended the priests. During a century or more, the Turks held no inconsiderable parts of Hungary, and the sultans favored the heretics. Buda was held by the Turks for more than a hundred years. All through Transylvania and Wallachia heretics abounded. The papal power was shorn of its strength beyond the river Theiss.

In 1541, John Sylvester published a complete Hungarian Testament, and dedicated it to the king's sons. They patronized it from political reasons, as it strengthened the nationality of their people. The Servians of the south received the Bible in their tongue in 1493. The Hungarian Bible was not completed till 1589. Many Hungarians were going continually to Luther and Melanchthon to study theology. These noble men often encouraged their remote pupils, after their return, by epistles.

In 1549, Ferdinand sent two royal commissioners to inquire into the religious state of Upper Hungary, and to discover the strength of Protestantism. The pastors, after the first timidity, took courage, and presented the "Five Cities' Confession," as it was called. It was drawn up in so mild a manner that they were not for the present disturbed. The last public act of Melanchthon was to ordain a new Hungarian pastor. His pupils always retained a deep love for him.

Ferdinand hoped that the Council of Trent would compromise matters with the reformers and heal the breach. He sent two Hungarian bishops to demand reforms. One of them, Dudith, wrote back to the king that there was no chance of reform in a Roman council; that the pope managed the whole business, through his puppets, for the aggrandizement of the papal see. Dudith soon became a Lutheran. Yet Ferdinand extorted from the pope the right of the laity to the cup, at least in Hungary. But the reformers were not satisfied. They covenanted together (as they did in Scotland) by an oath to support the truth, and signed their names to the document. This document is still preserved in a library at Presburg. When brought to trial, they declared the right of freedom of conscience, and their undaunted demeanor so influenced their judges that they were acquitted. Indeed, during the most of the century (1450-1550) the reformation flourished in Hungary.

In 1564, Ferdinand was succeeded by his son Maximilian. He tried to protect the Protestants by preventing the execution of the fulminations from Rome. He prevented the foisting of Jesuit teachers on towns, against their will. Yet he tried to suppress Zwinglianism. It is doubtful whether he were not really a Protestant at heart. In 1576, his son, Rudolph, succeeded, who was his exact opposite—like Philip II. of Spain.

Congregations, pastors, and even cities and private persons, published their several confessions of faith. This was to let the world know that they were not to be confounded with Socinians and others. Rome also pretended that the reformed were only disobedient children within her fold who needed chastisement. But when the Bishop Conrad demanded his church dues, he received, instead, a confession of their faith. Rudolph began, not open persecution on a large scale, but the harassing of congregations and pastors. There were nearly a thousand Protestant congregations in Hungary.

The approach of Mohammed III., with 150,000 troops, was considered by the reformed a blessing rather than an evil, as again these Moslems diverted the force of persecutors. Ten thousand papal troops, however, sent to help Rudolph, treated the reformed as if they were Turks. They harnessed them as beasts, and made them draw their forage. Protestant pastors were burned with Protestant books as fuel, or flayed alive. They declared their intention to put to the sword every adult who was not a Romanist. A fearful famine followed the war, so that even human

16

flesh was eaten. The king, resigning real power into the hands of papal tools, re-mained in indolence and debauchery in his palace at Prague.

The free cities, however, were nurseries of liberty, both civil and religious, and only yielded to actual force. From Kashaw the pastors were banished, and the churches handed over to priests. Leutshaw utterly refused to yield, and, force not being at hand, she for the present escaped.

In 1604, the king consented that a law should be forged, and added to the common laws, authorizing persecution, though in violation of his oath to the Hungarian con-stitution. This placed the lives of his best subjects at the complete disposal of the Romish clergy. He would not allow them even to remonstrate in future. But this act aroused the free blood of Hungary. A vigorous protest was presented, though with little effect. Matthias Botskay, a prince of the best blood, had long revolted against the barbarities of the Romanists. The king also refused him an audience, after a long journey. Returning home, he found the Romanists had committed great de-predations on the Protestants, and plundered two of his castles. With his vassals he pursued the Romish forces and defeated them. The cities which had suffered re-fused to shelter the soldiers of Rome. Another victory was gained over them. A gene-ral revolt against the king followed. The Turks helped the good cause. They offered to secure for Botskay the crown. But he refused, saying he was only fighting for freedom of conscience. Peace was concluded, 1606, the king being glad to compro-mise with Botskay. This was called "The Peace of Vienna." The forged law of persecution was annulled. All decrees against the Protestants were abolished. Freedom of conscience was secured; but with the added clause, "it was not to be interpreted in any way detrimental to the Roman religion"! Public and military appointments were open to all, irrespective of creed. But the pope protested, and his agents went to work. Botskay was poisoned. Oppressions began anew. But Rudolph was deposed. His brother was made king on his signing the articles of freedom. Afterward, Matthias II. had a brief reign of seven years, 1612-19. Bigotry depopulated Germany of at least a million of human beings.

Prince Bethlen Gabor, of Transylvania, was the Cromwell of Hungary. He devoted his life to the cause of religious liberty. With his small province he opposed the im-perial might of Ferdinand. Through the help of the Turks he conquered Presburg and gained possession of the crown of Stephen. He also helped the Bohemian Pro-testants. Then ensued the Thirty Years' War. In 1620, Ferdinand was willing again to guarantee religious freedom, except in Bohemia. He would not desert the cause in that province. It was proposed to make Bethlen King of Hungary. This he re-fused. Piety was the prevailing current of his life. Four days after the offer, he dissolved the diet. But the diet had decreed perfect equality in faith. Jesuits were banished. But Ferdinand would not agree as to Bohemia. He drove thence 12,000 Christians, (185 nobles,) who scattered to England, Flanders, or Switzerland. Bethlen made war on him. Peace was repeatedly made and broken. Bethlen died in 1629. Ferdinand died in 1637.

In 1633, the church in Hungary was the picture of suffering. Now Prince George Rakotzy, of Transylvania, raised the standard of revolt. He made alliance with Pro-testant Sweden and with Turkey. New promises were made. Rakotzy died in 1648. His son followed with his father's zeal.

The period 1657-1705 is called the golden age of the Jesuits.

It was during some such scenes that Matthias Corwin is supposed to have emi-grated to England, (though probably stopping in Germany,) and to have joined the Puritans there; but whether in the latter part of this period, as above presented, or at an earlier time, is not at present known.

APPENDIX E.

REVS. ANTHONY AND JOHN CORVINUS, OF GERMANY.

(From Herzog's Encyclopædia.)

"CORVINUS, ANTONIUS,'(Räbener,) a prominent reformer in the north of Germany, born February 27th, 1501, in Warburg, (not Marburg.) Little is known about his youth or conversion. In 1523, he was driven, on account of his adherence to Lutheranism, from the cloister Loccum, where, as in Riddagshausen, he received his training. From thence he went to Wittenberg, where he prosecuted his studies without coming into personal contact with the principal men of the reformation In 1526, he was at Marburg, when efforts were made toward founding the new university, but was never professor there. At the recommendation of Amsdorf, a leader of the reformation, in 1528, he was called to Goslar, as preacher at St. Stephen's, where he zealously labored to the close of 1531; when, owing to the unhappy turn of the reformation at Goslar and personal persecutions, he went to Witzenhausen. Here he served the ministry for a number of years, but was frequently at Marburg, and was employed by the landgrave in almost all the important transactions, as at the disputation between Melanchthon and Bucer, 1535, and the Smalcald Convention, 1537.

"In 1539, an important field of usefulness opened for him in Göttingen and Kalenberg. At the commencement of the reformation, Duke Erich, the elder, reigned there—a man of an upright disposition, but, as a fellow-soldier of the Emperor Maximilian, rough and severe; without taste for higher interests; averse to the reformation, as the friend of Charles V.; and, through the influence of his first wife, he strove to keep down the movement in his principality. His first wife died without issue. In 1525, he married Elizabeth, daughter of Joachim I., one of the bitterest foes of the reformation. Elizabeth, favorably impressed, perhaps, by the patience of her mother, more so by her brother, the Margrave, John of Küstrin, who visited her in Münden, 1538, when Corvin preached for the first time in Münden, formally passed over to the Evangelical Church in the same year. Erich, who remained in the Romish Church, did not hinder her, and ordered that Corvin, at the request of Elizabeth, and by the permission of Philip of Hesse, should come over, from time to time, to preach and administer the sacraments. From this time Corvin gained in influence. In 1539, he reformed Nordheim, and gave to the city a church government. Immediately after the death of Erich, July 26th, 1540, at the diet of Hagenau, Elizabeth having become regent for Erich II., the introduction of the new doctrine became the chief aim of her life. At her side stood Waldhausen, her chancellor, recommended by Luther; Burcard Mithob, her physician, a friend of Melanchthon, allied to Corvin; above all, Corvin himself, who at first remained at Witzenhausen, but in 1541 or 1542 removed to the province and engaged in her service, and was called as superintendent of Kalenberg-Göttingen.

"In the fall of 1540, the estates were consulted in Pattensen concerning the reformation, and consented to receive the word of God. An edict to this effect was published. The ceremonies at first remained unaltered, but the pure Gospel was preached. The change was first to be made internally; this accomplished, the outward soon followed. About Whitsunday, 1542, appeared the form of church government drawn up by Corvin, in German; then, November 4th, monastic rules; also rules for the church revenue. These were to be introduced by a visitation. This was undertaken by Corvin, with the assistance of selected clergy and laity, in Göttingen, 1542; in Kalenberg, 1543.

"Meanwhile, Corvin zealously assisted in carrying on the reformation at Hildesheim, whither he was sent, with Winkel and Bugenhagen, by princes of the Smalkaldian confederacy. He assisted in drawing up the ecclesiastical discipline and introducing the reformation into Braunschweig-Wolfenbüttel. Then he returned to his own country, and labored actively in furthering the good work. Although all opposition had not disappeared, yet the whole land seemed to be permanently reformed. Erich II., while engaged in the emperor's camp in fighting the Evangelicals, returned to the Roman Church. Many preachers had to fly. Corvin, with Walther Höcker, was seized, (November 1st, 1549,) in Pattensen, by Spanish soldiers, and taken to Kalenberg. He spent three years in prison. Elizabeth's intercession had no effect. He suffered patiently for the cause of the Gospel, which he had served so zealously. Toward the close of 1552, Erich II., through the influence of the Margrave Albrecht, of Brandenberg-Kulmbach, was regained for the side of the Evangelical party, and Corvin was released from his confinement. But the rigor of the prison consumed his energies. Already sick, he was brought to Hanover, where he died April 5th, 1553. He lies buried before the altar of the Church St. George and James.

"Corvin does not belong to the creative spirits of the reformation, but to the circle of men who assisted in proclaiming the word in its purity. More important than his writings, of which few (except his Postilla) reached several editions, was his talent for church organization. (See Baring, *Leben Corvin's*, Hannover, 1749. Schlegel, *Kirchen und Reformation-geschichte von Nord-deutschland*, ii., p. 141, etc.)
"ULHORN.—*Santee.*"

His principal work is the *Postilla in Evangelia et Epistolas*, or *Brief Notes on the Gospels and Epistles*. Ulhorn wrote *Ein Sendbrief von Antonius Corvinus, mit einer biographischen Einleitung*, (Göttingen, 1853,) or *A Synodical Letter of Anthony Corvin, with a Biographical Introduction*. Baring, as indicated above, wrote Corvin's Life, and Schlegel, in his *History of the Churches and Reformation in Northern Germany*, has occasion to allude to him. See also a brief notice in McClintock's and Strong's *Biblical Cyclopædia*.

REV. JOHN CORVINUS.

All that the writer has collected concerning this individual is incorporated in the Introduction.

APPENDIX F.

CORVINUS, JOHN AUGUSTUS. An engraver, born at Leipsic, 1682. He went to reside at Augsburg, where he engraved plates for several considerable works, published in that city; but his style, though neat, is stiff and without taste. He died at Augsburg, in 1738. Among his engravings are several of the plates for a work called, "Representatio Belli ob successionum Regno Hispanico," or, "A Representation of the War of Succession for the Kingdom of Spain;" most of those for the Bible of Scheuchtzer, and a collection of views of churches in Vienna, Strutt, and Heinecken. (*Rees's Encyclopædia.*)

APPENDIX G.*

RIVERHEAD, LONG ISLAND, N. Y., July 11, 1848.
TO REV. JASON CORWIN:
SIR: I am informed you have written to Mrs. Youngs, of this island, making inquiries as to the ancestors of the Corwins, in consequence of information derived

* In 1848, a Hungarian gentleman, named Denburgh, informed Charlton Jason Corwin that the collateral line of Corvinus in Hungary, had recently become extinct, and that a large property was awaiting claimants. The Hungarian Revolution soon breaking out, the above-mentioned Hungarian left the country, to assist his fellow-patriots. Of course, all such claims are Utopian, and unworthy of consideration.

from your son, living in New-York City, that an estate in Europe was without heirs, unless they could be found in this country, by the name of Corwin.

There are a great many Corwins upon Long Island and in other parts of this country, whose ancestor was Matthias Corwin, who came from England more than two hundred years ago, and who was among the first settlers of the town of Southold. There is, however, another family of Corwins, whose ancestor was Theophilus Corwin, who came to this country over one hundred years ago. He appears to have come from the continent of Europe, and I am taking pains to ascertain all the particulars in relation to him.

I would thank you to write to me on the receipt of this letter, and state the name of your son in the city of New-York, that you mentioned in your letter to Mrs. Youngs, as having given you information in relation to the property in Europe.

Respectfully, your obedient servant,   GEORGE MILLER.

MR. JASON CORWIN: NARROWSBURG, Sept. 8, 1848.

SIR : You will please pardon me the liberty I have taken in addressing these few lines to you, and the favor I ask of you. I am a son-in-law of Mr. Phineas Corwin, and he wishes me to ascertain the pedigree of the family on Long Island. I have been informed that you have a pedigree of the family, for a long time back. He wishes to know who the first Corwin was, and whom he married, for these reasons : It appears that there was once a Matthew Corwin, who lived in Hungary, and his eldest son emigrated to this country, and settled on Long Island. But soon after he left, his father died, and left a large estate, amounting to forty millions. It remained unclaimed for fifty years, when it went into the hands of the government. You will therefore do me a great favor in giving me what information you can in relation to the matter. My wife's father was born on Long Island, and he has a brother of the name of Jason, and their father's name was Matthew Corwin. The Corwin that this estate belongs to married a *Morton*. Your early attention to the above will be thankfully received.

Respectfully yours,   DANIEL W. GILLETT.

(This letter (it is here condensed) was missent, and fell into the hands of Elisha Corwin, of New-York City, by whom it was answered.)

MR. CORWIN :

SIR : By request of Mrs. Youngs, I write to inform you that our Long Island friends have ascertained that Matthias Corwin emigrated from Warwick, England, to Long Island; also that letters have been found written in Hungary, one hundred and fifty years ago, addressed to Theophilus Corwin, their brother. In examining my father's papers, I find a deed, which my great-grandfather, Theophilus Corwin, received in 1699. Our Long Island friends have written to Austria on the subject. If you have received any intelligence since I saw you, you will confer a favor by informing me.   Yours respectfully,   IRENA HUSTED.

ARCADIA, Oct. 16, 1848.

It will be observed that these rumors, and this excitement in portions of the family, about this Hungarian property, took place at the time that Kossuth visited this country, namely, in 1848. What foundation there was for this excitement, the writer can not positively say. He was, however, informed several years since by one of the family that Kossuth alluded in some of his speeches to the emigration of some of the descendants of Matthias Corvinus to America. That there were two distinct families on Long Island, descended, the one from Matthias, and the other from Theophilus, is no doubt an error, as the will of Matthias refers to a son Theophilus, and no documentary proof of a separate family descended from a Theophilus has been met with, in the preparation of this work.

The following "tree" gives a probable plan of the relation of the European Corvini.

THE EUROPEAN CORVINI.

Generations.
1440
1470
1500
1530
1500
1500

JOHN HUXTADI CORVIN,. (1376-1456.)
His son was

MATTHIAS CORVIN, (1441-1491.)
His sons were

LADISLAUS CORVIN,
(born about 1465.)

and

JOHN CORVIN.
(Born about 1470; yet living, 1540.)
John's teacher was Anthony Bonfinius.
John's son, probably,—

Rev. ANTHONY CORVIN.
(Born 1501; died 1553.)
He became a Protestant in 1526,
and a celebrated reformer, preacher, and author, in Germany.

A CORVIN was at the Council of Trent, 1540,
as Legate of the Pope;
(Born, probably, 1490-1500.)

(No names have been met with as belonging to this generation.)

JOHN CORVIN, (born about 1560.)
His son was

Rev. John Corvin, perhaps the same as

Matthias Corwin, (born about 1590, died 1658.)
The American Immigrant, 1638.
Possibly he was a brother or cousin of Arnold.

Arnold Corvin, (born about 1590.) He was an eminent lawyer,
and published Digests of the Law, in Aphorisms,
at Amsterdam, Holland, 1649.

1525-30. Many Hungarians fled to England.

1548. 120 Protestant Churches in Hungary.

1604. Many Protestants banished from Hungary.

1618. There was a Corvinus at the Synod of Dort.
Born about 1560-70.

APPENDIX II.

THE ENGLISH CULWENS, OR CURWENS.

Si je n'estoy.

HUTCHINSON's *History of the County of Cumberland, England*, and Camden's *Britannia*, are the sources of the information concerning this family. They state that John de Talbois, (who was brother of Fulk, Earl of Anjou, and King of Jerusalem, (in the crusades,) married Elgiva, daughter of the Saxon King, Ethelred, who reigned A.D. 866–871. John de Talbois was the first lord of the barony of Kendall, to which position his son Elred or Ethelred succeeded. Elred's son, Ketel, granted the church of Morland to the abbey of St. Mary, York, to which grant Christiana, his wife, was a witness. Ketel's eldest son was Gilbert, father of William de Lancaster, (who had a son of the same name,) from whom descended in a direct line the de Lancasters, barons of Kendall. Ketel's second son was William, and the third, Ormus or Orm. The latter married Gumilda, daughter of Gospatric, Earl of Dunbar and Northumberland. (Camden's *Brit.* iii. 423.) She was also sister of Waldieve, first lord of Allerdale.

Now, William the Conquerer had infeoffed Ranulphus de Meschines with the county of Cumberland, and Ranulphus infeoffed Waldieve in the barony of Allerdale. William, a brother of Ranulph, whom the king had made lord of Coupland, also infeoffed Waldieve, in the land between Cocar and Darwent, and in the townships of Brigham, Elysfield, Dene, Brainthwaite, Grisothen, the two Cliftons, and Stanebnrne. Ranulph also inherited the earldom of Chester, by the death of his brother Galfridus, on which he surrendered Cumberland back to the king, on condition that his tenants *in fee* should hold their land of the king *in capite*.

Waldieve now infeoffed Orm, when he married his sister Gumilda, in the manor of Seaton below Darwent, parcel of the barony of Allerdale, as also the towns of Camberton, Crayksothen, and Flemingby; and thereupon Orm was settled at Seaton. (Camden's *Brit.* iii. 9.)

The walls and ruins of the mansion-house of Seaton, Mr. Denton states to have been visible yet in his time.

Orm was succeeded by his second son, Gospatric, (named after his maternal grandfather,) to whom Allan, son of Waldieve, second lord of Allerdale, and his cousin-german, gave High Ireby, which remained vested in a younger branch of the Curwens, and which terminated in female heirs. This Gospatric was the first of the family, who was lord of WORKINGTON, having exchanged with his cousin, William de Lancaster, the lordship of Middleton, in Westmorland, for the lands of

THE CUMBERLAND COUNTY CURWENS, ENGLAND.

SEE HUTCHINSON'S CUMBERLAND, VOL. II., P. 143.

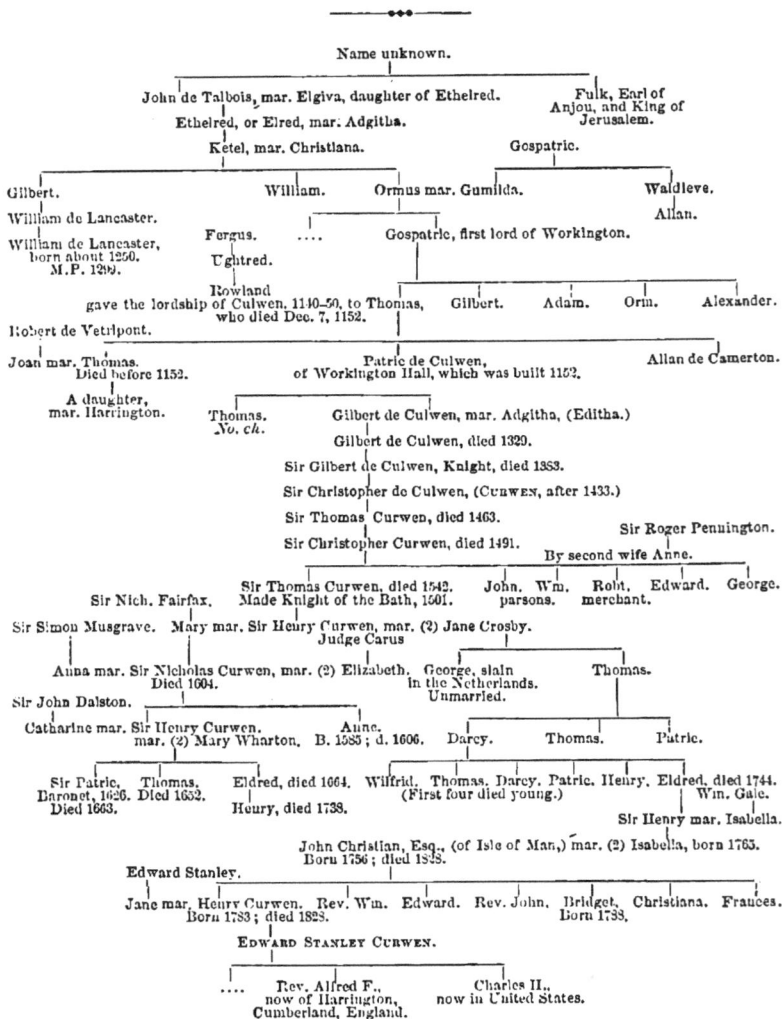

Name unknown.

John de Talbois, mar. Elgiva, daughter of Ethelred.   Fulk, Earl of Anjou, and King of Jerusalem.

Ethelred, or Elred, mar. Adgitha.

Ketel, mar. Christiana.   Gospatric.

Gilbert.   William.   Ormus mar. Gumilda.   Waldieve.

William de Lancaster.   Allan.

William de Lancaster, born about 1250. M.P. 1290.   Fergus.   ....   Gospatric, first lord of Workington.

Ughtred.

Rowland gave the lordship of Culwen, 1140–50, to Thomas, who died Dec. 7, 1152.   Gilbert.   Adam.   Orm.   Alexander.

Robert de Vetrlpont.

Joan mar. Thomas. Died before 1152.   Patric de Culwen, of Workington Hall, which was built 1152.   Allan de Camerton.

A daughter, mar. Harrington.

Thomas. No. ch.   Gilbert de Culwen, mar. Adgitha, (Editha.)

Gilbert de Culwen, died 1329.

Sir Gilbert de Culwen, Knight, died 1383.

Sir Christopher de Culwen, (CURWEN, after 1433.)

Sir Thomas Curwen, died 1463.   Sir Roger Pennington.

Sir Christopher Curwen, died 1491.   By second wife Anne.

Sir Thomas Curwen, died 1542. Made Knight of the Bath, 1501.   John.   Wm. parsons.   Robt. merchant.   Edward.   George.

Sir Nich. Fairfax.   Sir Simon Musgrave.   Mary mar. Sir Henry Curwen, mar. (2) Jane Crosby. Judge Carus

Anna mar. Sir Nicholas Curwen, mar. (2) Elizabeth. Died 1604.   George, slain in the Netherlands. Unmarried.   Thomas.

Sir John Dalston.

Catharine mar. Sir Henry Curwen. mar. (2) Mary Wharton.   Anne. B. 1585 ; d. 1606.   Darcy.   Thomas.   Patric.

Sir Patric, Baronet, 1626. Died 1663.   Thomas. Died 1652.   Eldred, died 1664.   Wilfrid. Thomas. Darcy. Patric. Henry, Eldred, died 1744. (First four died young.)   Wm. Gale.

Henry, died 1738.   Sir Henry mar. Isabella.

John Christian, Esq., (of Isle of Man,) mar. (2) Isabella, born 1765. Born 1756 ; died 1828.

Edward Stanley.

Jane mar. Henry Curwen. Born 1783 ; died 1823.   Rev. Wm.   Edward.   Rev. John, Born 1788.   Bridget, Christiana. Frances.

EDWARD STANLEY CURWEN.

....   Rev. Alfred F., now of Harrington, Cumberland, England.   Charles H., now in United States.

Lamplugh and Workington. In this bargain, de Lancaster retained for himself and heirs an annual rent charge of sixpence, to be paid at the fair of Carlisle, or a pair of gilt spurs, binding Gospatric and his heirs to homage, and to discharge his foreign service, as of the castle and barony of Egremont.

Gospatric gave two parts of the fishery in Derwent to the abbey of Holm Cultram, except Waytcroft, which he gave to the priory of Carlisle ; which was granted over by the priory to Thomas, son of Gospatric, upon a reserved rent of 7s.

Thomas was called, after the fashion of the times, "son of Gospatric." To him one Rowland, son of Ughtred, son of Fergus, gave the lordship of Culwen, in Galloway. Thomas confirmed the grant of Flemingby made by his father to the convent of Holm Cultram, and gave to that house the whole fishing of Derwent. His eldest son, Thomas, died before his father; and Allan, his third son, acquired by gift from Patric the lands of Camerton, taking that surname, from whom descended the Camertons. Thomas, son of Gospatric, gave Lamplugh to Robert de Lamplugh, and his heirs, to be holden by yearly presentment of a pair of gilt spurs. To his second son, Patric, he had given, while his eldest son was yet living, the lordship of Culwen in Galloway, and Patric, marrying the heiress of Culwen, (Camden, iii. 423,) assumed the surname, and was thenceforward known as Patric de Culwen,* (about 1140–50.) [Culwen is in the county of Kirkcudbright, on the south-west coast of Scotland.]

But Patric's elder brother, Thomas, dying without male heirs, Patric succeeded to the entire estate, and removed to Workington. His eldest son Thomas left no children, and the estate again passed to the line of the second son, Gilbert. Gilbert was returned to Parliament by the county of Cumberland in 1373, 1374, and 1376. Sir Christopher also represented the county in several Parliaments. He was also sheriff in 1423 and 1427, under the name of Culwen, and in 1433 under the name of CURWEN, to which cognomen the family have ever since adhered.

"The family of Curwens settled here (Workington) is of great ancestry. The name is local, derived from Culwen, and by a corruption, which first appeared in the public records in the reign of King Henry VI., the family name was changed to Curwen, one of them being returned in the sixth year of that king's reign, as sheriff of the county, by that name." (Hutchinson's *Cumberland*, ii. p. 143.)

In 1542, by an inquisition of knight's fees in Cumberland, it was found that Sir Thomas held the manor of Workington of the king, by knight's service, as of his castle of Egremont, by the service of one knight's fee, 45s. 3d., cornage, 4s. seawake, and puture of two sergeants. He also held the manor of Thornthwaite, one third of the manor of Bothill, the manors of Seaton and Camberton, divers tenements in Gilcrouse, Great Broughton, and Dereham. Sir Henry had the honor of affording an asylum to Mary Queen of Scots, when that unhappy princess sought the protection of England. She landed near the mouth of the Derwent in 1568, after her escape from the castle of Dunbar. Taking refuge at the Curwen house, she was hospitably entertained till the pleasure of Elizabeth was known, when she was removed first to Cockermouth castle, and then to Carlisle. The chamber in which she slept is still called the Queen's chamber. (Camden, iii. 437.) Sir Henry represented Cumberland in parliament in 1552 and 1558. Sir Nicholas, his son, was sent to parliament in 1621. Sir Henry's son, Patric, was made a baronet in 1626; but dying without heirs, the title became extinct, and the estate passed to his half-brother, Eldred, who dying in a year, was succeeded by his son, Sir Henry, with whom that branch of the family expired. The estate now passed over to Elred, of the younger branch descended from Sir Henry Curwen and Jane Crosby. He was

* The prefix *de* used before Culwen for a while shows the Norman French influence, but does not, of course, permit us to assign a French origin to the word. The preposition was used for about three centuries.

sheriff in 1729. He also represented Cockermouth in parliament. His son and successor, Sir Henry, represented Carlisle in parliament in 1762, and stood for the county of Cumberland, at the general election in 1768, and, after an unprecedented contest, was victorious by a considerable majority. He left an only daughter, Isabella, heiress of his estate, who was married very young to her paternal first cousin, John Christian, Esq., of Unerigg Hall. The family tablet of the Christians is given in Hutchinson's *Cumberland*, vol. ii. p. 146–8. The record extends back to 1422. John Christian of Milntown, Unerigg and Workington, Esq., was baptized July 13th, 1756. He was M.P. for the city of Carlisle in two parliaments. He assumed the surname and arms of the Curwens, by virtue of the king's sign-manual, dated March 1st, 1790. Isabella was born October 2d, 1765, and was married at Edinburgh, October 5th, 1782.

It was remarkable of this lady that she was the last and only living child of a great number, her mother having had fifteen children, all either still-born, or who died in a few moments after birth. John Christian, (Curwen,) Esq., had previously been married to Miss Jaubman, of the Isle of Man, who left at her death an only son, John Christian, Esq., of Unerigg Hall, one of the dempsters of the island. By the heiress of the Curwens, he left five sons and two daughters. Henry, born December 5th, 1783, was heir of his mother; William entered the ministry, and became rector of Harrington in Cumberland; Edward, of Bell Grange; John in the ministry, rector of Harrington; Bridget married Charles Walker, Esq., in Ashford court, county of Saluph; Christiana, and Francis, of Upington, in Shropeshire.

Sir Henry was sheriff in 1784. In 1786, he was returned to parliament. He became knight of that shire, for Cumberland, and so remained till his death. Mr. Curwen was succeeded in his own estate (on the Isle of Man) by his eldest son, John Christian, and in those of the Curwens, by his second son, Henry, as above said, who died December 9th, 1828. Henry's son, Edward Stanley Curwen, is now possessor of the estate. Of the sons of the latter, Rev. Alfred F. Curwen is now rector of Harrington; Charles H. Curwen is now in the United States. (*See Introduction.*) The eldest son's name the writer has not learned.*

"The Curwen family is a very ancient and respectable one. Their principal residence has long been at Workington Hall, in the county of Cumberland, where they had large possessions in landed property and coal mines. The last gentleman of that name and family was Henry Curwen, Esq., late member of the county. It was chiefly by his interest that Sir James Lowther, late Earl of Lonsdale, lost his parliamentary influence in the famous contested election for Cumberland, in the year 1768, when Henry Fletcher, Esq., now a baronet, first obtained a seat in the House of Commons, with Mr. Curwen, who sat in the preceding parliament for the city of Carlisle." . . . . . "The mansion-house is a large quadrangular building, which still bears marks of great antiquity, (1152,) notwithstanding various alterations and improvements which have been made during the last thirty years. The walls were so remarkably thick that they were able a few years since, in making some improvements, to excavate a passage, sufficiently wide, lengthways, through one of the walls, leaving a proper thickness on each side of the passage to answer every purpose of strength." . . . (Camden's *Britannia*, iii. 437.)

"The manor house of the family of Curwens, called Workington Hall, stands upon a fine eminence on the banks of the river Derwent. It is an elegant mansion,

---

* In 1870, the writer wrote to England, directing his letter to "The Representative of the Curwen Family at Workington." The letter was handed to the Rev. Alfred F. Curwen, who kindly answered it, suggesting a probable origin of the Salem Curwens. He said that the church records were much injured in the time of the Commonwealth, but that he would have them searched; but probably nothing was found, as the writer has received no further communication on this subject. Two interesting letters, however, have been received from Mrs. Edward Stanley Curwen, in the mean time, who also kindly sent a photographic *carte* of Workington Hall, a lithograph of which is presented in this volume.

surrounded by excellent lands. The house commands a prospect of the town, the river, and its northern banks, and the western ocean for a considerable tract. Here is a park, with beautiful cattle." (Hutchinson's *Cumberland*, ii. 142.)

"The river Derwent, in one unbroken stream, throws itself into the ocean at Wirkington, famous for a salmon fishery. It is now the seat of the ancient knightly family of the Curwens, who derived their descent from Gospatric, Earl of Northumberland, and took their surname by agreement from Culwen, a family of Galloway, whose heir they married. They have here a noble mansion, like a castle, and from them, if I may be allowed to mention it, without the imputation of vanity, I derive my descent from my mother's side." (Camden's *Britannia*, iii. 423. Quoted, also, in Hutchinson's *Cumberland*, ii. 137.)

"Also on the west syde of Darwent is a prety creke, where as shyppes cum to, wher ys a lytle prety fyssher toun, cawled Wyrkenton, and ther is the chefe howse of Sir Thomas Curwyn." (*Lel.* vii. 71.)

"Of this family was Hugh Curwen, Archbishop of Dublin, who having sat twelve years, (and in the mean time having been constituted one of the Lords Justices of Ireland,) old age growing heavy upon him, he took care to be translated to Oxford, and lingering one year in that see, he died at Sumbroch, near Bruford, and was there buried in the parish church, November 1st, 1568. He was made Bishop of Oxford in 1567.

"Curran, the celebrated Irish advocate, was born at Newmarket, in the county of Cork, Ireland, July 24th, 1750, and died at Amelia Place, Brompton, October 14th, 1817. He was for several years a member of parliament, but was far more successful in the forum than in the senate. It is said that his family originally bore the name of Curwen, and that his paternal ancestors came over to Ireland from the north of England with one of Cromwell's leaders." (*Fragment of an obituary notice.*)

SOME OTHER CURWENS NOW IN ENGLAND.

(From recent Directories of London and Liverpool.)

LIVERPOOL:

John Curwen, 11 Hope street, Berkenhead.
Mossop " 3 Water street, Egremont.
Robert " 13 Clifton road, Clifton.
Park " Berkenhead.
Thomas " 4 Parker street, "

LONDON :

Robert Curwen, 10 Wood street, E. C.
Henry F. " 23 Old Square, W. C.
Randle " 306 New North-road, Hoxton, N.
Thomas " 3 Bartholomew lane, E. C.

APPENDIX I.

WORKINGTON, Cumberland co., England, is a market-town and sea-port, situated on the left bank of the river Derwent, about a mile from its entrance to the sea, in St. George's Channel. The population in 1851 was 5837. The living is a rectory in the Archdeaconry of Richmond and Diocese of Chester. Workington is indebted for its prosperity chiefly to the collieries in its neighborhood, which furnish the principal article of export. Timber and flax are imported to a considerable extent. Ship-building, rope and sail-making, and block-making, employ some of the inhabitants. A straw plait, in imitation of leghorn, is made here. There are also extensive iron foundries, hat-works, breweries, malt-kilns, dye-works, chemical works, temper-yards, nail-works, and flour-mills. The river Derwent is here crossed on a stone bridge of three arches, built in 1763. Workington possesses a safe and capacious harbor, with a breakwater and extensive quays. The custom-house and commodious warehouses are situated on the quays. The number of vessels registered belonging to this port in 1854 was 3 under 50 tuns and 92 over 50 tuns, with 1 steam vessel of 18 tuns.

St. Michael's, the parish church, was rebuilt about 1780, in semi-gothic style. St. John's Chapel, erected in 1825, is of the Tuscan order of architecture. The Wesleyan Methodists, Independents, English Presbyterians, and Roman Catholics also have church edifices here. There are also national, British, infant, and Roman Catholic schools, a school of industry for girls, and a savings-bank, a mechanics' institute, a subscription library and news-room, a theatre, assembly-rooms, and a dispensary. A lock-up and justice's rooms have been recently built. The principal market for corn and provision is held on Wednesday, and a less important market on Saturday. Fairs, held in May and October, have lately been revived. Races are held annually. From Workington there is a branch railroad to Cockermouth.

----•◆•----

APPENDIX J.

PHILIP CORWINE.

THE Conversion of Philip Corwine, a Franciscan Fryar, to the Reformation of the Protestant Religion, anno 1589, formerly written by John Garvey, sometime Primate of all Ireland, being a copy of the Original, remaining amongst James Usher, late Primate of the same, his papers; and now entred amongst Sir James Ware's Manuscripts. Published for the good of the Protestant Church of England by R. W., Gent, Dublin. Printed by Jos. Ray, at Colledge Green, for a Society of Stationers. 1681.

To the Reader. Zeal, the mother of all Religions, caused the devout (and otherwise honorable) women to persecute St. Paul: the same stirred up St. Paul to persecute Christ, before he had knowledge of him. Thus was the Zeal of this Convert, which a Reverend Father of the Protestant Church wrote to be preserved for future memory.

This narrative I shall lay before Protestant and Papist upon two accounts: First, That all men may behold the Danger of immoderate Zeal against those of another Judgment. Secondly, To show how that Zeal and Knowledge ought to walk together.

1. Immoderate Zeal, saith Nazianzen, was in his time the cause of great Broyls and Troubles; insomuch thet Truth it self hath been stretched too far; so that by a vehement dislike of Errour on one side, men have run into errour on the other, as Dionysius and Alexandrinus did, being too fervent against Sabellius; for these two first laid the ground of Arrianisme.

2. Zeal without Knowledge may well be compared to Phaeton in the Poet, who took upon him to drive the Chariot of the Sun, but yet through his inconsiderate Rashness set the world in a flame. St. Bernard hits full on this point in these words: Discretion without Zeal is slow-paced, and Zeal without Discretion is strongheaded: let therefore Zeal spur on Discretion, and Discretion reign in Zeal.

Thus the zeal of this Convert, having joyned with the knowledge of the holy Scriptures, became a lively Faith in Christ, not choosing Saint or Angel to plead his Cause, but the best of Advocates, Christ Jesus, who is the only Mediator between God and Man.   R. WARE.

The Conversion of Philip Corwine, a Franciscan Fryer, to the Reformation of the Protestant Religion, *anno* 1589.

Philip Corwine, Nephew unto *Hugh Corwine,* late Archbishop of *Dublin,* supposing our Predecessor *John Long,* late Archbishop of *Armagh* and Primate of all *Ireland,* had been then living, wrote this Epistle following; which came to my hands, thinking it had been to me directed, my name being *John,* and his Successor, within a little time after the Expiration of our said Predecessor, I opened it.

After my long and tedious travels, wandering like a Jew finding no settlement, (ever since the Reformation of the Church of England revived, and was appointed to be observed by our Gracious Sovereign the Queen and Her High Court of Parliament,) I departed from *Ireland* into *England,* from thence into *Spain,* and so into *Italy;* being ambitious to visit *Rome,* purposely to behold her Modes and Forms: but being not satisfied with the Passages and cruel Objects which I there beheld, I returned into Spain, and took up my lodging in St. *Francis* his Monastry in the City of *Sevill,* where I have hitherto remained, expecting your Gracious acceptance, and encouraging of a Prodigal Son to come home to the church of Christ. For were it no other then the Tyranny of the *Romish* Church, and the large indulgences which she hath sent abroad into most Nations, for to massacre those whom she esteems Heretical, purposely to advance her coffers, and to raise up her luxurious appetite, it would be for a rational man a sufficient president to lay before him what an Idol, or rather what a devouring Dragon, he and his ancestors have hitherto worshiped.

I acknowledge I have feared this monster beyond my Creator or Redeemer, and depended on his sinful Indulgences more than on the better Passions of my Saviour Jesus, for which I only crave mercy from above, whose Power is most infinite.

Also I have from my youth upward wandred in the paths of Idolatry, worshiping of Stocks and Stones framed into human shapes by the art of man, whose eyes behold neither my bowing or my kneeling, whose ears heard not my petitions, nor their mouths able to return me an answer: therefore I can not but lament, both for the precious time which I have lost, and for the happiness of which I have all this time been bereaved; all thus justly happening unto me through my perverseness in not embracing those comfortable offertures which my Gracious deceased Uncle, *Hugh,* late Archbishop of *Dublin,* proposed unto me, in case I had adhered unto the Protestant faith, as it is now, according to the Apostolick manner, established.

Yet knowing your Fatherly Clemency and pious Inclinations, by your former correspondence with my deceased Uncle, I shall returne with the Prodigal Son, and come over, if I may be so happy to receive your encouragement, and so end the residue of my days to God's Glory and mine own Salvation.

Sevill, March 29, 1589.

To the Reverend Father in God
John, *Archbishop of* Armagh.

Your trusty and obedient Servant,
PHILIP CORWINE.

Upon the receipt of this epistle,* rejoicing to embrace especially the reformation of so understanding a man as this convert was, I returned an answer. Yet beforehand, I showed this, his epistle, unto our judicious and most learned brother in Christ, Adam, now Archbishop of Dublin and Chancellor of Ireland, with whom I consulted, and then we gave him all the encouragement imaginable, and sent him this answer following, by a Portuguese merchant, then going from Dublin to Waterford, and from thence being bound for Spain:

---

* The above was copied *verbatim et literatim,* and I intended thus to copy the whole pamphlet, which I found in the Public Library of Philadelphia; but my time falling short, and being unable quickly to find any one to do the work for me, I was obliged to copy the rest rapidly in phonography, even some. times giving only an abstract.—E. T. C.

To Mr. Philip Corwine, in St. Francis his Convent, at Seville.

Mr. Corwine: Your epistle is welcome, though not coming to the hands of that Rev. Father in God, John, our predecessor, who lately departed this life to obtain an everlasting one in the kingdom of heaven. Although he was deprived of this happy *offerture* of yours, (by his death,) yet was it as acceptable to me, his successor, who, with the rest of our brethren in Christ, do rejoice at your recantation, (in a manner,) as the angels and saints of heaven do at the repentance of a sinner, which causeth his soul to enter therein. Whatever you expected from our predecessor shall be granted by his successor, your humble servant, who embraces any one who desires to become a member of Christ and one of his flock. My weak endeavors shall not be wanting, neither shall the ability of our church here be slack, to assist you accordingly. I and our brother, Adam, now Archbishop of Dublin and Chancellor of this kingdom, have unanimously embraced your learned epistle, and in token thereof have sent you this small testimony of our affection toward your journey expenses. Therefore, let this be unto you an assurance of your kind reception by us, the clergy here, and for the future brotherly love and correspondency between you and us, your brethren and clergy of this nation.

Dublin, June 27th, 1589.   Your true and loving friend,

                                                        JOHN ARMACHANUS.

Upon the receipt hereof, this Philip, beholding the clergy's invitation together with the sum (being 50 pounds, English) which Adam, my brother aforesaid, and myself had sent him for his supply in his journey, upon the first conveniency, . . . toward Ireland, first coming into the Netherlands, where, meeting with a vessel bound for Waterford, he there landed, being civilly entertained by our brother, Thomas Whitherhead, the then Bishop of that diocese. He landed on October 22d following, after the date of this our epistle, where this Philip tarried for some certain days, discoursing with the Bishop wherefore he came over hither; as also showing him our encouragement, our brother encouraged him (sending a guide along with him) to Drogheda. Of which I, having noticed, sent word to the mayor and corporation to receive him kindly, which was accordingly performed.

Some few days after this his kind reception at Drogheda, this Philip Corwine being minded to make an open manifestation of his true and hearty conversion, desired me to give notice beforehand to the town and corporation that he intended (God willing) to make a recanting speech, and to declare therein the reasons why he fell from the Church of Rome, and the causes of his adherence to the Protestant faith. Which speech was declared in St. Peter's Church, where he chose this following text for his discourse:

Acts 9:4, 5: "*And he fell to the earth, and heard a voice saying unto him, Saul, Saul, why persecutest thou me? And he said, Who art Thou, Lord? And the Lord said, I am Jesus, whom thou persecutest. It is hard for thee to kick against the pricks.*"

He then compares himself to Saul. He left England, on the accession of Elizabeth, to go to Rome, desiring to find ways to persecute the Church of England. His uncle, Hugh, offered him excellent inducements if he would stay, but he would not. As he traveled through southern Europe, he compared the low condition of the people with those of England. He also could not help contrasting the wickedness of the popes with the purity of Christ. While in Rome, eight persons were condemned to death for saying that Christ was their only mediator. Their resolution and faith *pinched* his conscience, and he wondered how Rome could be so cruel. He tried to divert himself by reading the Bible, and opened by the passage, "Saul, Saul, why persecutest thou me?" Again, at another time, with similar purpose, he opened by Psalm 94, "O God, the Avenger, show Thyself clearly." He heard the voice of the murdered crying for vengeance. He thought that by belonging to the Church of Rome he was guilty of her cruelties. At last, hoping to change the current of his thoughts, he left Rome, going to Seville. Here he heard in the streets the cries of heretics, as they were arrested. The Spanish inquisition was just as bad as the Roman. A poor man (Leopold Dinsever) said that he thought the inquisition made more heretics than it did Romanists. He was imprisoned, and his goods confiscated. He ventured to rebuke these persecutors, and was silenced by the retort, "We believe you will be such another." Further reflection made him still more dissatisfied with Rome. Opening the Scriptures again, at Ezekiel 18 : 27—"*Again, when the wicked man turneth away from his wickedness that he hath committed, and doeth that which is lawful and right, he shall save his soul alive*"—his heart was changed, his reason and understanding were changed. He fell into a passion of weeping. Two coming up to him at the time, asked him why he wept. He answered, For the multitude of his sins. He concealed his conversion for a time, and then he wrote the letter above. He confessed the goodness of God to him in opening his eyes, and his heart was filled with gratitude.

The author of this pamphlet continues: I entertained this Mr. Corwine for ten months as one of my chaplains; but he suddenly died of a flux, which took him on September 3d, 1590, having received the Lord's Supper, and continuing faithful to the end.

His conversion was delightful, and he said the following things about the late Council of Trent:

1. That the Council had in their litany, that on certain days all heretical kings should be cursed, among whom was named the Queen of England and all her obedient subjects.

2. That Her Majesty's dominions are to be given to any prince conquering them, especially to Spain, since the King of Spain had married the Queen's sister. But Ireland should be independent, because of her fidelity.

3. That every Roman Catholic in England, of any quality or estate, having two or more sons, should train them up in either of these callings, being most convenient for the aid of Rome—viz., lawyers, physicians, clerks, vintners, inn-keepers, apothecaries, grocers, brewers, victuallers.

4. That all deeds held by Roman Catholics, or other ecclesiastical lands, be reposed privately under the signet of Rome, the party enjoying the same until the Catholic faith be restored. Then, such deeds appearing on record, the possessor, if a Catholic, to be imbursed out of heretics' estates.

5. That all oaths on the Protestant Bible are null.

6. That no Catholic marry any heretic, unless it will be a great advancement to the Catholic, and then only under promise to a confessor to advance only the Roman faith.

7. That dispensations be granted to the wife, and to learned Catholics, for oaths, that they may hold place or office. But he shall confess three or four times a year that he is in heart a Romanist.

8. That they shall give all proper information to the Court of Rome.

9. That excommunication or a perpetual curse shall light on the families that will not assist, in every way, Mary, Queen of Scots, to the throne of England.

10. That the Roman Catholics be obedient to the bishops or priests set over them, paying them taxes, etc.

11. That they try and induce Elizabeth to marry a Catholic, or secure, in some way, a Romish prince for England.

Thus spake Philip Corwine, before his death, about what the Council of Trent had done, and he was the cause of the conversion of the Mayor of Drogheda, his wife, and several others. But his sudden death was thought to be the effect of poison.*

* This volume is found in the Philadelphia Library, 947, Q. 9.

APPENDIX K.

NICASIUS DE SILLE.

(*See* p. 49, *note, near bottom of page.*)

NICASIUS DE SILLA, or Sille, who was first counselor in the administration of Governor Stuyvesant of the affairs of New-Netherlands, was a man of considerable attainments in literature and science. He was probably a descendant of a person of the same name who filled several important positions under the government of the United Provinces of Holland, as well as of the municipality of Amsterdam, in the latter part of the sixteenth century. The identity of their names is, however, all the evidence in our possession to justify the opinion in favor of this relationship, unless, indeed, his title of " Well-born," renders the supposition more probable. The elder de Sille was a native of Malines or Mechlin, in the Belgian provinces, which he abandoned on account of religious persecution, at the time when Balthazar de Moncheron, Peter Plancius, and other of his countrymen, for the same reason, fled to Holland, where they became doubly entitled to the rights and considerations of citizens of the Dutch Republic by reason of the part they took in the struggle of their adopted country for independence of the imperial power of Spain.

The American Nicasius de Sille was a native of Arnheim, and came to New-Netherlands in the summer of 1653, bringing with him a commission from the West-India Company as first counselor of the colonial government. In this commission he is declared to be an able and experienced statesman and soldier, and is instructed to reside at Fort Amsterdam, to deliberate with the governor on all questions of war, police, or public affairs; to promote alliances of amity and commerce; to assist in the administration of justice, both civil and criminal, and to advise generally on all matters which might transpire in the colony. He appears to have enjoyed the confidence of the governor, whom he accompanied in his expedition to the Delaware against the Swedes. He succeeded Van Zenhoven, in 1656, as attorney-general. He was at the same time appointed Sheriff of New-Amsterdam. In this office it was his duty, among other important labors, to make nightly tours around the town. It marks the simplicity of the times to read his complaints on one occasion to the burgomasters and schepens of the city, of the dogs making dangerous attacks on him while performing that service, of the hallooing of the Indians in the streets, and the boys playing *hoecje*, that is, hide and seek, around the hooks or corners of the streets, to the prejudice of quiet and good order.

He became one of the proprietors of New-Utrecht, on Long Island, where, in 1657, he built the first house erected in that town. It was standing until within a few years past. It was inclosed by high palisades, to protect it from attacks by the Indians. He resided there in 1659, in 1674, and probably until his death, which is unknown as to time and place.

He began the records of the town of New-Utrecht, which are not only in his hand-writing, but evince, in different respects, his literary acquirements. It is to these writings that we are indebted for the few specimens of his verse which make him one of the earliest poets of New-Netherlands.

De Sille's closing days were clouded by domestic troubles, and he seems to have needed all the consolations of that religion which breathes throughout his poetry. He left three children, all by his first wife. A son, Lawrence, from whom are descended many bearing his name, now generally abbreviated into Sill ; and two daughters, Gerdientje, who married Gerritse Van Cowenhoven, of Brooklyn, and Anna, who became the wife of Hendrik Kip, of New-Amsterdam.

(From Murphy's *Anthology of New-Amsterdam*, which also gives a woodcut of de Sille's house at New-Utrecht, and specimens of his poetry.)

KIP FAMILY. (See p. 49.)

[From Holgate's *American Genealogies.**]

AMONG the members of the Association entitled "The Company of Foreign Companies," as having been organized in 1588 for the purpose of exploring a north-east passage to the Indies, around the coast of Asia, was an individual named Hendrick Kype, ancestor of a highly respectable family in the State of New-York. He left Amsterdam with his family and came to New-Netherlands in 1635. Returning to Holland, however, soon after, he died there. His sons, remaining in this country, seem to have had considerable enterprise; for we find them securing large tracts of land and holding prominent stations in the government of the colony, as far down as the time of its conquest by the English in 1664. In 1647–49, when Governor Stuyvesant, to remodel the government, organized a popular assembly, composed of nine men, chosen from among the people, who should coöperate with him and his counsel in the administration of the government, Hendrick Kip, one of these sons, was among those selected for the purpose. Jacobus (James) Kip, another son, was secretary to the Council of New-Netherlands, and obtained a grant of land on the island of Manhattan, about two miles above the City Hall, on what was afterward called Kip's Bay. Here, in 1641, he erected a house of bricks, imported from Holland, a part of which was afterward rebuilt in 1670 and 1696. Five generations of the family were born here. A few years since, on the opening of Thirty-fifth street, on the line of which it stood, the house, at that time the oldest on the island, was taken down. In the following generation, we find the family purchasing from the Esopus Indians, on the east side of the Hudson, where Rhinebeck now stands, a tract of land extending four miles along the river and several miles inland. The original deed, which is still preserved, is dated July 28th, 1686, and signed by three Indian chiefs, Ankony, Anamaton, and Callicoon. Two years after, a royal patent dated June 2d, 1688, was granted by His Excellency Thomas Dongan, Governor of the province of New-York, under the name of the Manor of Kipsburgh, in confirmation of the Indian title. One fifth part of this manor was afterward sold to Colonel Henry Beckman, through whose granddaughter, the mother of Chancellor Livingston, it passed into the Livingston family. About three hundred acres of this property are still held by the Kip family, under the original deed. It seems that, while a part of the family adhered to the British cause during the Revolution, Jacobus Kip, of Kip's Bay, was a staunch Whig, his son having joined the American army. His residence for a short time was Washington's head-quarters. It will be recollected that on Sunday, September 15th, 1776, the British, under Sir William Howe, landed at Kip's Bay, and, after a skirmish with the Americans in the rear of Mr. Kip's house, they took possession of it, and, for several years, it was occupied by British officers as their head-quarters.

Jacob Kip, of Kipsburg, became a captain in the British infantry; and of another member of the family, we find the following account in Bolton's *History of Westchester County,* vol. ii. p. 254:

"The command of the loyalist rangers afforded Colonel de Lancey facilities for communicating with his old associates in this section of the country, and was the means for inducing some of the landed gentry to take an active part in the contest. This was particularly the case with Samuel Kip, Esq., of a family which, from the first settlement by the Dutch, had possessed a grant of land at Kip's Bay, and in other parts of New-York island. Having been always associated with the government, and from their landed interests wielding an influence in its affairs, they were naturally predisposed to espouse the royal cause. In addition to this, Mr. Kip's estate was near to that of Colonel de Lancey, and a close intimacy had always existed

* See also Lossing's *Field Book of Revolution,* ii. 597, for similar remarks and a woodcut of the Kip house ; Duyckinck's *Cyc. of Am. Lit.,* ii. 551, and *Hist. Notes of the Kip Family,* (by Bishop Kip.) Published by Munsell. Albany. 1871.

17

between them. He was therefore easily induced to accept a captain's commission from the royal government, and embarked all his interests in this contest. He raised a company of cavalry, principally from his own tenants, joined the British army with the colonel, and from his intimate knowledge of the country was enabled to gain the reputation of an active and daring partisan officer. For this reason he was for a time assigned to a command in the loyalist rangers. In one of the severe skirmishes which took place in Westchester County, in 1781, Captain Kip, while charging a body of American troops, had his horse killed under him, and received a severe bayonet wound. He survived, however, several years after the war, though, like his friend de Lancey, a heavy pecuniary sufferer from the cause he had espoused."

The following is a list of the municipal offices the family have held in the city of New-York, under the different forms of government, during the last two hundred years, taken from the corporation records : ·

Schepens, (*or justices*,) Hendrik Kip, 1656; Jacobus Kip, 1659, 1662, 1663, 1665, 1673, 1674. *Common Councilman*, Johannes Kip, 1694. *Aldermen*, Johannes Kip, 1685, 1687, 1691–93, 1696, 1697; Jacobus Kip, 1709–28; Leonard Kip, 1820, 1821. *Assistant Aldermen*, Samuel Kip, 1729–31; Samuel Kip, 1807, 1808; Leonard Kip, 1817–19.

Most, if not all, of the Kips' baptisms and marriages, of the parties who remained in New-York City, are recorded in the records of the Dutch Church, and are copied in Valentine's Manual of the Common Council, New-York City.

The author has considerable material relating to the descendants of Hendrik Kip, (brother of Isaac and James,) of New-Amsterdam, N. Y., as has also Henry Kipp, now of Hackensack, N. J. The records of the various Dutch churches in Bergen and Passaic counties, N. J., such as Hackensack, Passaic, Schraalenberg, Paramus, etc., contain much material also relating to the Kipps, though often in a perishing condition.

PREAKNESS, N. J.

The tract of country now called Preakness (N. J.) was originally obtained by letters patent from the crown of Great Britain. The first, perhaps, who possessed any special claim over that region was Thomas Hart, of Enfield, in the county of Middlesex, (England,) who was a merchant and proprietor of several tracts of land in New-York and East New-Jersey. By will dated December 19th, 1704, he bequeathed two third parts of said lands to his sister Patiente Ashfield, and one third part to Mercy Benthall, wife of Walter Benthall, all of London. His sister, Patiente Ashfield, was made by said will sole executrix of said lands. These lands were inherited by Priscilla and Mercy Benthall, daughters of Walter and Mercy Benthall, and by Richard Ashfield, grandson of said Patiente Ashfield. Priscilla and Mercy Benthall, by will, dated July 20th, 1721, appointed Rip Van Dam, Sr., and Rip Van Dam, Jr., attorneys of New-York, to dispose of their real estate and its appurtenances in New-Jersey, which had been bequeathed them by the will of Thomas Hart. These attorneys sold it to GEORGES DU REMOS and CORNELIUS KIP, December 4th, 1723, for £270, current money of New-York. The descendants of these two persons have held parts of this section until the present day. The name of the district is in the deed spelled Perekenos. The tract of land sold was thus described :

" Beginning at the south end of the land of Mdmle. Brackhuls, at a beech-tree standing near by the run ; from thence W. by N. 20 chains, to a stake there standing ; from thence N. 7¼° E. along the marked trees to a stake there standing, 90 chains ; thence W. 7¼° N. along David Danielson's line, as the trees are marked, to a white-oak tree, marked on four sides, near by the run 112° . . . . . degrees south, ′ along the run 30 chains ; then N.N.E. along the run 23° to the place where it first begun ; containing 600 acres or thereabouts, with all," etc., etc., etc. The original deed is held by Nicholas Kipp, late of Preakness, now of Chicago, Ill.

GENERAL INDEX.

— • • —

This index includes the names of parties intermarried, which are designated by *m.*; of descendants of other surnames than Corwin, which are designated by—see; of localities of residence, of sketches, incidents, etc. Those who have spelled their surnames Curwin, Curwen, Corwine, Curran, or in any other way, are also specially indexed.

* This list is not complete, but it contains what material on the subject the author had on hand.

18

* This list is imperfect, but it contains all the author has secured.

Paving, *m.* Amos Blinn, 3.
Paris, see Bourbon Co., Ky.
Paris, see Anna, 42; Elizabeth, 64; Jacob, 6; James, 57. *m.* Julia Blinn, 18. See Laura, 7; Linnie, 1; Lucinda, 10; Thomas, 24; Walter, 8.
Parker, *m.* Moses, 2.
Parkman, *m.* Rev. George, (4.)
Parshall, *m.* Phineas, 1.
Parsons, *m.* Aminda Goble, 5.
Partridge, Stella Baker, 1.
Patents, David, 4.
Eugene, 1.
Jesse, 3.
Paterson, see Passaic Co., N. J.
Patty, *m.* Elizabeth, 1½.
Payne, Edward, 3. *m.* Mary Whitney, 18¾, see last page.
Peake, (Sir William,) George, (1.)
Pease, *m.* Alonzo, 1.
Peck, *m.* George, 13.
Peconic, see Suffolk Co., N. Y.
Peken, John, 2.
Pellett, see Elizabeth, 65; Gabriel, 8; Harriet, 23. *m.* Mary, 24.
Pembroke, see Genesee Co., N. Y.
Pence, *m.* George Blinn, 56.
Pendleton, see Madison Co., Ind.
Penfield, see Monroe Co., N. Y.
Penney, *m.* Annie, 12. See Anna, 32; David, 28; Deborah, 10; Esther, 6. *m.* Ezra, 2; Helen M., 3. See Huldey, 6; Franklin, 1; Jesse, 11; Lewis, 12; Mary, 164. *m.* Sarah, 3, ?. See Thomas, 23.
Pennington, see Mercer Co., N.J.
Pennsylvania.
Joab H., (1.)
Joseph, 10.
Mary, 7.
Rebecca, (2.)
Sophia Smith, 3.
Alleghany Co., Pittsburg.
Jemima, 8½.
Henry, 23½.
David P., ...
Armstrong Co., Brady's Bend.
Lydia, 6½.
Bradford Co., Troy.
Benjamin, 19.
Clearfield Co., Curwinsville.
John, (b.)
Dauphin Co., Harrisburg.
John, (c.)
Erie Co., Erie.
Rev. Ira, 4.
Fayette Co.
David, 8.
Ichabod, 1.
Jesse, 2.
Matthias, 5.
Lebanon Co., Lebanon.
William H., 36.
Lehigh Co., Catasauqua.
Charles, 7.
George S., 12.
Luzerne Co.
Gabriel, 4.
George C. Wetherby, 59.
Henry, 8.
Richard, 8.
William, 50.
Greenfield, (township.)
Anna, 17.
Eliza, 3.
Green Grove.
James Wetherby, 49.
Providence.
Isaac, 5.
Scott.
Eliza, 3.
Wilkesbarre.
Seth, 1.
Lycoming Co., Williamsport.
Edward Watkins, 17.
Ferdinand Crassons, 1.
McKean Co., Norwich.
Edward, 3.
Southport.
Ghordis, 1.
Montgomery Co.
John, (c.)
Philadelphia Co., Philadelphia, p. xxxii.
Aaron, (1.)
George, (a.)
John E., 45.
John, (a.)
Sullivan Co., Shunk.
Anne, 17.
Susquehanna Co., Great Bend.
Rev. Jason, 4.
Herrick.
James Wetherby, 49.
Montrose.
Albert, 1.
Alsop, 1.
Silas, 2.
Susan, 3.
New-Milford.
Benoni, 1.
George, 14.
Horace Little, 3.
Rosetta, 1.
Silas, 2.
William, 13.
Tioga Co., Blossburg.
Joel L. Davis, 1.
Dagget's Mills.
Margaret Reed, 7.
Liberty.
John, 68.
Tioga.
Amos, 1.
Venango Co., Oil City.
Charles, 42.
Pleasantville.
Benjamin, 20.
Washington Co., Washington.
Mary Matthews, 103.
Stephen, 4.
Wayne Co., Honesdale.
Asa, 4.
Penn Yan, see Yates Co., N. Y.
Peoria, see Peoria Co., Ill.
Persecution, Lucinda, 4.
Peru, see Miami Co., Ind.
Peters, (Hugh,) George, (1;) John, (1.)
Peters, see Eleazar, 3; George, 69; Horace, 2, 4. *m.* Mary, 30, see 115, 159. See Matilda, 3; Sarah, 84; William, 80.
Petit, see Charles, 41; Charlotte, 12; Emma, 14. *m.* Hannah, 12. See Isabella, 7.
Petty, *m.* Jacob, 2; Sarah Howell, 94.
Phelan, *m.* Malcolm, (A.)
Phelps, see Ontario Co., N. Y.
Philadelphia, see Philadelphia Co., Pa.
Philips, *m.* Ardon, 1. See Eliza, 20; George, 55. *m.* Hannah, 3. See Jesse, 12. *m.* Joshua, 4; Linea, 1. See Mary, 105. *m.* Rhody, 1. See Sarah, 63; William, 69.
Phipps, (Sir William,) George, 3.
Physicians.
Bela Colgrove, 1.
Caroline, 8.
Daniel, 8.
Clinton Colgrove, 1.
Elias Corwin, 6.
Emelin, 5, *m.* Dr. Van Harlingen.
Gabriel S. Corwin, 5.
Hannah, 3, *m.* Dr. Elijah Philips.
Henry H. Corwin, 25.
Irena Tuthill, 1, *m.* Dr. E. L. Blakeley.
Isaac L. Millspaugh, 17.
James B. Colgrove, 45½.
John, (c.)
Joseph, 16.
Marcus H., 2.
See Mary, (7.)
Pickard, see Eugene, 1. *m.* Margaret Halstead, 10.
Pickle, *m.* Nathaniel, 2.
Pickman, *m.* George, (5.)
Pierce, (President,) Amos, (2.) *m.* Helen M., 5½.
Pierrepont, see St. Lawrence Co., N. Y.
Pierson, *m.* Martha, 2.
Pike, *m.* Dorothy, 5; Isabella, 2. See Le Roy, 1; Mary, 149; Olivia, 5; Orsamus, 1; Rollin, 1.
Piketon, see Pike Co., O.
Pine Bush, see Orange Co., N. Y.
Pinkerton, John, 2½, (supplement.)
Plainfield, see Windham Co., Ct.
Plattekill, see Ulster Co., N.Y.
Pleasantville, see Venango Co., Pa.
Plymouth, see Hancock Co., Ill.
Plymouth, see Plymouth Co., Mass.
Plymouth, see Wayne Co., Mich.
Pollion, *m.* Abel, 2.
Pollifly, see Bergen Co., N. J.
Pompey, see Onondaga Co., N.Y.
Pompton Plains, see Morris Co., N. J.
Porter, see Adelia Horton, 1; Albert B., 7. *m.* Elizabeth, 9. George, 40. See Jane, 12; Levisa, 2.
Port Jervis, see Orange Co., N.Y.
Portraits, George R., (8.)
Portsmouth, see Scioto Co., O.
Potter, *m.* Herbert Baker, 4; Mary Galloway, 131.
Post, *m.* George Smith, 74.
Prattsville, see Ohio.
Prawl, *m.* Mary Matthews, 103.
Preakness, see Passaic Co., N. J.
Preble Co., see Ohio.
Price, *m.* Annie, 5; Julia, 8; Noah, 2.
Prince, *m.* Hervey, 1.
Providence, see Luzerne Co., Pa.
Providence, see Providence Co., R. I.
Prudens, *m.* Nancy Woodhull, 4.
Purdy, *m.* Mehetable Warner, 13.
Putnam, *m.* Rosina, 1.

Quakers, p. xxxi.
Quincy, see Adams Co., Ill.
Quinton, *m.* Richard, (6.)

Ramsay, *m.* Daniel, 1; Theophilus, 2.
Rand, *m.* Daniel Campbell, 29.
Randall, *m.* Rebecca, 2.
Randolph, see Orange Co., Vt.
Ransom, *m.* William Carr, 86.
Ransomville, see Niagara Co., N. Y.
Rapalje, *m.* John, 30.
Raritan, see Somerset Co., N. J.
Raritan Landing, see Somerset Co., N. J.
Raynor, *m.* Asa, 2. Barnabas Reeves, 4; Henry, 19.
Raze, *m.* Frederic, 6.
Reading, see Schuyler Co., N.Y.
Rebellion.
Albert Baldwin, 13.
Amos, (2.)
Charles, (½,) supplement.
Clinton Colgrove, 1.
David, 25.
Elmore, 1.
Frederic M. Fountain, (2.)
George, 40.
George Smith, 78.
Henry, 20½.
Horace H. Peters, 2.
James, 20, 21.
James B. Colgrove, 45½.
Jesse Blinn, 13.
John, (10½.)
Jonathan, 2.
Joshua, 9.
Mary, 33.
Oliver, 3.
Perry, 1.
Quincy, 1.
Richard M., (6.)

* These are a few wills of the family which the author has consulted, or met with, in the preparation of this work. Of course there are many more, but the material needed was more easily obtained from other sources. All the ante-Revolutionary wills of the State of New-York, south of Albany County, are now deposited in New-York City.

SUPPLEMENT.

ADDITIONS AND CORRECTIONS.

Only so much is here inserted as is necessary to make the correction or addition.

———•◦•——

1 AARON,[7] (*Gershom*, 3,) p. 1, and Gershom, 3, below.

1 ABIGAIL, (*John*, 1,) etc. See p. 2.

(1) Abigail, etc.
 M. Eleazar Hawthorne, etc.   P. 3.

1 ABNER, etc.
 Ch. (additional) Cynthia, Hannah.

1½ ABNER, (*Barnabas*, 1,) etc.   P. 3.

¼ ABRAHAM,[7] (*Joseph*, 10¼,) b. in Rockland Co., 1805–15, N. Y., yet
 living.
 M.
 Ch.   (*Morgan Co., Ill.*)

1 AMAZIAH ; perhaps another child, Elizabeth.   P. 9.

2 AMY,[7] (*Joseph*, 10¼,) b. April 10th, 1810.

½ ANGELINE,[7] (*Seth*, 1,) b. March 18th, 1813.
 M. Sylvester Downs.
 Ch. Amelia, (*m.* Hiram Carter,) Frank, Sarah J., (*m.* E. F. Squires,)
 Edward, Lewis, Benjamin, Allison.

(2) ANTOINETTE,[8] (*Stephen O.*, 8½,) b. ——, 1845.

2½ BETHIA,[6] (*Gilbert*, 1,) b. ——, 1790–1800.
 M. —— Gardner.

7½ BETHIA,[7] (*Matthias*, 6½,) b. ——, 1810–20.

6½ CAROLINE, (*David*, 18.) P. 24.

1½ CHARITY,[7] (*Joseph*, 10¼,) b. December 23d, 1811.
 M. Amaziah B. Clinton.
              (*Paterson, N. J.*)

53½ Charles Daniel Evans, (*Sarah E. Corwin*,) b. February 24th, 1856.

[1] CHARLES II. CURWEN,[1] (*Edward Stanley Curwen*, of England,) b. ——,
 1830–40.   Came to America about 1868.

3½ CYNTHIA,[7] (*Abner*, 1,) b. about 1780.
 M. David Gildersleeve.

P. 48, note †.   Olferts Suert, also written Olphert Sjoerts.

He was from Hedronvoon, Holland, and married
(1) Margaret Cloppers, of New-York, September 9th, 1682;
(2) Hilligont (or Hellegond) Lucas, May 22d, 1702 or 3.

In his family were one male, one female, five children, one negro, one negress, three negro children.   He had also children, (additional to those mentioned in note,) Adyltje, bapt. September 2d, 1688, Maritje, bapt. January 7, 1686.

Also a Margaret Shourt and Richard Newbold were licensed to be married, (New-York City,) 1758, and a Johannes Cloppers married Marybem Sourt, in New-York City, July 2, 1684.

Adolph Soert and Margaret E. Ekkert had a child Hendrik baptized at Hackensack, N. J., April 27th, 1735.

P. 49, 9th line of note from top, read,—the children of Leah Kipp and Christian Shurte were, Mary Ann, b. March 15th, 1809, who married Edward C. Corwin, No. 5.

15th line from top, read, Christian Shurte was drowned March 5th, 1822.

17th line from top, read, to Jasper Dodd, May 8th, 1829.

Note †, 10th line, read, Isaac, widow, married, in 1653, Catalina Hendrik de Suyers; and in next line, for "Maria Vermilyea," we sometimes find "Maria Vornelse," widow of John de la Montaign.

Isaac Kip was then of Kill von Kull, N. J., and she of New-Haerlem.

The De Kype family was settled for a long time near Alençon, Bretagne, France.   Ruloff was a Romanist, and in the triumph of the Protestants, under Condé, in 1562, his chateau was taken and burned.   He and his three sons lived in Holland under an assumed name.   In 1569, Ruloff, with his son Henri, reëntered France, and fell in battle on the banks of La Charente, near Jarnac, March 13th.   His son Jean Baptiste, a priest, secured his burial in a neighboring church, where an altar tomb was erected to his memory, both of which were destroyed in the French Revolution. The inscription mentioned him as Ruloff De Kype, Ecuyer, (entitled to coat armor,) and was surmounted by his arms, with two crests, one a gamecock, the other a demi-griffin, holding a cross.   The arms were pictured in the stained glass window of the Dutch Church in New-York City, built in 1640.   They were also carved in stone over the door of the Kips Bay house, which was erected in 1655.   The motto was, "*Vestigia nulla retrorsum.*"

Of Ruloff's three sons, Henri entered the army of an Italian prince, where he spent his life, and died unmarried.   Jean Baptiste was a Roman priest, and Ruloff settled at Amsterdam, Holland, and became a Protestant.   He was born in 1544, and died in 1596, and left one son, Hendrik, born 1576. He married Margaret de Marneil.   After the failure of the effort to find a north-eastern route to Asia, (pp. 49, 257,) the company, of which he was a member, employed Hendrik Hudson in 1609, who, in the Half-Moon, discovered Hudson River.

P. 49, note, 6th line from bottom, read, "Nicasius de Sille of Aernhorn, Holland, and Maria," etc.   (De Sille married a second time, Tryntje Congers, from the Hague, May 26th, 1655.) ,

    4th line from bottom, after "Antje Bryant," read (daughter of Cornelius Breyendt and Margaret Simese.)

    3d line from bottom, after "now very numerous," read, This Nicasie Kip bought 220 acres of land of Garret Lydekker, adjoining the land of Henry Kip, in New-Barbadoes, (near Hackensack,) bordering on the river.

    2d line from bottom, read, James, born December, 1702, who removed, etc.

P. 50, note, 4th line from top, read, Leah Mandeville, (b. February 18th, 1740, d. June 7th, 1802.)

    7th line, Hester Johnson, (b. April 6th, 1786, d. May 22d, 1859.)

P. 50, note, 5th line from bottom, read, Nicholas Kipp, of Geneva, b. / November 5, 1806, married Mary Freshour, (daughter of Henry Freshour and Catharine Clice, of Maryland,) February 7th, 1832 ; Mary, b. December 17th, 1808, married James Jones, February 13th, 1825 ; John, b.'October 30th, 1812, married Esther Burt, January 31st, 1843 ; he died February 25th, 1863 ; Leah, b. December 12th, 1816, married Frederic Backenstose, November 24th, 1836 ; (Geneva, N. J.;) Cornelius, b. November 9th, 1818, d. December, 1818.

½ Elijah Woolsey,[7] (*Joseph*, 10¼,) b. October 25th, 1806, in Rockland Co., N. Y.

    M. Jane Stults, of Paterson, N. J., August 16th, 1828.

    Ch. Mary Ann, and three, who died young.   (*Calusa, Cal.*)

2¼ Elizabeth,[6] (*Gilbert,* 1,) b. ——, 1790, living.

    M. David Gesner.   (*Nyack, N. Y.*)

11¼ Elizabeth,[7] (*Joseph*, 10¼,) b. in Rockland Co., N. Y.

    M. William Filmer.   (*Polk City, Iowa.*)

28½ Elizabeth,[7] (Joseph,   ,) b. September 7th, 1808.

1¼ Ella,[8] (Stephen O., 8⅓,) b. about 1850.

2¼ Ellen,[6] (*Stephen*, 8⅓,) b. about 1850.

4 Elvira,[8] (*Ira*, 4¼,) b. June 5, 1852.   (*La Salle, Ill.*)

½ Emma,[6] (*Joseph,* .)

    M. J. E. Ware.   (*Paterson, N. J.*)

1 Evan Corwin Evans,[8] (Sarah E. Corwin,   ,) b. November 13th, 1863.

1 Garret,[7] (*Matthias*, 6⅓,) b. ——, 1820–30.

2 Gershom,[8] (*Matthias*, 3,) b. about 1742, d. about 1792.

    M.

    Ch.

He was a seafaring man, and was taken prisoner by a British cruiser, and carried to England. He escaped to France, and subsequently joined the American flotilla under John Paul Jones, and assisted in the capture of the Serapis, and other vessels. After peace returned, he located at Hudson, Columbia Co., N. Y., and died there. See p. 82.

3 GERSHOM,[6] (*Gilbert*, 1,) b. about 1775.
M. Margaret ——.
Ch. Aaron? Lucy?
Received a grant of 450 acres, etc. See p. 82.

4 GERSHOM,[7] (*Matthias*, 6½,) b. 1810–20.

Note on p. 82, 1 Gershom, etc., is right as it stands. 2 Gershom, is confounded with Gershoms 3 and 4, all being supposed, at first, to be identical. In these corrections, they are properly discriminated.

1 GILBERT,[6] (*Matthias*, 3,) b. at Southold, February, 1745, d. April 11th, 1810.
M. Amy Knapp, etc., b. about 1750, d. 1807.
Ch. Joseph, Elizabeth, Gershom, Mary, Matthias, Bethia.
He went to Nantucket, etc. P. 83.

4½ GILBERT,[7] (*Joseph*, 10½,) b. August 15th, 1803, d. in Paterson, N. J.

4½ HANNAH,[6] (*Abner*, 1.)
M. Eleazar Kellum.

(1) HARBERT,[4] etc., etc.

½ HARRIET,[6] (*Joseph*, 10½,) b. in Rockland Co., N. Y., 1823, d. in Quincy, Ill., June, 1834.

20¾ HENRY WISNER,[8] (*Stephen O.*, 8½,) b. about 1851.

4 Hester Kate Evans,[8] (*Sarah E. Corwin*, —,) b. March 17th, 1858.

2¼ JACOB,[7] (*Matthias*, 6½,) b. ——, 1820–30.

36¼ JAMES,[7] (*Matthias*, 6½,) b. ——, 1820–30.

38¾ JAMES BUCHANAN,[8] (*Stephen O.*, 8½.)

3¼ JANE,[7] (*Joseph*, 10¼,) b. September 23d, 1804.
M. —— Giddings.

7½ JESSE H.,[8] (*Lewis J.*, 2½,) b. ——, 1840–50.

5½ JOHN,[5] (*Jedediah*, ½,) b. about 1740.

36¾ JOHN W.,[7] (*Joseph*, 10½,) b. ——, died at St. Louis, Mo., 1861.
No ch.

37 JOHN TOTTEN, (*David*, 18,) etc.

62½ JOHN Y.,[8] (*Lewis J.*, 2½,) b. ——, 1840–50.

23¾ JOSEPH B.,[8] (*Lewis J.*, 2½,) b. ——, 1840–50.

8¾ LYDIA,[8] (*Henry W.*, 23½,) b. and d. March 28th, 1854.

13½ MARY,[6] (*Gilbert*, 1,) b. ——, 1785–95.
M. —— Sarvant.
Ch. Garret, etc. (*Nyack, N. Y.*)

31½ MARY S.,[8] (*Lewis J.*, 2½.)

81¾ MARY ANN,* (*Elijah*, ½,) b. December 20th, 1835.

   M. George W. Ware, March 28th, 1859.   (*Calusa, Cal.*)

1 Matthias, (p. 160,) is mentioned in a document (in Clerk of Court's Office, Essex Co., Mass.) as an owner of real estate in Topsfield, in that county, in 1649. (G. R. C.)

6⅓ MATTHIAS,⁵ (*Gilbert*, 1,) b. ——, 1780–90, d. ——, 1851.

   M. Margaret Amerman, of Haverstraw, N. Y.

   Ch. Jacob, Garret, Gershom, Richard, James, Bethia, Matthias.

                (*Friendship, Alleghany Co., N. Y.*, 1819.)

10⅓ MATTHIAS,⁷ (*Matthias*, 6⅓,) b. ——, 1810–20.

8½ NATHAN F.,⁸ (*Lewis J.*, —,) b. ——, 1840–50.

7⅓ RICHARD,⁷ (*Matthias*, 6½,) b. ——, 1810–20.

10⅓ Richard Jennings Evans,* (*Sarah E. Corwin*, —,) b. October 15th, 1866.

54¼ SARAH,* (*Joseph R.*, 10¼,) b. ——.   (*San Francisco, Cal.*)

68½ Sarah Caroline Evans,* (*Sarah E. Corwin*,) b. May 28th, 1860.

A family of Corwins has just been heard from, living formerly in South-Carolina, but more recently in Indiana and California, and apparently of the Salem stock, probably having descended from John, (2,) and John, (4.) They are placed here separately, in order of time.

   (2) *JOHN,⁵ [*Bartholomew*, (1),] b. February 22d, 1722, d. September. 26th, 1744.

   M. ——.

   Ch. John? and perhaps Richard. (See p. 127.)

   (2½) *JOHN,⁶ [*John*, (2), ?] b. about 1743? d. ——, 1772, in York District, S. C.

   M. Mary Barron. She afterward married a Mr. Pinkerton, who was a Revolutionary soldier. He served under Gens. Morgan and Green.

   Ch. Elizabeth, James.

   (2½) *RICHARD,⁶ [*John*, (2), ?] b. ——, 1745–50?

   M.

   Ch.   (*New-Jersey.*)

   (3⅓) *ELIZABETH,⁷ [*John*, (2½), ?] b. about 1770.

   M. —— Clark, 1790–1800.

   They moved from York District, S. C., to Georgia, about 1816.

   (⅓) *JAMES,⁷ [*John*, (2½), or perhaps *James*.] b. March 22d, 1775, in York District, S. C., d. May 22d, 1852, in Warwick Co., Ind.

   M. Lucina Haynie, in South-Carolina. She was born October 14th, 1787, in Pendleton Co., S. C., was of Scotch descent, and d. April 1st, 1857, in Warwick Co., Ind.

   Ch. John, Charles, Rev. James, Elizabeth, Henry, Samuel, Rev. William S., Mary, Lucina, Louis M.

He removed from York District, South-Carolina, to Warwick County,

               * Corwine.

Indiana, in 1815. Being an anti-slavery man even at that early time, he did not wish to train up his children in the South. His father's family records were destroyed in the Revolutionary War, (*see Bartholomew*, 1, p. 18;) and James's father dying in 1772, when James was only two years old, the descendants have lost the traditions of the family, only being able to say that they emigrated from New-Jersey about the beginning of the Revolution, and that this James had an uncle, Richard Corwine, remaining in New-Jersey; also that the mother of James was a Mary Barron, (or widow Rowan,) who long survived her first husband, marrying a second time a John Pinkerton, in York District, S. C. Mr. P. was also a Revolutionary soldier under Generals Morgan and Green. Hence James is probably of the Hunterdon Co., N. J., Corwine family, which went to New-Jersey from Massachusetts in 1717. Now, there was a John, No. (2), p. 127, of this New-Jersey family, whose descendants moved to Baltimore, Md., at an unknown date. This John (4) the writer would suggest as the probable father of James, (‡.) (See *Curwen's Journal and Letters*, Edition of 1864, p. 4.)

The following children of James (‡) are divided, some spelling their name Corwine and some Corwin, but the writer does not know which follow the one spelling or the other:

(11‡) JOHN,[8] [*James*, (‡),] about 1800, in York District, S. C., d. ——.
    M. (1) Catharine Leright; (2) Susannah Sprinkle.
    Ch. William Franklin, John Q. Adams, Minor Leright, James; Elizabeth A., Eliza Ann, Sarah, James, Catharine Gaines, Louisa Anna, David S., Carolina.

(‡) CHARLES,[8] [*James*, (‡),] b. about 1800–1810, in York District, S. C.
    M. Mary Duncan.
    Ch. Amanda, Jonas, (killed at battle of Corinth,) William Spencer, James, Martha Jane, Harriet. He moved with his father to Indiana, in 1816.   (*Monmouth, Ill.*)

(‡) Rev. JAMES,[8] [*James*, (‡),] b. ——, 1800–10, in York District, S. C.
    M. Susan Ann Shaw.
    Ch. Edson, (died at age of nine months.)
Of the Methodist Episcopal Church. Removed to Indiana with his father, 1816; to California, 1849.   (*Los Angeles, Cal.*)

(5‡) ELIZABETH,[8] [*James*, (‡),] b. about 1812, in York District, S. C.
    M. Alonzo D. Johnson, who died 1870.
    Ch. Three sons and three daughters.   (*Warwick Co., Indiana.*)

(‡) HENRY,[8] [*James*, (‡),] b. in York District, S. C., about 1814.
    M. Mary Martin.
    Ch. Sarah.   (*Liberty, Mo.*)

(12‡) SAMUEL ROWAN,[8] [*James*, (‡),] b. about 1818, in Warwick Co., Ind.
    M. Louisa J. Taylor.

Ch. Mary, Lucina Jane, James Taylor, William Spencer.

(*Evansville, Ind.*)

(1) Rev. WILLIAM SPENCER,* [*James*, (¼),] b. in Warwick Co., Ind., November 16th, 1822.

 M. Eliza Ann, daughter of Zabina and Experience Lovejoy, November 12th, 1845.

 Ch. Minerva-Helen, (b. November 12th, 1846; *m.* George H. Bean, in Solano Co., Cal., August 22d, 1869 ; now of Marion Co., Cal.,) Mary Lucina Elizabeth, (b. July 19th, 1849; *m.* Francis Marion Shepler, in Solano Co., Cal., January 31st, 1869 ; now of Sacramento, Cal.)

He removed to California in 1853. Of the Methodist Episcopal Church.   (*Elk Grove, Cal.*)

(3½) MARY,* [*James*, (¼),] b. about 1824, in Indiana.

 M. Anthony Walters Lockwood.

 Ch. Three sons and three daughters.   (*Evansville, Ind.*)

(1) LUCINA,* [*James*, (¼),] b. about 1826 in Indiana.

 M. William L. Haynie.

 Three Ch.   (*Warwick Co., Ind.*)

(1) LOUIS MARION,* [*James*, (¼),] b. September 26th, 1830, in Warwick Co., Ind.

 M. Annie De Arcy, October 24th, 1859. She was born in Galway County, Ireland, October 24th, 1837.

 Ch. James-Henry, born October 19th, 1870.

He removed to California in 1860, and has been County Recorder of Napa County since 1868.   (*Napa City, Cal.*)

(1) MINOR L.,* [*grandson of James*, (¼).]   (*Cincinnati, O.*)

P.S.—Any facts or arguments in possession of any one, in reference to the origin of the Long Island family, either in harmony or in conflict with the statements and theories herein presented, or any corrections, additions, or remarks upon the subject, will be gratefully received by the author, and preserved for incorporation in a future edition, if such be ever published.

E. T. C.

www.ingramcontent.com/pod-product-compliance
Lightning Source LLC
Chambersburg PA
CBHW021450210326
41599CB00012B/1018